# *Economics of the Environment*

# *Economics of the Environment*

## *Selected Readings*

Edited by ROBERT DORFMAN
HARVARD UNIVERSITY

and NANCY S. DORFMAN
NORTHEASTERN UNIVERSITY

W · W · NORTON & COMPANY · INC · NEW YORK

To Peter and Ann


FIRST EDITION
Library of Congress Cataloging in Publication Data

Dorfman, Robert, comp.
Economics of the environment.

Includes bibliographical references.
1. Pollution—Economic aspects—Addresses, essays, lectures. 2. Environmental policy—Addresses, essays, lectures. I. Dorfman, Nancy, joint comp. II. Title.
HC79.P55D65 333.9 72–3848
ISBN 0–393–05275–3

Published simultaneously in Canada
by George J. McLeod Limited, Toronto
PRINTED IN THE UNITED STATES OF AMERICA

1 2 3 4 5 6 7 8 9 0

# Contents

# *Preface*

This collection is a response to the widely felt concern over the profligate way in which mankind, and particularly Americans, have been depleting their natural resources and exploiting their environment. Economists have a double concern in the matter: that the viability and healthfulness of the environment be preserved, and that the measures taken to protect the environment be effective and not reduce the flow of ordinary useful goods and services any more than is necessary. This second concern is the economist's special province, and for that reason his adrenalin begins to flow when he hears some of the more alarmist proposals for meeting the "environmental crisis." Is there really an environmental crisis, or only a problem that we can deal with soberly? Is the environmental problem so severe that the growth in the output of goods and services must be brought to a halt while millions in this country and hundreds of millions throughout the world still live in poverty? Irrespective of how severe the problem is, how can we best maintain the flow of ordinary real income without abusing the environment? These are among the crucial questions that are addressed by the papers in this collection.

The collection consists of twenty-six papers divided into five groups. The first group blocks out the problem by assessing its nature and severity and pointing out the main economic issues. The second group is the most technical: it consists of five essays in which the concepts and methods of economic analysis are developed with a view to their application to environmental problems. The third group, containing seven papers, argues the pros and cons of the policies for environmental protection that emerge from the analyses in the preceding group. The fourth group explores the reasons for the rapid increase in the burdens imposed on the environment during the last thirty years or so. Finally, the fifth group of papers illustrates the methods and difficulties encountered in trying to make quantitative assessments of environmental damage or the costs of abating it. All together, the papers survey the key facets of the problem from an economic point of view.

There are both advantages and drawbacks in trying to gain insight into a difficult social problem by studying a collection of source materials such as this. The main drawback is that the collection cannot be as neat, tidy, and coherent as a textbook or treatise. Some matters are taken up again and again; others, equally important, fall in the gaps between the papers. The papers are not even consistent with one another. Sometimes the authors consciously contradict each other, sometimes they disagree out of ignorance. The authors do not always use the same assumptions, and they often do not make their assumptions clear in the passages that we have incorporated.

All these characteristics of a collection inject a certain amount of confusion into the reading. But they also reflect the liveliness, the heat of controversy, the earnestness of firsthand testimony that a placid text or treatise cannot convey. On balance, there is much to be said for joining the struggle with the authors and not smoothing over the rough edges and unresolved issues.

So in this volume we are allowing some of the main contenders to speak with their own voices, and by the same token we are allowing the reader to draw his own conclusions from their assorted evidence and arguments. In the introduction, however, we shall try to sort out the principal premises and postulates that divide the authors—to sort these out, in the hope not of resolving the issues, but of clarifying how the various discussions and viewpoints relate to each other.

R. D.
May 1972 N. S. D.

# *Introduction*

The essays in this collection are concerned with the economic aspects of current environmental problems. There are other aspects—moral, medical, biological, chemical, and more—but they are treated here only to the extent necessary for understanding the economic aspects. Even with this limitation, the environment remains too large a concept and word to be analyzed in a single volume. We shall therefore confine ourselves to the aspects of the environment that people seem to have in mind when they talk of "the environmental crisis." In that context the subjects of concern are the purity of air and public waters, the plentifulness and vitality of natural landscapes, fauna, and flora, and the integrity of certain other natural features such as beaches. All of these are comprehended in the *ecosphere*—the living space shared by all living creatures (including man) and the creatures themselves.

Man has been tampering with the ecosphere for a very long time; one might almost define man as the animal that modifies its environment consciously, not instinctively. The great transformation from primitive hunting and gathering to settled civilization occurred when man learned to convert forests and savannahs into farms and to breed domesticated varieties of plants and animals. This was the most radical change in the environment that mankind has undertaken to this very day. Over the course of time this transformation altered the ecology of entire continents, and there is good reason to believe that ancient civilizations rose and fell as a result of its progress.

Not all of the environment changes engineered by man have been for the worse, though some writers presume that they have been. Quite apart from the enormous increase in food production, malarial swamps have been drained, deserts have been made habitable, and much more has been achieved. (Lewis W. Moncrief pursues this question in Selection 17.) But in the last few generations mankind's propensity to change his environment has itself been transformed, as symbolized by the contrasts between the whaleboat and the radar-equipped factory ship, the waterwheel and the nuclear power plant, or the country road and the interstate highway. The power to use and adapt has become the power to destroy abruptly. Barry Commoner catalogs some alarming particulars in Selection 16. Our understanding of the environment has by no means kept pace with our capacity to alter it, and our ability to control our impact has fallen far behind. Therein lies the current threat.

The visible and impending environmental impacts of our newly acquired powers have forced us to recognize that the environment consists of scarce and exhaustible resources. That is where economics enters, for economics is the science of allocating scarce resources among competing ends. What has happened, obviously, is that the economic institutions that sufficed when ecological side effects were mild and gradual have abruptly become inadequate. It is as if we were still trying to control automobiles with bridles. The task for economics is to perceive just why the old institutions cannot control the new forces, and to devise new methods of control that can. The papers in this collection are addressed to that task. First we shall examine what they say about the adequacy of the economic system we have inherited.

## Environmental Resources and Private Property

Several of our authors (Coase in Selection 6, J. H. Dales in Selection 19) lay great emphasis on the concept of private property and its role in directing the use of resources. Their point is twofold. First, the institution of private property and our other economic institutions have evolved together so that our economic system is well attuned to securing the efficient use of things that are owned. This follows from the familiar argument, going back to Adam Smith's "invisible hand," that competitive markets guide resources into the uses in which they will produce the things that consumers want most. But it should be noticed, first, that this argument applies only to resources that are privately owned and to commodities that consumers buy individually and use as they see fit. Second, if any resource is not privately owned, or if a consumer wants a commodity that he cannot procure and use individually, then the invisible hand doesn't work. On the contrary, ordinary economic institutions do not provide any incentives for furnishing such resources or commodities, or for using them efficiently if they occur naturally. In particular, resources in the environment, which no private person owns, tend to be used heedlessly, with results that are all too obvious.

One explanation, then, of our environmental problems is that many vitally important resources are not owned by anyone and consequently lack the protection and guidance that a private owner normally provides.[1] This state of affairs was tolerable when the unowned resources

1. There is no implication that resources that are owned are automatically used in the public interest. For example, if the owner of a resource is a monopolist, he will not use it in a socially efficient way.

were plentiful, and careless use had virtually no effect upon them, but now we can no longer be so relaxed about it.

It does not follow that more extensive private ownership is the solution to our environmental problems. There are good reasons that explain why it is either impractical or undesirable to confer private titles to certain resources. Since these are the underlying causes of the misuse of the environment, and any corrective measures must take them into account, it pays to examine them closely.

The resources that make up the environment are unsuitable for private ownership because they lack the "excludability property." That is, it is typically not practical to exclude people from these resources or to prevent people from benefiting from them, either because of physical impossibility, or because controlling access would be inordinately expensive or cumbersome, or because limiting access would be socially unacceptable. This is the case with the atmosphere, and with most public roads, waters, beaches, and so on. It is also the case with flood-protection works, sewage-treatment plants, and many other facilities that improve the condition of the environment. There isn't much point in owning anything from which other people cannot be excluded.

Nonexcludability is not the whole story. Another peculiarity of environmental resources is that there are likely to be enormous economies in the joint consumption or use of the resource as contrasted with individual use. This second characteristic is illustrated by the contrast between housing and streets. It is economical to divide living space into family-size lots and devote each of them to housing one or a few families, but it would be fantastically wasteful to provide each family with its private road to the central business district. Accordingly, most houses are private property, but streets are public property. Most resources in the environment are analogous to streets, not houses, and are therefore used in common by substantial numbers of people. The difficulty that this common use creates is that each user may interfere with the others, reducing the serviceability of the resource to them; moreover, he has little incentive not to do so. Typical examples are road congestion and the use of the atmosphere and waters for discharging waste products. Resources that are most economically used in common frequently are privately owned—ski slopes, for example. But frequently also, it is deemed undesirable to have such resources controlled by private individuals. It is not much help to notice that resources from which people cannot be excluded or which it is very economical for them to share would be better managed if they were ordinary, private economic goods. Their very nature precludes that solution.

Although the characteristics of nonexcludability and shared, mutually interfering usage are especially prevalent among environmental goods, they do occur elsewhere. For example, fire protection and scientific discoveries have the nonexcludability property. Public and university libraries are used in common, and the users interfere with each other. In fact, the economic theory of resources that exhibit these characteristics has been worked out in other contexts. One way to define public goods in general is to say that they are resources or commodities from which potential users cannot be excluded, and a fairly complete theory of the provision and use of public goods has been developed. The mutual interference of the users of a shared resource is a special case of the general phenomenon of externalities: an externality occurs whenever the activities of one person affect the welfares or production functions of other people who have no direct control over that activity. The general theory of externalities is also well developed.

It follows that the analysis of environmental problems is to a large extent an application of the general principles of public goods and of externalities. This is just the approach used in the analytic papers in the collection.

The general principles for the management of public goods are easily explained.[2] Since access to them cannot be controlled, by definition, the only decision is how much of the goods to provide. If the good occurs entirely naturally, as do oceanic shipping lanes, again there is no decision to be made. But if it has to be created at some trouble and expense, as does a flood-protection dam or a lighthouse, then no single individual is likely to find providing it worthwhile, since he cannot enforce a charge for its services. A social decision therefore has to be made about whether to provide the good and, if so, on what scale. The theory of public goods is concerned with those decisions.

To be concrete, consider a community consisting of three individuals, A, B, and C. They are all bothered by the smoke from the power plant that supplies their electricity, and the public good in question is a smoke-precipitator that might be installed in the power plant's stack. None of them is willing to pay for the precipitator singlehandedly, but each would accept a moderate increase in his electric bill (or taxes) to defray the expense. Whether a precipitator should be installed and, if so, how large a one depend on its cost and on its value to each of the beneficiaries.

The relevant data are shown in Figure 1. The size of the precipitator, measured in terms of the percentage of smoke removed from the discharge plume, is plotted horizontally; the vertical scale is a scale of dollars. The line labeled $P$ is the marginal-cost curve. The smallest conceivable precipitator costs $4 per year to own and operate, and would achieve virtually no reduction in smoke emission. If 20 percent of the smoke is being precipitated, it would cost an additional $7.50 a year to raise the proportion to 21 percent. If 80 percent is precipitated, it would cost an extra $18 a year to raise the proportion to 81 percent.

2. Not so easily applied. But of that, more later.

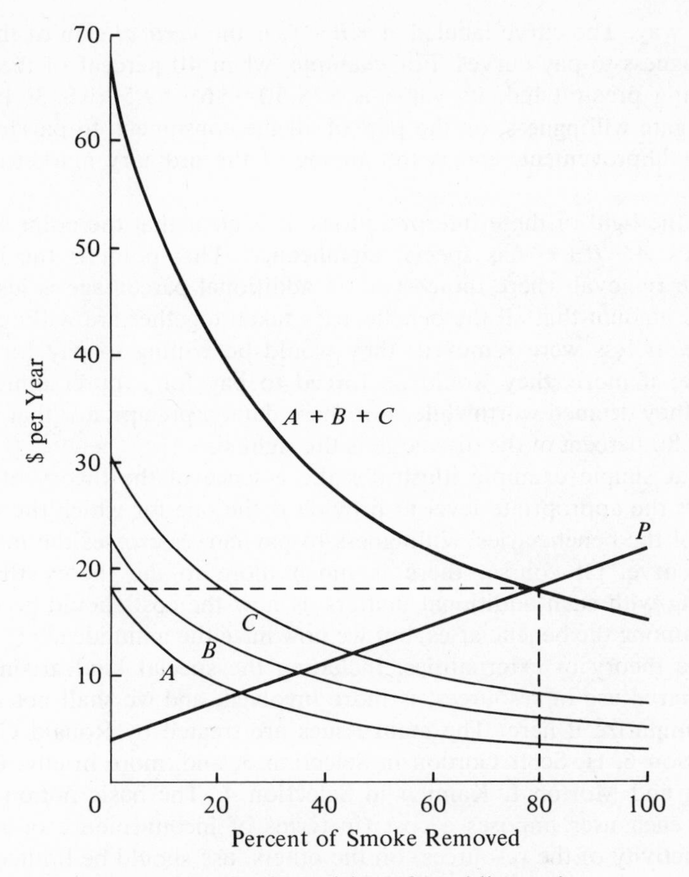


FIG. 1. Optimal provision of a public good.

The other points are similarly interpreted. This line is effectually the supply curve for smoke removal. For example, if the consumers were willing to pay $180 for each percentage point of smoke removed, the power plant would find it worthwhile to remove 80 percent of its smoke from its discharges.

The curves labeled *A, B,* and *C* are willingness-to-pay curves, one for each beneficiary. The curve for A shows how much he would be willing to pay for incremental improvements in the atmosphere. If only 10 percent of the smoke were being removed, he would be willing to pay $12 a year to have an additional 1 percent removed, and so on. If A were able to buy smoke removal as an individual, this would be his demand curve for it, telling how much he would buy at each stated price. Of course, he cannot do that, since the precipitator is a public good, and its services cannot be supplied to him without providing the same amount to others.

The willingness-to-pay curves for B and C are interpreted in the same way. The curve labeled $A+B+C$ is the *vertical* sum of the three willingness-to-pay curves. For example, when 40 percent of the smoke is being precipitated, its value is \$28.50 = \$6 + \$9.50 + \$13. It is the aggregate willingness, on the part of all the consumers, to pay for additional improvement, and is the analog of the ordinary market-demand curve.

In the light of these interpretations, it is clear that the point where $P$ crosses $A+B+C$ has special significance. That point is the level of smoke removal where the cost of an additional percentage is just equal to the amount that all the beneficiaries taken together are willing to pay for it. If less were removed, they would be willing to pay for an increase; if more, they would be forced to pay for a purer atmosphere than they deemed worthwhile. For these data, a precipitator that will remove 80 percent of the discharge is the right size.

That simple example illustrates the essence of the theory of public goods: the appropriate level to provide is the one for which the vertical sum of the beneficiaries' willingness-to-pay curves crosses the marginal-cost curve. Of course, there is much more to the theory that this, dealing with such additional matters as how the cost should be distributed among the beneficiaries, but we now have the main idea.[3,4]

The theory of externalities, including the special kind arising from the shared use of resources, is more involved, and we shall not attempt to summarize it here. The main issues are treated by Ronald Coase in Selection 6, H. Scott Gordon in Selection 5, and, more briefly, Otto A. Davis and Morton I. Kamien in Selection 4. The basic notion is that since each user imposes a cost (in terms of inconvenience or reduced productivity of the resources) on the others, use should be limited to the point where the utility or benefit to an additional user just counterbalances the total cost that he imposes on all the others.

The main point of these concepts and theories for us is that they explain why environmental resources, which partake of these difficulties, tend to be overused, misused, and abused unless special measures are taken. They show that each individual's incentives induce him to use environmental resources more heavily and to contribute less to protecting them than is socially desirable. It should be emphasized that no malign intent or even blameworthy negligence is involved. It would be silly to stay home from Coney Island just because our presence would make

3. For brief and clear discussions of the theory of public-goods provision, see R. A. Musgrave, *The Theory of Public Finance* (New York: McGraw-Hill Book Co., 1959), pp. 74–78 or L. Johansen, *Public Economics* (Chicago: Rand-McNally, 1965), pp. 129–140.

4. The obvious ideal would be to charge each beneficiary in accordance with his willingness to pay, but if this were the policy, the consumers could not be expected to disclose their preferences with sufficient candor. (Contrast this with behavior with respect to ordinary commodities.) For discussion of this and other difficulties, see the recommended further readings.

it even more congested. No individual can reduce the prevalence of smog by putting a smog-suppressor on his car. The only meaningful solutions are collective solutions.

It follows that we have to make collective decisions about the use of the environment, and that one reason for its current state is that we have thus far defaulted on this obligation. We have to decide both what we want to achieve and what means to use for achieving it. Many of the papers, especially those in Part III, deal with this second problem but, since there is no systematic consideration of the first, we shall take it up in the next section.

## Criteria for Economic Performance

So far we have taken it as accepted ground that the environment is not being used properly at present. We are putting too much DDT and phosphates in the water and too much nitrous oxide and lead in the air, taking too much nitrogen out of the soil, despoiling our scenic heritage, and much more. Although it is obvious that what we are doing is wrong, it is by no means obvious what would be right. One thing is certain: we have to go on using the environment, and using it in common. People have to live and congregate somewhere, dispose of waste products, and even use depletable resources. These activities cannot be abolished, though they can be controlled. Controlling them means finding the proper balance between the utility of these activities to the individual and the disutility they impose, via the environment, on others.

It is easier to talk about the proper balance than to define it. Economists have hammered out useful formulations—useful in the sense that they can be applied to specific problems and decisions—only after long and earnest effort. And still, these formulations are not all that might be desired, as we shall see.

The critical difficulty lies in diversity of interest. What is good for Mrs. Goose is not necessarily good for Mr. Gander, and both of them have to be weighed in the proper balance. Reconciling divergent interests is what politics is all about, but economists have something to say, too.

Because of diversity of interest, it is very hard to say which of two policies or situations is better than the other unless, perchance, everyone happens to agree. An appeal to majority rule is a cop-out. It is a way to reach a decision in concrete instances, but it will not serve as a definition of the "right" decision, and no one maintains seriously that it can be relied on to produce the "right" decisions. Our task at the moment is to define, as well as we can, what we mean by a right decision as a foundation for policies for using the environment.

After generations of hard thought, economists have arrived at five criteria for judging policies or decisions. So many criteria are needed because none of them is entirely satisfactory, for one reason or another. Four of the criteria relate to the efficiency of the economic system, the fifth to equity. We shall discuss them all, since all are invoked in the papers that follow.

The four efficiency criteria consist of two pairs. The first, and fundamental, pair relate to the success of the economy in promoting welfare or satisfaction. We shall refer to these as the "utility criteria." The second, and more tractable, pair relate to the success of the economy in producing goods or other physical results. We shall refer to these as the "productivity criteria."

Within each pair there is one fully specific criterion that purports to pick out the best single decision or situation. Unfortunately in both cases the device that singles out the *optimum optimorum* is questionable. So there is a kind of fall-back criterion that identifies the class of decisions or situations in which the best must lie, if there is a best, but does not compare or evaluate the decisions in that class. That is how the four efficiency criteria are related logically. Now we can define them, beginning with the two utility criteria.

## *The Broad Utility Criterion: Pareto Optimality*

The task of an economy is to produce the combination of goods and services that will promote the welfare of the members of the community as much as possible with the resources and production techniques available. The welfare of the community is some resultant of the welfares of its individual members; these welfares are generally called their utilities. The level of utility of each member of the community is presumed to depend on two things: his own consumption of private goods and services, and the environmental conditions to which he is exposed. We can compare the desirability of two or more different modes of operation of the economy by noting the amount of utility that each affords to each member of the community, and drawing conclusions on the basis of these individual utilities.

A simple diagram is frequently used to illustrate these concepts and make their application more concrete. Suppose that a community consists of only two individuals, Mr. Able and Mr. Baker. The entire output of the economy is divided between them, and both are affected by whatever public goods or environmental conditions result from the operation of the economy. Then we can depict the welfare results of the economy on a diagram like Figure 2, where Mr. Able's utility is measured along the horizontal axis and Mr. Baker's is measured vertically. Point *A*, for example, represents the results of some mode of operating the economy in which Able's utility level is 100 units and Baker's is

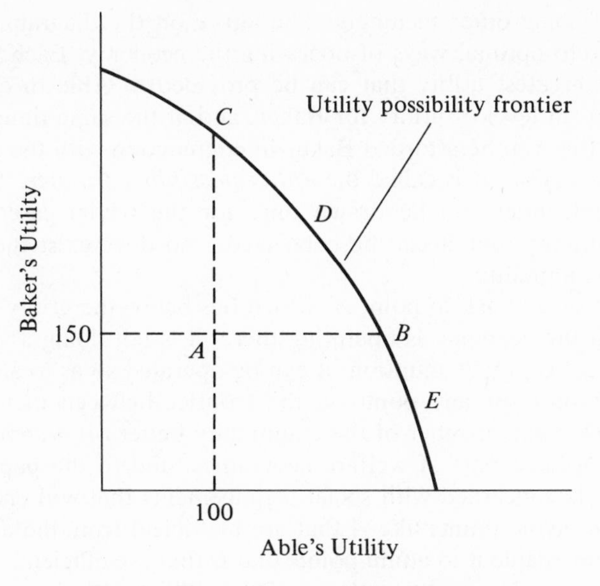


FIG. 2. The utility-possibility frontier.

150 units. It may be possible to change the pattern of economic activity so as to increase Able's welfare without diminishing Baker's. For example, Baker may enjoy playing his stereo equipment at a resounding volume that interferes with the yoga meditations to which Able is devoted. Able might be willing to pay Baker as much as $100 to use earphones instead of stereo speakers (i.e., to turn over to Baker $100 worth of the private goods and services to which he would otherwise be entitled), and Baker might be willing to comply for as little as $50. In this case, if Able paid Baker $50, Baker would lose some utility by resorting to earphones but would gain it back by consuming additional private commodities and be just as well off as before, whereas Able's utility would be increased $100 worth by the reduction in the noise level and would be diminished only $50 worth by his reduced consumption of other commodities, so that he would be better off than before. This result is shown in the figure by point *B*. Point *B* is indubitably a better point for the operation of the economy as a whole: one of the members of the community has benefited and none has suffered. Coase analyzes this type of social arrangement at some length in Selection 6.

Let us suppose that there are no further possibilities for increasing Able's utility without reducing Baker's. Then the output of the economy corresponding to point *B* is said to be *Pareto optimal* or *Pareto efficient*. As a formal definition, the operation of the economy is Pareto optimal if there is no way to change it that will make some member or members better off by increasing their utilities without reducing the utilities of some other members. The curve on the diagram represents all the Pareto-optimal ways of operating the economy. Each point on it shows the greatest utility that can be provided to Able in conjunction with the given level of utility for Baker, and at the same time the greatest utility that can be afforded Baker in conjunction with the given level of utility for Able. It is called the *utility-possibility frontier*. The utility-possibility frontier can be drawn only for the trivial case of a two-man community, but it can be conceived, and does exist theoretically, for every community.

Now let us go back to point *A,* which lies below the utility-possibility frontier. If the economy is operating there, it is not doing as good a job as possible. For, by assumption, it can be operated so as to attain points such as *B* or *C* or any point on the frontier between them that will make at least one member of the community better off without harming any one. A large part of welfare economics, and of the papers in this collection, is concerned with social arrangements that will enable a real economy to avoid points like *A* that are inefficient from the utility point of view, and enable it to attain points like *D* that are efficient.

One method for resolving the conflict of interest between Able and Baker would be for the community to require Baker to use earphones. This would increase Able's utility even more than if he had to pay for silence, but it would reduce Baker's. The result might be point *E,* which is drawn on the utility-possibility frontier. This assumes that this social arrangement is Pareto optimal, as we defined it. But, to illustrate a terminological pitfall, the *change* from *A* to *E* is not Pareto optimal, because a member of the community is harmed thereby.

The comparison between points such as *C* and *E,* or between social arrangements such as free bargaining between Able and Baker as against government regulation, raises important social issues. As we have drawn them, both social devices lead to Pareto-optimal points, but there is good reason to believe that free bargaining is more likely to produce this result than governmental decrees. We shall consider that issue when we take up various practical methods of control. For the moment we note that the Pareto-optimality criterion does not enable us to discriminate between such points. The next criterion does so.

## *The Sharp Utility Criterion: Social Welfare*

We noted above that the welfare of a community is some resultant of the welfares of its individual members. Suppose we could agree on precisely what that resultant was. Then we should have a *social-welfare function,* that is, a rule by which we could evaluate the welfare of an entire community if we knew the welfares, or utilities, of its individual members. For example, the social-welfare function might state that the welfare of the community is the simple sum of the welfares of its individual members. A slightly more complicated social-welfare function would be the sum of the individual utilities minus one-half the difference between the greatest utility and the lowest one.[5] This social-welfare function would pay attention not only to the aggregate of utility but to the amount of inequality in the society. Any real social-welfare function would undoubtedly be very complicated indeed, and in fact, so far as we know, no satisfactory social-welfare function dependent on individual utilities has ever been constructed. There is good reason to believe that the social-welfare function is a philosopher's stone and does not exist in any real society. This matter is pursued at greater length by Dorfman and Jacoby in Selection 14.

In spite of the difficulties with the concept of a social-welfare function, it is very appealing because it is the only way to arrive at clear-cut, unambiguous social evaluations. Look again at our diagram. Point *D* is clearly socially superior to point *A,* since everyone benefits from a movement from *A* to *D*. But is point *E* socially superior to *A?* It is Pareto optimal, while A is not; but, nevertheless, Baker is likely to feel that a regimented society such as *E,* where people are not allowed to play their loud speakers, is a poorer place to live in. And Baker is half of this society. Without a social-welfare function it is not even clear that a Pareto-optimal operation of the economy is superior to an inefficient one; and it is even more impossible to compare the merits of a number of points on the utility-possibility frontier. So some people conclude that since social decisions are made in practice, some social-welfare function must be implicit in the social-decision process.

If there were a social-welfare function, we could portray it like an individual consumer's utility map. This is done in Figure 3. Figure 3 shows the same data as Figure 2 plus a number of "social-indifference curves," distinguished by Roman numerals. The value of the social-welfare function is constant along any social-indifference curve and is higher on higher indifference curves. Since points *A* and *F* are both on indifference curve *I,* they are equally desirable socially according to the social-welfare function depicted. Point *E* is above that social-indifference curve and is superior to either of them. Points *C* and *B* are better still. Point *D* is the most desirable point on the utility-possibility frontier. The points on social-indifference curve *IV* are even more desir-

5. One restriction has to be mentioned. The value of the social-welfare function has to increase whenever the utility of any individual increases, other utilities remaining the same. For example, the following is *not* a social-welfare function: the sum of individual utilities minus twice the difference between the greatest and the lowest. The reason is that this function would actually decrease if the happiest individual became still happier, all others remaining the same. Because of this restriction, it is very dangerous to try to include such considerations as altruism or envy in a social-welfare function. Strange paradoxes can result—for example, a recommended distribution of income in which the most envious people receive the highest incomes and the most generous the lowest.

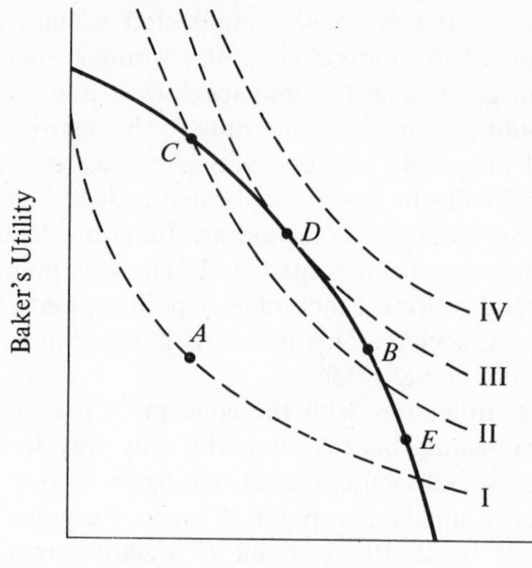


FIG. 3. The utility-possibility frontier with social indifference curves.

able, but they are unattainable. From this diagram one can conclude that according to this social-welfare function it would be better to let Able and Baker bargain than to forbid Baker to use his loud speakers, but that it would be better still to adopt the social arrangements that lead to point *D,* whatever they are. As the example illustrates, if there is a social-welfare function, the social desirability of any two configurations of individual utilities, Pareto optimal or not, can be compared.

But we have already cast doubt on the existence of such functions.[6] Indeed, there is even good reason to doubt that individual utilities can be defined or measured, either in principle or in practice. So criteria are needed for assessing economic performance without appeal to these dubious concepts.[7] The two productive-efficiency criteria are intended to fill this need.

## *The Broad Productivity Criterion*

The broad productivity criterion is a more hardheaded analog of Pareto optimality. Instead of looking for that possible will-o'-the-wisp, utility,

6. For a more profound and decisive critique see Kenneth J. Arrow, *Social Choice and Individual Values,* second edition (New York: John Wiley & Sons, 1963).

7. The Pareto-optimality criterion does not depend on having a social-welfare function or on being able to measure individual utilities. All it requires is that we be able to tell for each individual which situation in any pair he prefers. But even this is excessively demanding for the most practical purposes.

it concentrates on measurable outputs of goods and services and, possibly, environmental quality. Otherwise it is similar. Specifically, an economy is said to be productively efficient if it is producing as much of every good and service (rather than utility) as is technically possible, given the outputs of all other goods and services and the amounts of resources used.

As an example, consider an economy that produces only two commodities (there may be any number of consumers). Suppose, to be concrete, that the two commodities are lumber and parkland. The productive possibilities for the community are shown in Figure 4, with acres of parkland plotted horizontally and output of lumber vertically. The key feature of the diagram is the arc PP′, called the *production-possibility frontier*. It shows the greatest amount of lumber that can be produced in conjunction with a specified acreage of parkland, and the greatest amount of parkland consistent with a specified output of lumber.

Any combination of lumber and parkland that is on or below the production-possibility frontier is technically feasible (for example, the combinations shown by points *A, B,* and *C*), but combinations that lie above the frontier are physically impossible. A productively-efficient economy will produce an output right on the frontier (*B* or *C,* not *A*). A badly managed economy could find itself at *A,* perhaps by turning its most productive stands of timber into a park. If those stands were turned over to lumbering and an equal, less heavily wooded area turned

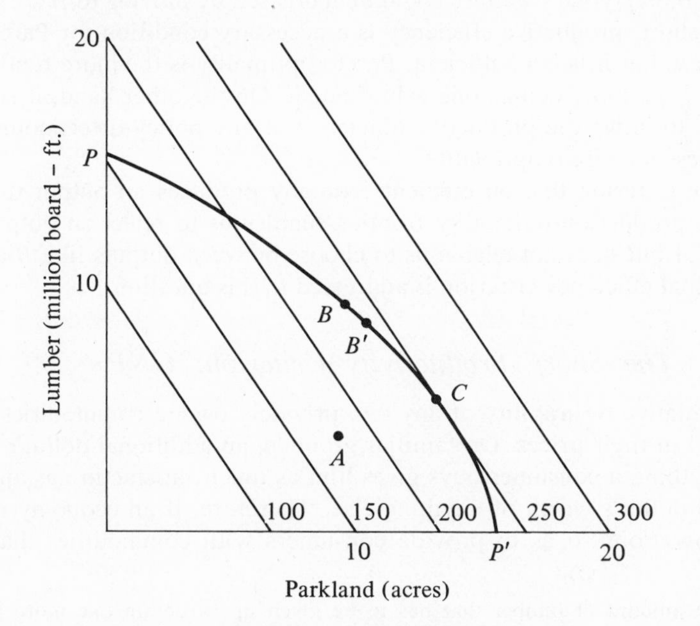


FIG. 4. The production-possibility frontier.

into a park, more lumber could be produced without diminishing the amount of parkland, as at efficient point *B*. An important criterion of economic efficiency, therefore, is whether the rules and procedures of an economy tend to lead it to produce at a point like *A* or at one like *B*.

The relationship between productive efficiency (i.e. producing on the production-possibility frontier) and Pareto optimality is clear, but not simple. On the simple assumption that everyone would like both more lumber and more parkland, an economy cannot be operating Pareto optimally unless it is productively efficient. For if it were at a point like *A*, it would be possible to change its operations so as to produce more lumber without sacrificing any parkland. There would then be more lumber, and more utility, to distribute; and some people could be made better off without harming anyone. On the other hand, it is quite possible for an economy to be productively efficient without being Pareto-optimal. That is, it could produce at point *B* on Figure 4 and yield the utilities of point *A* of Figure 3. This would happen if it produced, however efficiently, the wrong combination of commodities. Notice that, as the figure is drawn, if the economy were producing at *B* it could have about 1.5 acres more parkland by giving up 1.5 million board-feet of lumber (thus moving to point *B′*).[8] If there were a thousand families in the community and each would be glad to reduce its consumption of lumber by 1,000 board-feet or more in order to expand the parks by an acre, then it would not be Pareto optimal for the economy to produce at point *B;* everyone's welfare could be increased by moving to *B′*.

In short, productive efficiency is a necessary condition for Pareto optimality, but it is not sufficient. Pareto optimality is the more fundamental and the more demanding achievement. On the other hand, it is much easier to judge the productive efficiency of any policy or economy than to assess its Pareto optimality.

The criterion that an efficient economy produces an output that lies on its production-possibility frontier enables us to reject an output like point *A* but does not enable us to choose between outputs like *B* and *C*. The final efficiency criterion is addressed to this question.

## *The Sharp Productivity Criterion: GNP*

The relative desirability of any two privately owned commodities is reflected in their prices. On familiar grounds, an additional dollar's worth of anything a consumer buys gives him as much satisfaction as an additional dollar's worth of anything else. Therefore, if an economy revises its operations so as to provide consumers with commodities that they

8. The amount of lumber that has to be given up to obtain one more acre of parkland is known as the *marginal rate of transformation* of lumber for parkland.

value more highly, in dollars-and-cents terms, it is providing them with more utility (is moving closer to the utility-possibility frontier) even though it must reduce the quantities of some commodities to provide more of others. This remark discloses the social significance of prices as guides to economic activity, as well as suggesting how to use prices to choose among the points on the production-possibility frontier. It suggests that the best point on the frontier is the one at which the value of the goods and services produced is as great as possible, using market prices for the private goods and services.[9] This gives rise to the criterion that gross national product, the market value of goods and services produced, should be as great as possible.

This criterion also is illustrated in Figure 4. If lumber is worth $10 per million board-feet and a year's use of parkland is worth $15 per acre, then

$$\text{GNP} = \$10 \times \text{millions of board-feet} + \$15 \times \text{acres of park}.$$

The slanting lines are lines of constant GNP, according to this formula. The output at point *C* is worth $250, and this is greater than the value of output at any other point on or under the production-possibility frontier. According to the GNP-maximization criterion, point *C* should be chosen.

The GNP criterion is the most practicable one of all, in fact the only one that can be applied with much assurance. We have followed a long road to arrive at it, and it is time to reconsider what we have learned. The ideal is the sharp utility criterion, which maximizes the society's social-welfare function. But a society may not have a social-welfare function, and in any event we do not know it, so that is impracticable. Next best is the broad utility criterion, but that one is indecisive among many options and, besides, depends upon unascertainable data for the most part. The GNP criterion does select a decision that satisfies both the broad utility criterion (as far as private goods and services are concerned) and the broad productivity criterion, and this is its main justification. Its weakness is that it may pick out the wrong decision from the point of view of equity. We have yet to discuss equity criteria but still should note here how the GNP criterion may lead us astray.

First, the market prices used in computing the GNP result from the market-demand curves which, in turn, reflect the current, and perhaps inequitable, distribution of income. If there were a better distribution of income there would be different market prices (on the plausible assumption that people in different income brackets have different tastes), and GNP would be maximized at a different point on the production-possibility frontier. Thus, accepting the GNP criterion amounts to en-

9. We discuss below the prices to be used to evaluate non-owned goods and services.

shrining the current distribution of income. By this token, the GNP criterion is valid only if the current distribution of income is regarded as satisfactory; otherwise the point of maximum GNP has no special claim over the other points on the production-possibility frontier.

Second, the position of the economy on the production-possibility frontier not only responds to the distribution of income; it strongly influences it. A movement, say, from point *B* to point *C* will enrich some people at the expense of others. If the change in the income distribution is adverse, the movement may not be desirable, even though it increases GNP and attains Pareto optimality.

Both of those are serious critiques. They entail the idea that the GNP criterion cannot stand on its own but has to be used with a judicious eye to the implications of different measures for the distribution of income. In other words, the GNP criterion has to be paired with some equity criterion, and we have to be prepared to sacrifice some amount of GNP in the interests of equity.

Another frequently-raised objection to the GNP criterion is that it ignores environmental impacts as well as externalities in general. The GNP estimates that are published as indicators of the level of economic activity can be faulted on that ground, but the truth is that GNP is a poor measure of economic welfare for many reasons.[10] On the other hand, the measure of GNP that is used to assist in making environmental decisions is immune to this objection. To see this, as well as to descend from our high and abstract level of discussion, we have to consider how these criteria are used in practice.

## *Application of the Criteria: Benefit-Cost Analysis*

The four criteria express the objectives of environmental improvement or other public programs as matters of general principle. Their real significance emerges when they are applied in practice. We shall now sketch the task of implementation, emphasizing, for concreteness, the GNP criterion.

The context is always the same. There is some specific pollution or environmental problem, and some specific measure or measures are proposed to combat it. The question is, Which of the measures should be adopted, if any, and which specific design should be chosen? To answer the question, fortunately, it is not necessary to compute the entire gross national product but only to estimate the effect that the various alternatives will have on it. The alternative that has the most favorable effect

10. This is not the place to discuss the limitations and proper interpretation of GNP estimates. For one thing, they are "gross"; they ignore depreciation of the capital stock. National income is a better measure of economic welfare, but even it has severe limitations.

(which may be to do nothing) is the one that maximizes GNP as far as this problem is concerned.

The effect of any measure on GNP is the resultant of its favorable effects, called benefits, and its unfavorable effects, called costs. Thus the task of estimation is called benefit-cost analysis. In outline, it goes as follows. For an environmental-protection measure, the benefits are the value of the improvement in the state of the environment. These benefits are sometimes easy to calculate, but most often difficult. (See Selections 24 and 25). For example, if an environmental measure will reduce the concentration of nitrous oxide in the atmosphere around Los Angeles, one of the favorable effects is a reduction in atmospheric damage to crops. That is an easy benefit to measure: since the crops are sold, they have prices, and the value of the crops saved can be computed.[11] Another benefit is a reduction in the incidence of respiratory disease. That is harder to evaluate. One method is to estimate the decrease in the number of work days lost because of respiratory illness. Since those days have a monetary value, so does the saving of them. That device is not very satisfactory, since it ignores pain and suffering, though it is often used in the absence of other "hard" data. One way to incorporate the value of pain and suffering averted is to find out how much people are willing to pay to avoid similar disabilities. The amount that people are willing to pay for antihistamines or allergy-preventives may provide an adequate indication. If we know how many days of illness the measure will avert and how much people are willing to pay to avoid a day of similar illness, we can estimate this benefit.

Those few examples should suffice.[12] One goes through the list of good things that may be expected to result from the measure and, by hook or crook, estimates the values of those good things and adds them up. The result is the total benefit or favorable effect on GNP.

The estimation of the costs is similar but usually easier. The most important costs are generally the diversion of resources from other uses in which they would contribute to monetary GNP. The monetary costs of the resources used by the measure are therefore a good estimate of the gross reduction in GNP that it entails. The resources in question

11. One critically important technicality must be mentioned. Any environmental measure is likely to have favorable effects, such as reducing crop damage, for a number of years. The value of crops saved in a single year (for example) is therefore not an adequate measure of its benefit in this regard. We must estimate the time-stream of benefits, year by year, and then aggregate them by applying a discount factor to each year's benefits and adding up the stream of discounted benefits. Costs, to be mentioned below, also give rise to time-streams and have to be treated in similar fashion. This method for dealing with time-streams is exactly the same as the methods used to evaluate ordinary investments. The selection of the discount factor to be used is very important, but it would be too much of a digression to discuss it here.

12. For further illustration see Selections 24 and 25.

may be used either by the government in administering the measure or building requisite facilities, or by business firms or consumers who have to do costly things in order to comply. For example, the cost of smog-suppressors is the major component of the cost of combating automobile-generated air pollution, and represents a genuine reduction in the value of other goods and services available to consumers. Sometimes there are nonmonetary costs, and these must be estimated by the same devices used for estimating nonmonetary benefits.

When the total benefits and total costs of a particular alternative have been evaluated by these methods, the net benefits can be calculated as total benefits minus total costs. The GNP criterion is satisfied by the alternative whose net benefits are greatest. If total costs exceed total benefits for all the measures under consideration, the recommendation is to do nothing, for the net benefits of doing nothing are zero by definition. From this brief sketch of an elaborate and refined procedure, it is evident that the calculations required to apply the GNP-maximization criterion are difficult and far from precise, but still entirely feasible.

The broad productivity criterion avoids many of the difficulties just described: it does not require the monetary evaluation of benefits, or summing them, or discounting the benefits that accrue in different years. Its shortcoming is that it is too likely to be indecisive. To apply it, we estimate the physical results of each alternative measure in each year of operation, just as before, but stop short of assigning values to the nonmarket costs and benefits. Then we compare these physical results: if one alternative produces more of every kind of benefit in every year than any other, and no more of any kind of cost, then it is the best alternative. But since such a clear-cut and decisive advantage is very unlikely to occur, this criterion has limited usefulness. It can sometimes be used, however, to rule out definitely uneconomical proposals by showing that there are alternatives that dominate them in every respect.

The two utility criteria are entirely different, since they are addressed not to physical results, but directly to individual welfare. Because it is out of the question to consider the welfare of every individual in a society, these criteria have to be applied by aggregating people into broad groups of similarly circumstanced individuals and considering the welfares of representative members of each group. (This is the approach followed in Selection 14.) If a measure advances some group's welfare without harming any group, there is no real question about its advisability. But virtually all real-life proposals advance some people's interests at the expense of others'.[13] That is when a real issue arises. To re-

13. Fervent proponents typically deny this, claiming that their proposals are in the "general interest." The correct response is "That's what *they* say." Sometimes advocates try to deprecate their opponents by asserting that the opponents are acting in "their selfish interests." This is generally true, but if the opponents are members of the society, their selfish interests are legitimate social interests.

solve such conflicts within the framework of the broad utility criterion, economists have developed the "compensation principle." According to this principle, if the advocates of a measure are willing to pay sufficient compensation to the opponents so that the opponents are willing to accept the measure and the compensation, then the measure should be adopted and the compensation paid. The justification is a strict application of Pareto optimality: everyone will be better off, the advocates because they have their measure without excessive cost, and the opponents because they are adequately compensated. The result is a movement toward the utility-possibility frontier.

In practice, the amounts that the gainers would be willing to contribute and that the losers would have to be paid can be estimated by the same methods used to evaluate results for the GNP criterion. The practical difficulty is that there is usually no way to arrange for the compensatory payments. (However, it is often possible to compensate people injured by environmental measures.) To meet this objection it has been urged, but less convincingly, that the compensation test should be applied even when the compensation will not actually be paid. The argument is that if the gains to the gainers are greater than the losses to the losers—which is what the compensation test reveals—then the overall welfare of the community is advanced by the measure. In effect, this is analogous to the GNP criterion, except that the sum to be maximized is the sum of utilities instead of the sum of commodity values. But the notion that individual utilities can be compared or summed was abandoned long ago.

Another way to view the compensation principle without compensation is to regard it as an application of the social-welfare criterion in which everyone's welfare is counted equally and the effect of a proposal on a man's welfare is measured by either the amount he is willing to contribute to obtain it or the amount he must be indemnified to agree to it. We generally endorse that everyone's welfare does count equally, but as generally deny that monetary payments are a valid indication of welfare changes in a world of wide disparities in income. So it is generally adjudged that the compensation principle doesn't stand up unless the compensation is to be paid.

There is, finally, the social-welfare criterion. It is easy to see how to apply it if the social-welfare function is known, but this is never the case.

The upshot of this list is that the GNP-maximization criterion, with all its shortcomings, is the only one that is much used or of practical importance. The others serve chiefly to justify it and to illuminate its significance and limitations. Perhaps it is well to repeat that the GNP criterion, as applied, does incorporate environmental impacts. Its defect in this regard is that the valuations placed on environmental impacts, which are obtained by a difficult process of estimation, are likely to be crude and disputable.

## *Equity, the Distribution of Welfare*

Efficiency and equity are the two desiderata of economic performance. We have now defined efficiency at great length, for that is where economists are skilled. But they are no more skilled than anyone else when it comes to equity, a concept that good men have been trying to define since before Aristotle, and with only modest success.

The lack of an adequate definition of equity is a great embarrassment to economists,[14] since their efficiency criteria, except for social-welfare maximization,[15] are not decisive without one. We have already encountered the problem in the Able-Baker example. Any resolution of their conflict will result in some point in Figure 2. Only points on the utility-possibility frontier are efficient, but on the frontier the better the point is for Able, the worse for Baker. Which point on the utility-possibility frontier should be chosen? Efficiency criteria cannot answer that question, for the issue is equity or relative deserts, and we have no touchstone for answering it.[16]

The same kinds of questions always arise in more realistic environmental decisions, for several reasons. First, there is the problem of allocating the costs of environmental protection fairly. Everyone favors a better environment, but few people want to see it attained largely at their expense. Environmental measures are expensive: they result in higher taxes, higher costs of production for industries required to introduce new methods of waste disposal or nonpolluting kinds of raw materials, higher prices for the consumers who use those products. The question of equity is, Who shall bear what share of those costs?

Second, an environmental measure inevitably changes the allocation of welfare; it is inherently redistributive. This is so partly because some people care more than others about any particular aspect of the environment, so that a protection measure redistributes society's resources in favor of the advocates or neighbors who are most affected. More basically, it is so because the environment consists of natural resources to which people are accustomed to have free access. But it is exhaustible, and the exercise of each man's right is likely to interfere with others'. The question of whose right shall prevail—Able's or Baker's, bird-watchers' or snowmobilers', swimmers' or tanners' or shoe-wearer's—is a question of equity. Any protection measure will constrict or revoke some people's long-standing right to use the environment as they see fit—and thus reduce their welfare for the benefit of other people.

14. This is probably the least of the evil consequences of this lack.
15. The social-welfare criterion incorporates equity along with everything else, which is probably why it is so impracticable.
16. This very example has been analyzed at length by R. B. Braithwaite in *Theory of Games as a Tool for the Moral Philosopher* (Cambridge: Cambridge University Press, 1955).

So the question of fair sharing of society's resources, including the environment, is an inexorable part of any environmental decision. Also, as we have noted, a very difficult part.

There aren't any generally accepted dicta. In each case the conflict of interest seems to be resolved ad hoc, in the light of the prevailing climate of public opinion and the cries of the advocates and opponents. The general tendency is to conceal the fact that there is a conflict of interest. Proponents staunchly maintain that their measure is in everyone's interest—but the very existence of opponents casts doubt on this claim. Even the papers in this collection acquiesce in this conspiracy of silence; the questions of equity and distribution are scarcely mentioned. In a few of the papers (say, Selection 6) its pertinence is clearly denied.

We are thus in the uncomfortable position of knowing what is efficient but not what is equitable, but having to know both to arrive at a well-founded policy. To avoid this impasse, economists used to recommend that equity and efficiency be considered in isolation from each other. In other words, that environmental and other social questions be decided by applying one of the efficiency criteria, and that any resultant inequities be rectified by instituting suitable taxes and transfer payments. That disjunction is now regarded as unworkable, however, for two reasons. On the practical side it is usually impossible to discover or legislate the required taxes and subsidies. On the theoretical side, the taxes and subsidies, if arranged, would be almost sure to upset the economic efficiency of the original proposal. This is true because taxes and subsidies almost inevitably distort the price structure and interfere with the efficient allocation of resources.[17] So when we make an environmental decision, we should be aware that we are going to have to live with its distributional consequences. The equity issue has to be faced from the outset.

## Controlling the Use of the Environment

Like the lilies of the field, economists neither toil nor spin. Their practical task, rather, is to design social instruments and institutions that will guide toiling, spinning, and all other economic activities in accordance with the criteria that have just been described. The entire apparatus of economic analysis has been devised with this end in view: to help us predict how people will react to different possible social arrange-

17. For example, sales taxes and taxes on commodities drive a wedge between the prices of taxed commodities and the marginal cost of the resources needed to produce them. Efficiency requires that the price be equal to this marginal cost. However, a tax on a polluting commodity may actually be beneficial, as we shall see below.

ments and incentives so that we can propose arrangements that will elicit desirable behavior.

We have seen why ordinary economic market arrangements, which are efficient guides for the use of private resources, do not lead to the efficient use of environmental resources. New instruments have to be devised. In this section we shall discuss the principal options in the light of the criteria of economic performance.

Let us consider a standard environmental problem. There is some environmental resource—an airshed, a beach, or a waterway. It is used jointly by a substantial number of users—consumers or producers or both. The various users interfere with each other, that is to say, the more it is used by some, at least, of the users the less productive or serviceable it is for the others. Some method is needed for controlling the use of the resource in the general interest.

There are three broad classes of methods. One is the present state of affairs, the status quo, in which each potential user has free and unrestricted access to the resource. We already know that this policy is likely to be highly unsatisfactory, and we consider it only a basis for comparison. Any alternative policy must do better than this one. The second class is governmental regulation. It is a very broad class. It includes prohibition of certain uses—for example, open incineration. Also, limiting certain uses, as by setting limits to the amounts of effluent that factories can discharge into a stream. It also includes prescribing protective measures such as installing smog-suppressors on cars or subjecting all liquid effluent to "secondary treatment." All are similar in invoking the force of legal sanctions to require or prohibit specified acts that affect the resource. The final class is economic or financial inducement: imposing taxes or charges, or offering subsidies to discourage or encourage different ways of using the resource, and leaving the users free to decide how best to respond to the inducement. This, too, is a very broad class, and we shall not attempt to list the variants.

The basic social problem is to choose among these methods of control and, indeed, to choose the specific methods within the class selected. The principles of choice are the criteria discussed in the last section—in fact, that is the reason we discussed them. In this section we shall pull the threads of the argument together by seeing how those principles of choice apply to the broad classes of controls that we have just defined. Three classes of controls times five criteria give fifteen cases to be discussed. That would be a tedious and repetitious task. Rather than undertake it exhaustively, we shall pick and choose, concentrating on the cases that are most illuminating.

*The Status Quo* The status quo exposes the resource to all the dangers and abuses we have discussed earlier. Each potential user is invited to use the resource at his convenience and at no cost, disregarding the effects of his use on other firms and people. In fact, he has no way of knowing what those effects are. Our criteria should show that this situation leads to an inefficient use of the resource, and they do.

It is best to begin with the GNP criterion. To be concrete, consider the power-plant example of Figure 1, for it is typical. In the status quo there is no inducement for the power plant to install a precipitator and, since it is costly, the plant will not do so. But this decision foregoes a potential increase in GNP. For example, if a small precipitator, capable of removing 20 percent of the smoke, were installed, the consumers would be willing to pay at least $820 a year for its services ($20 \times \$41 = \$820$). Actually, this is an understatement. It assumes that the value of the precipitator to the consumers is the percentage of smoke removed multiplied by the amount they would be willing to pay for an additional 1 percent. In fact, the value is the area under the aggregate-demand curve up to the percentage removed, or, in this case, $1068 per year. The marginal-cost curve shows that the annual cost of a 20-percent-removal precipitator is $115. So GNP could be increased by $953 by installing such a precipitator. Our previous analysis showed that even this is far from optimal. An 80-percent-removal precipitator would increase GNP by $1760 as compared with no precipitator at all.

This instance follows the general rule that, in the status quo, individuals act so as to reduce GNP by ignoring the effects of their actions on each other.

The GNP criterion is closely related to the broad utility criterion, the one that requires the economy to generate a set of individual utilities that lies on the utility-possibility frontier. In fact, that is the basic justification for employing the GNP criterion. If the GNP criterion is not satisfied, it is possible to rearrange the pattern of production so as to provide consumers with things that they value more highly than those being currently produced. Since consumers' valuations measure the amounts of utility they derive, this will make it possible to increase the utilities enjoyed by some consumers without reducing those of any.

In the present example everyone's utility would be increased if the consumers were assessed $39 each ($= \$115 / 3$) and the proceeds were used to defray the cost of a 20-percent precipitator. It would be even better to assess each consumer his marginal willingness to pay for an 80-percent precipitator, if that could be determined, and to use the proceeds to cover the cost of a precipitator of that size. The status-quo policy clearly fails according to this criterion.

It fails also, and automatically, according to any social-welfare criterion. A social-welfare criterion simply picks out the most preferred point on the utility-possibility frontier. Since we have just found that the status-quo policy does not attain any point on that frontier, it cannot satisfy any social-welfare criterion.

The remaining efficiency criterion is productive efficiency, the requirement that it not be possible to increase the output of any desired good or service without reducing the output of some other one. In the example of the power plant, this requirement appears to be met. It is not possible, within the bounds of that example, to reduce the amount of smoke emission (i.e., increase the quality of the atmosphere) without either reducing the amount of power generated or devoting to smoke precipitation resources that could be used to produce other valued outputs. That is so in the current instance, but the status quo policy does not assure that even this mild version of efficiency will be attained. Gordon presents a contrary instance in Selection 5.

All in all, the status-quo policy shows up very badly, as we expected. It does so because it induces producers and consumers to ignore the adverse consequences of their activities upon the usefulness of the environment to others. The other two policies try, in different ways, to modify this behavior. We turn to them now.

*Government Regulation* In principle the government can determine the social optimum using any of the criteria, and ordain it. In practice it can do no such thing. Governmental efforts to control the use of environmental resources are beset by a number of difficulties that we can classify as (1) inadequacy of information, (2) crudeness of regulatory instruments, (3) problems of enforcement, and (4) haphazardness of burdens. In assessing the efficiency of governmental regulation in the light of the criteria, then, it is irrelevant to analyze performance under ideal conditions. The task is to consider how the practical difficulties of regulation in a complex world affect performance.

The example that we used above is a bit too simple to illustrate all the perplexities that have to be confronted, but we can base our discussion on it, enriching it as needed.

The first decision is to establish what is called an environmental standard, that is, a target level of environmental quality. We have already ascertained that the level that maximizes GNP corresponds to an 80-percent reduction from the current level of smoke emission, but the government cannot know that. It does not know the marginal-cost curve for smoke reduction, for only the power company knows its operations intimately enough to make such estimates. Still less can it know the willingness-to-pay curves of the affected population. So it estimates these data as best it can by benefit-cost analysis. Its decision is likely to be influenced by political considerations such as those discussed in Selection 14. It may settle on a 60-percent reduction, or a 90-percent. In either case it will fail to maximize GNP and therefore fail in the light of the utility criteria. But it is most unlikely to miss nearly as badly as the status-quo policy.

Two other kinds of ignorance reduce the efficiency of governmental regulations. The first kind depends on the form of regulation chosen.

There are essentially two options: the government may simply require the power plant to reduce its emissions by, say, 80 percent, or it may prescribe some specific measure such as installing a precipitator of specified size. The trouble with the first option is that it is difficult to enforce—it requires continuous monitoring, and provisions for dealing with occasional, and perhaps excusable, infractions. The difficulty with the second option is that it is likely to be inefficient. Only the power company can know the most economical way to reduce its emissions—precipitator, change in the type of fuel, change in operating temperatures, modernization of the boilers, or whatnot. On the whole, efficiency recommends that detailed decisions be made by people with the best access to information, which argues for the first option; but difficulties of enforcement may outweigh this consideration. The second form of ignorance enters when there are several sources of pollution, as is always true except in simplified examples. Then even the broad productivity criterion requires that emissions be controlled in such a way that the marginal cost of further reductions be the same for all sources of pollution. Requiring every polluter to cut back by the same 80 percent is sure to be inefficient. But the government has no way of knowing the marginal-cost curves of the individual emitters.

These inefficiencies are exacerbated by the crudeness of regulatory instruments. If the government orders a cutback, it is very hard not to impose the same level of reduction on all sources, and we know that to be inefficient. If the government chooses the route of prescribing antipollution measures, it has virtually no choice but to issue a blanket regulation binding on all sources of pollution, and any one measure is sure to be inappropriate for some of the sources. The resultant inefficiencies are not merely theoretical possibilities. One study found that it would cost about $11 million a year to bring water quality in the Delaware Estuary up to a reasonable standard by means of blanket requirements, of which at least 38 percent could be saved by tailoring the requirements to the individual circumstances of each polluter.[18]

There is, finally, the problem of equity. Governmental regulation imposes burdens that have little relationship to the benefits received from the resultant improvement in the environment—burdens on consumers, on workers who may lose their jobs, on stockholders, and on others. It is almost inevitable that some people's utilities will be diminished as a result of the regulation, even substantially, and those people are justified in protesting that they should not be made to bear the cost of improving the environment for the benefit of others. The incidence of the costs of regulation is hard to foresee and still harder to rectify.

The most severe drawback of governmental regulation is its inevita-

18. Edwin L. Johnson, "A Study in the Economics of Water Quality Management," *Water Resources Research, vol. 3* (Second Quarter, 1967), pp. 291–306.

ble and substantial inefficiency in terms of any of the four criteria. The use of economic incentives is increasingly advocated as a method of environmental control that is largely free of this defect. We turn to it next.

*Economic Inducements* We shall concentrate on the use of effluent charges, taxes, and the like, and shall neglect subsidies, which are not as widely advocated and entail special problems. A wide variety of tax-like devices for controlling the use of environmental resources is available, and they all share the characteristic of striving for efficiency by remanding a maximum amount of discretion to the individual user or polluter. In effect they are surrogates for the ordinary market prices that control the use of privately owned goods and resources, and are designed to induce the same kind of efficiency in resource allocation. But artificial markets are not real markets, and do not achieve quite the same result, as we shall see.

Taxes and charges assist in controlling the use of environmental resources by bringing home to the user the costs that he imposes on other users, which he would otherwise disregard. The standard phrase is "they internalize the externalities." The variety of taxes and charges is limited only by the ingenuity of the controlling officials and their advisors, and we shall mention only a few of the leading proposals.

The leading contender is probably effluent charges, strongly recommended by Larry E. Ruff (Selection 1) and Allen V. Kneese and Blair T. Bower (Selection 9) among others. We can illustrate this device by using our power-plant example again.

If the pollution-control authorities assess a charge on the power plant in proportion to the amount of smoke it emits, the electric company will have an inducement to reduce its discharges, by installing a precipitator or by more economical means if available. Thus if the pretax level of emission is 100 units and a tax of $7.50 per unit is levied, then, according to Figure 1, it will be worthwhile for the power plant to reduce its emissions by 20 units. Furthermore, it will be induced to do this in the most efficient way. If there are several sources all subject to the same tax, they will all react similarly. They will not all reduce their emissions by the same 20 units, they will do even better: each will reduce its emissions to the point where the cost of a further unit reduction is $7.50, which is just what economic efficiency requires. No regulatory device is flexible enough to achieve this result.

In terms of efficiency criteria, a unit tax or charge on emissions will induce an improvement in the quality of the atmosphere and will attain this improvement in the most economical way possible. It will thus lead to a point on the production-possibility frontier, satisfying the broad productivity criterion. But this is not enough to satisfy the GNP criterion. That would require, in addition, that the tax be set at the level that will induce emitters to cut back their discharges to the point where the amount that sufferers would be willing to pay for a further reduction is just equal to the tax. In terms of our diagram, the GNP-maximizing tax would be $18 a unit of emission. This is a complicated requirement to state, and even more complicated to satisfy, since it requires estimating both the amount of reduction that any given tax would elicit and the amount that sufferers would be willing to pay for further reduction. Kneese and Bower argue, in Selection 9, that the correct unit tax can be found by trial and error. It is true that a tax found to be erroneous can be changed readily, but the responses may not be so flexible. Once the power plant has installed the precipitator appropriate for the $7.50 tax, enlargement or reduction of the device is likely to be prohibitively expensive if the tax rate should be changed. We can take for granted that the tax rate will be more or less in error and that the GNP criterion will not be satisfied fully. To the extent that the GNP criterion is violated, so will be the broad utility criterion, of which it is the most concrete indicator.[19]

If the amount of the tax were set correctly, the GNP criterion and the broad utility criterion would be satisfied. A number of proposals attempt to achieve this result by having the tax set by market-like forces instead of by administrative decision.[20] If, for example, rights to discharge pollutants (or go to the beach) were auctioned off, potential polluters would bid for the rights, their bids revealing the social value of making the discharges, and various consumer groups or representatives could also bid with the intent of holding the rights they buy off the market. The correct number of rights to sell—i.e., the correct amount of pollution to permit—is the one where the marginal usefulness as revealed by dischargers' bids is equal to marginal obnoxiousness as disclosed by the consumer groups' bids. If such a scheme were practical (it is not, for various reasons), it would achieve the GNP criterion. Dales's proposal is less ambitious, but similarly motivated.

We cannot here discuss the many possible variant tax schemes: preset

19. A different example of the very same principle: A million people go to Jones Beach on a hot Sunday if the admission charge is 50¢. At that level of congestion, each person annoys each other person to an extent valued at 0.00005¢ (i.e., each person would be willing to pay one cent to have 20,000 other users disport themselves elsewhere). If the admission charge were reduced to 20¢, an additional 500,000 people would come. We now argue that this would be socially harmful. Each of the additional users is willing to pay less than 50¢ but at least 20¢ for his visit, say 35¢ on the average. So the additional use is worth about $175,000. On the other hand, the new throng discommodes each of the original users to the amount of 500,000 / 20,000 = 25¢, or $250,000 in toto. The social costs outweigh the social gains according to the GNP criterion. If the charge were rescinded entirely, the loss would be still greater, though the recission might be politically popular. Moral: It doesn't always pay to give things away. The figures were cooked up, of course, to give this result: $50¢ = 10^6 \times 0.000{,}05¢$ = number of users × unit congestion cost.

20. Dales has made the leading proposal along these lines. See Selection 10.

taxes versus market-determined taxes, taxes on effluents versus taxes on pollution-generating products, and so on. Nor can we discuss the very important question, bearing on the equity criterion, of what should be done with the taxes collected—in particular, whether they should be used to compensate the sufferers from the pollution permitted.[21] We have said enough to indicate how the effluent-charge approach qualifies in the light of the fundamental criteria. On the whole, it is the only approach that satisfies the productivity criteria. Economists, accordingly, are strongly attracted to it.

---

Pollution is a dirty word. One's automatic response on hearing it is to say that it must be avoided at all costs. But that automatic response, by mentioning costs, triggers a more considered response. It makes one wonder what the costs are, and whether they might not be excessive.

And indeed they might be. Pollution is a by-product of living. Sewage systems pollute streams, but they also are essential for controlling communicable diseases. Pesticides are notorious pollutants, but, particularly in poor countries, they are essential in averting famines. Of course, not all polluting activities have such high payoffs as these examples, but these instances make the point that some pollution is inevitable, and more than the barest minimum may be socially worthwhile. We are not about to stop polluting at all costs, for the cost would be removing ourselves and our depredations from the planet.

This being so, the genuine questions are, How much pollution and other disturbance of the environment shall we indulge in? What kinds? and What controls shall be imposed? To answer these questions sensibly, we have to assess in the light of reasonable criteria the benefits and the costs of activities that impose on the environment, and we have to judge the probable consequences of proposed protective and control measures. The proper use of environmental resources is more a matter of economics than of morals.

This is the point of view taken throughout this volume. The introduction has presented the best-established criteria for evaluating economic policies, along with some background for applying them. The twenty-six selections in the volume will move on from this background to analyze various aspects of the environmental problem, and will examine some specific proposals for dealing with it.

21. Ralph Turvey treats some aspects of this in Selection 7.

# I

# *AN OVERVIEW*

This opening section provides perspective on the role that economics can play in finding rational solutions to environmental problems. It also supplies some necessary background on the nature and extent of these problems and on the relationships among measures for contending with them.

In the very first paper Larry E. Ruff announces the main theme that will be repeated and elaborated in the rest of the volume: Under normal circumstances, people have to pay for the resources and commodities they use, and are thereby led to use only the amounts that will yield results, or benefits, that are worth the cost. But we do not have to pay for using the environment—for venting fumes and noise into the atmosphere, for pouring pollutants into the lakes and rivers, and so on. So we treat the environment as if it were a free good and clutter it with our wastes even when the benefits or savings to ourselves are far less than the "external" costs we impose on other people. The price system fails when there are "external effects" because we don't apply it.

The most straightforward corrective is to prohibit or regulate abuses of the environment by government decree. Ruff, along with most economists, argues that instead the government should "put a price on pollution" and thus incorporate the environment into the normal operation of the price system. His point is that charging for the use of the environment will lead to the same relatively efficient use of its capacities that charging for ordinary commodities induces for their use. Government regulation, in contrast, would protect the environment at an excessive cost in terms of the production of ordinary goods and services.

This issue—the sensitiveness of price allocation versus the bluntness of government regulation—will arise repeatedly throughout the volume. But setting and charging prices for the use of the environment is more easily said than done. Later papers will examine the complexities hidden in this ostensibly simple solution.

The second paper, by Allen V. Kneese, is a broad, factual survey of the most pressing aspect of the environmental problem: the degradation of the environment by inserting into it large volumes of the waste products that result from human activity. Kneese emphasizes that the generation of wastes is inevitable. Unless they are recycled, all have to go somewhere. He summarizes the magnitudes of the major waste streams and the current state of knowledge about their deleterious consequences. Finally, he points out that the degradation of the land, air, and water components of the environment are not separate problems but are closely interconnected, since the by-products of economic and other human activities that are not received by one medium must unavoidably be deposited in some other one.

Clifford S. Russell and Hans H. Landsberg deal with worldwide and international aspects of the pollution problem. All nations of the world help contaminate the oceans and the atmosphere, but there are many more subtle ways in which one nation's activities impose environmental damage on other nations. These less obvious mechanisms, which are Russell and Landsberg's main concern, range from the destruction of the world's historic and artistic, as well as natural, heritage to the "exporting" of pollution in which more affluent countries sometimes engage when they attempt to restrict the extent of damage inflicted on their own local environments.

# 1

LARRY E. RUFF

# *The Economic Common Sense of Pollution*

*Larry E. Ruff is Chief of the Systems Evaluation Branch of the Implementation Research Division of the Environmental Protection Agency.*

We are going to make very little real progress in solving the problem of pollution until we recognize it for what, primarily, it is: an economic problem, which must be understood in economic terms. Of course, there are *noneconomic* aspects of pollution, as there are with all economic problems, but all too often, such secondary matters dominate discussion. Engineers, for example, are certain that pollution will vanish once they find the magic gadget or power source. Politicians keep trying to find the right kind of bureaucracy; and bureaucrats maintain an unending search for the correct set of rules and regulations. Those who are above such vulgar pursuits pin their hopes on a moral regeneration or social revolution, apparently in the belief that saints and socialists have no garbage to dispose of. But as important as technology, politics, law, and ethics are to the pollution question, all such approaches are bound to have disappointing results, for they ignore the primary fact that pollution is an economic problem.

Before developing an economic analysis of pullution, however, it is necessary to dispose of some popular myths.

First, pollution is not new. Spanish explorers landing in the sixteenth century noted that smoke from Indian campfires hung in the air of the Los Angeles basin, trapped by what is now called the inversion layer. Before the first century B.C., the drinking waters of Rome were becoming polluted.

Second, most pollution is not due to affluence, despite the current popularity of this notion. In India, the pollution runs in the streets, and advice against drinking the water in exotic lands is often well taken. Nor can pollution be blamed on the self-seeking activities of greedy capitalists. Once-beautiful rivers and lakes which are now open sewers and cesspools can be found in the Soviet Union as well as in the United States, and some of the world's dirtiest air hangs over cities in Eastern Europe, which are neither capitalist nor affluent. In many ways, indeed, it is much more difficult to do anything about pollution in noncapitalist societies. In the Soviet Union, there is no way for the public to become outraged or to exert any pressure, and the polluters and the courts there work for the same people, who often decide that clean air and water, like good clothing, are low on their list of social priorities.

In fact, it seems probable that affluence, technology, and slow-moving, inefficient democracy will turn out to be the cure more than the cause of pollution. After all, only an affluent, technological society can afford such luxuries as moon trips, three-day weekends, and clean water, although even our society may not be able to afford them all; and only in a democracy can the people hope to have any real influence on the choice among such alternatives.

What *is* new about pollution is what might be called the *problem* of pollution. Many unpleasant phenomena—poverty, genetic defects, hurricanes—have existed forever without being considered problems; they are, or were, considered to be facts of life, like gravity and death, and a mature person simply adjusted to them. Such phenomena become problems only when it begins to appear that something can and should be done about them. It is evident that pollution had advanced to the problem stage. Now the question is what can and should be done?

Most discussions of the pollution problem begin with some startling facts: Did you know that 15,000 tons of filth are dumped into the air of Los Angeles County every day? But by themselves, such facts are meaningless, if only because there is no way to know whether 15,000 tons is a lot or a little. It is much more important for clear thinking about the pollution problem to understand a few economic concepts than to learn a lot of sensational-sounding numbers.

## Marginalism

One of the most fundamental economic ideas is that of *marginalism,* which entered economic theory when economists became aware of the differential calculus in the 19th century and used it to formulate economic problems as problems of "maximization." The standard economic problem came to be viewed as that of finding a level of operation of some activity which would maximize the net gain from that activity, where the net gain is the difference between the benefits and the costs of the activity. As the level of activity increases, both benefits and costs will increase; but because of diminishing returns, costs will increase faster than benefits. When a certain level of the activity is reached, any further expansion increases costs more than benefits. At this "optimal" level, "marginal cost"—or the cost of expanding the activity—equals "marginal benefit," or the benefit from expanding the activity. Further expansion would cost more than it is worth, and reduction in the activity would reduce benefits more than it would save costs. The net gain from the activity is said to be maximized at this point.

This principle is so simple that it is almost embarrassing to admit it is the cornerstone of economics. Yet intelligent men often ignore it in discussion of public issues. Educators, for example, often suggest that, if it is better to be literate than illiterate, there is no logical stopping point in supporting education. Or scientists have pointed out that the benefits derived from "science" obviously exceed the costs and then have proceeded to infer that their particular project should be supported. The correct comparison, of course, is between *additional* benefits created by the proposed activity and the *additional* costs incurred.

The application of marginalism to questions of pollution is simple enough conceptually. The difficult part lies in estimating the cost and benefits functions, a question to which I shall return. But several important qualitative points can be made immediately. The first is that the choice facing a rational society is *not* between clean air and dirty air, or between clear water and polluted water, but rather between various *levels* of dirt and pollution. The aim must be to find that level of pollution abatement where the costs of further abatement begin to exceed the benefits.

The second point is that the optimal combination of pollution control methods is going to be a very complex affair. Such steps as demanding a 10 per cent reduction in pollution from all sources, without considering the relative difficulties and costs of the reduction, will certainly be an inefficient approach. Where it is less costly to reduce pollution, we want a greater reduction, to a point where an additional dollar spent on control anywhere yields the same reduction in pollution levels.

## Markets, Efficiency, and Equity

A second basic economic concept is the idea—or the ideal—of the self-regulating economic system. Adam Smith illustrated this ideal with the example of bread in London: the uncoordinated, selfish actions of many people—farmer, miller, shipper, baker, grocer—provide bread for the city dweller, without any central control and at the lowest possible cost. Pure self-interest, guided only by the famous "invisible hand" of competition, organizes the economy efficiently.

The logical basis of this rather startling result is that, under certain conditions, competitive prices convey all the information necessary for making the optimal decision. A builder trying to decide whether to use brick or concrete will weigh his requirements and tastes against the prices of the materials. Other users will do the same, with the result that those whose needs and preferences for brick are relatively the strongest will get brick. Further, profit-maximizing producers will weigh relative production costs, reflecting society's productive capabilities, against relative prices, reflecting society's tastes and desires, when deciding how much of each good to produce. The end result is that users get brick and cement in quantities and proportions that reflect their individual tastes and society's production opportunities. No other solution would be better from the standpoint of all the individuals concerned.

This suggests what it is that makes pollution different. The efficiency of competitive markets depends on the identity of *private* costs and *social* costs. As long as the brick-cement producer must compensate somebody for every cost imposed by his production, his profit-maximizing decisions about how much to produce, and how, will also be socially efficient decisions. Thus, if a producer dumps wastes into the air, river, or ocean; if he pays nothing for such dumping; and if the disposed wastes have no noticeable effect on anyone else, living or still unborn; then the private and social costs of disposal are identical and nil, and the producer's private decisions are socially efficient. *But if these wastes do affect others, then the social costs of waste disposal are not zero. Private and social costs diverge, and private profit-maximizing decisions are not socially efficient*. Suppose, for example, that cement production dumps large quantities of dust into the air, which damages neighbors, and that the brick-cement producer pays these neighbors nothing. In the social sense, cement will be over-produced relative to brick and other products because users of the products will make decisions based on market prices which do not reflect true social costs. They will use cement when they should use brick, or when they should not build at all.

*This divergence between private and social costs is the fundamental cause of pollution of all types,* and it arises in any society where decisions are at all decentralized—which is to say, in any economy of any size which hopes to function at all. Even the socialist manager of the brick-cement plant, told to maximize output given the resources at his disposal, will use the People's Air to dispose of the People's Wastes; to do otherwise would be to violate his instructions. And if instructed to avoid pollution "when possible," he does not know what to do: how can he decide whether more brick or cleaner air is more important for building socialism? The capitalist manager is in exactly the same situation. Without prices to convey the needed information, he does not know what action is in the public interest, and certainly would have no incentive to act correctly even if he did know.

Although markets fail to perform efficiently when private and social costs diverge, this does not imply that there is some inherent flaw in the idea of acting on self-interest in response to market prices. Decisions based on private cost calculations are typically correct from a social point of view; and even when they are not quite correct, it often is better to accept this inefficiency than to turn to some alternative decision mechanism, which may be worse. Even the modern economic theory of socialism is based on the high correlation between managerial self-interest and public good. There is no point in trying to find something—some omniscient and omnipotent *deus ex machina*—to replace markets and self-interest. Usually it is preferable to modify existing institutions, where necessary, to make private and social interest coincide.

And there is a third relevant economic concept: the fundamental distinction between questions of efficiency and questions of equity or fairness. A situation is said to be efficient if it is not possible to rearrange things so as to benefit one person without harming any others. That is the *economic* equation for efficiency. *Politically,* this equation can be solved in various ways; though most reasonable men will agree that efficiency is a good thing, they will rarely agree about which of the many possible efficient states, each with a different distribution of "welfare" among individuals, is the best one. Economics itself has nothing to say about which efficient state is the best. That decision is a matter of personal and philosophical values, and ultimately must be decided by some political process. Economics can suggest ways of achieving efficient states, and can try to describe the equity considerations involved in any suggested social policy; but the final decisions about matters of "fairness" or "justice" cannot be decided on economic grounds.

## Estimating the Costs of Pollution

Both in theory and practice, the most difficult part of an economic approach to pollution is the measurement of the cost and benefits of its abatement. Only a small fraction of the costs of pollution can be estimated straightforwardly. If, for example, smog reduces the life of automobile tires by 10 per cent, one component of the cost of smog is 10 per cent of tire expenditures. It has been estimated that, in a moderately polluted area of New York City, filthy air imposes extra costs for painting, washing, laundry, etc., of $200 per person per year. Such costs must be included in any calculation of the benefits of pollution abatement, and yet they are only a part of the relevant costs—and often a small part. Accordingly it rarely is possible to justify a measure like river pollution control solely on the basis of costs to individuals or firms of treating water because it usually is cheaper to process only the water that is actually used for industrial or municipal purposes, and to ignore the river itself.

The costs of pollution that cannot be measured so easily are often called "intangible" or "noneconomic," although neither term is particularly appropriate. Many of these costs are as tangible as burning eyes or a dead fish, and all such costs are relevant to a valid economic analysis. Let us therefore call these costs "nonpecuniary."

The only real difference between nonpecuniary costs and the other kind lies in the difficulty of estimating them. If pollution in Los Angeles harbor is reducing marine life, this imposes costs on society. The cost of reducing commercial fishing could be estimated directly: it would be the fixed cost of converting men and equipment from fishing to an alternative occupation, plus the difference between what they earned in fishing and what they earn in the new occupation, plus the loss to consumers who must eat chicken instead of fish. But there are other, less straightforward costs: the loss of recreation opportunities for children and sportsfishermen and of research facilities for marine biologists, etc. Such costs are obviously difficult to measure and may be very large indeed; but just as surely as they are not zero, so too are they not infinite. Those who call for immediate action and damn the cost, merely because the spiney starfish and furry crab populations are shrinking, are putting an infinite marginal value on these creatures. This strikes a disinterested observer as an overestimate.

The above comments may seem crass and insensitive to those who, like one angry letter-writer to the Los Angeles *Times,* want to ask: "If conservation is not for its own sake, then what in the world *is* it for?" Well, what *is* the purpose of pollution control? Is it for its own sake?

Of course not. If we answer that it is to make the air and water clean and quiet, then the question arises: what is the purpose of clean air and water? If the answer is, to please the nature gods, then it must be conceded that all pollution must cease immediately because the cost of angering the gods is presumably infinite. But if the answer is that the purpose of clean air and water is to further human enjoyment of life on this planet, then we are faced with the economists' basic question: given the limited alternatives that a niggardly nature allows, how can we best further human enjoyment of life? And the answer is, by making intelligent marginal decisions on the basis of costs and benefits. Pollution control is for lots of things: breathing comfortably, enjoying mountains, swimming in water, for health, beauty, and the general delectation. But so are many other things, like good food and wine, comfortable housing and fast transportation. The question is not which of these desirable things we should have, but rather what combination is most desirable. To determine such a combination, we must know the rate at which individuals are willing to substitute more of one desirable thing for less of another desirable thing. Prices are one way of determining those rates.

But if we cannot directly observe market prices for many of the costs of pollution, we must find another way to proceed. One possibility is to infer the costs from other prices, just as we infer the value of an ocean view from real estate prices. In principle, one could estimate the value people put on clean air and beaches by observing how much more they are willing to pay for property in nonpolluted areas. Such information could be obtained; but there is little of it available at present.

Another possible way of estimating the costs of pollution is to ask people how much they would be willing to pay to have pollution reduced. A resident of Pasadena might be willing to pay $100 a year to have smog reduced 10 or 20 per cent. In Barstow, where the marginal cost of smog is much less, a resident might not pay $10 a year to have smog reduced 10 per cent. If we knew how much it was worth to everybody, we could add up these amounts and obtain an estimate of the cost of a marginal amount of pollution. The difficulty, of course, is that there is no way of guaranteeing truthful responses. Your response to the question, how much is pollution costing *you,* obviously will depend on what you think will be done with this information. If you think you will be compensated for these costs, you will make a generous estimate; if you think that you will be charged for the control in proportion to these costs, you will make a small estimate.

In such cases it becomes very important how the questions are asked. For example, the voters could be asked a question of the form: Would you like to see pollution reduced $x$ per cent if the result is a $y$ per cent increase in the cost of living? Presumably a set of questions of this form could be used to estimate the costs of pollution, including the so-called "unmeasurable" costs. But great care must be taken in formulating the questions. For one thing, if the voters will benefit differentially from the activity, the questions should be asked in a way which reflects this fact. If, for example, the issue is cleaning up a river, residents near the river will be willing to pay more for the cleanup and should have a means of expressing this. Ultimately, some such political procedure probably will be necessary, at least until our more direct measurement techniques are greatly improved.

Let us assume that, somehow, we have made an estimate of the social cost function for pollution, including the marginal cost associated with various pollution levels. We now need an estimate of the benefits of pollution—or, if you prefer, of the costs of pollution abatement. So we set the Pollution Control Board (PCB) to work on this task.

The PCB has a staff of engineers and technicians, and they begin working on the obvious question: for each pollution source, how much would it cost to reduce pollution by 10 per cent, 20 per cent, and so on. If the PCB has some economists, they will know that the cost of reducing total pollution by 10 per cent is *not* the total cost of reducing each pollution source by 10 per cent. Rather, they will use the equimarginal principle and find the pattern of control such that an additional dollar spent on control of any pollution source yields the same reduction. This will minimize the cost of achieving any given level of abatement. In this way the PCB can generate a "cost of abatement" function, and the corresponding marginal cost function.

While this procedure seems straightforward enough, the practical difficulties are tremendous. The amount of information needed by the PCB is staggering; to do this job right, the PCB would have to know as much about each plant as the operators of the plant themselves. The cost of gathering these data is obviously prohibitive, and, since marginal principles apply to data collection too, the PCB would have to stop short of complete information, trading off the resulting loss in efficient control against the cost of better information. Of course, just as fast as the PCB obtained the data, a technological change would make it obsolete.

The PCB would have to face a further complication. It would not be correct simply to determine how to control existing pollution sources given their existing locations and production methods. Although this is almost certainly what the PCB would do, the resulting cost functions will overstate the true social cost of control. Muzzling existing plants is only one method of control. Plants can move, or switch to a new process, or even to a new product. Consumers can switch to a less-polluting substitute. There are any number of alternatives, and the poor PCB engineers can never know them all. This could lead to some costly mistakes. For example, the PCB may correctly conclude that the cost of installing effective dust control at the cement plant is very high and hence may allow the pollution to continue, when the best solution is for the cement plant to switch to brick production while a plant in the desert switches from brick to cement. The PCB can never have all this information and therefore is doomed to inefficiency, sometimes an inefficiency of large proportions.

Once cost and benefit functions are known, the PCB should choose a level of abatement that maximizes net gain. This occurs where the marginal cost of further abatement just equals the marginal benefit. If, for example, we could reduce pollution damages by $2 million at a cost of $1 million, we should obviously impose that $1 million cost. But if the damage reduction is only $1/2 million, we should not and in fact should reduce control efforts.

This principle is obvious enough but is often overlooked. One author, for example, has written that the national cost of air pollution is $11 billion a year but that we are spending less than $50 million a year on control; he infers from this that "we could justify a tremendous strengthening of control efforts on purely economic grounds." That *sounds* reasonable, if all you care about are sounds. But what is the logical content of the statement? Does it imply we should spend $11 billion on control just to make things even? Suppose we were spending $11 billion on control and thereby succeeded in reducing pollution costs to $50 million. Would this imply we were spending too *much* on control? Of course not. We must compare the *marginal* decrease in pollution costs to the *marginal* increase in abatement costs.

## Difficult Decisions

Once the optimal pollution level is determined, all that is necessary is for the PCB to enforce the pattern of controls which it has determined to be optimal. (Of course, this pattern will not really be the best one, because the PCB will not have all the information it should have.) But now a new problem arises: how should the controls be enforced?

The most direct and widely used method is in many ways the least efficient: direct regulation. The PCB can decide what each polluter must do to reduce pollution and then simply require that action under penalty of law. But this approach has many shortcomings. The polluters have little incentive to install the required devices or to keep them operating properly. Constant inspection is therefore necessary. Once the polluter has complied with the letter of the law, he has no incentive to find better methods of pollution reduction. Direct control of this sort has a long history of inadequacy; the necessary bureaucracies rarely manifest much vigor, imagination, or devotion to the public interest. Still, in some situations there may be no alternative.

A slightly better method of control is for the PCB to set an acceptable level of pollution for each source and let the polluters find the cheapest means of achieving this level. This reduces the amount of information the PCB needs, but not by much. The setting of the acceptable levels becomes a matter for negotiation, political pull, or even graft. As new plants are built and new control methods invented, the limits should be changed; but if they are, the incentive to find new designs and new techniques is reduced.

A third possibility is to subsidize the reduction of pollution, either by subsidizing control equipment or by paying for the reduction of pollution below standard levels. This alternative has all the problems of the above methods, plus the classic shortcoming which plagues agricultural subsidies: the old joke about getting into the not-growing-cotton business is not always so funny.

The PCB will also have to face the related problem of deciding *who* is going to pay the costs of abatement. Ultimately, this is a question of equity or fairness which economics cannot answer; but economics can suggest ways of achieving equity without causing inefficiency. In general, the economist will say: if you think polluter A is deserving of more income at polluter B's expense, then by all means give A some of B's income; but do *not* try to help A by allowing him to pollute freely. For example, suppose A and B each operate plants which produce identical amounts of pollution. Because of different technologies, however, A can reduce his pollution 10 per cent for $100, while B can reduce his pollution 10 per cent for $1,000. Suppose your goal is to reduce total pollution 5 per cent. Surely it is obvious that the best (most efficient) way to do this is for A to reduce his pollution 10 per cent while B does nothing. But suppose B is rich and A is poor. Then many would demand that B reduce his pollution 10 per cent while A does nothing because B has a greater "ability to pay." Well, perhaps B does have greater ability to pay, and perhaps it is "fairer" that he pay the costs of pollution control; but if so, B should pay the $100 necessary to reduce A's pollution. To force B to reduce his own pollution 10 per cent is equivalent to taxing B $1,000 and then blowing the $1,000 on an extremely inefficient pollution control method. Put this way, it is obviously a stupid thing to do; but put in terms of B's greater ability to pay, it will get considerable support though it is no less stupid. The more efficient alternative is not always available, in which case it may be acceptable to use the inefficient method. Still, it should not be the responsibility of the pollution authorities to change the distribution of welfare in society; this is the responsibility of higher authorities. The PCB should concentrate on achieving economic efficiency without being grossly unfair in its allocation of costs.

Clearly, the PCB has a big job which it will never be able to handle with any degree of efficiency. Some sort of self-regulating system, like a market, is needed, which will automatically adapt to changes in conditions, provide incentives for development and adoption of improved control methods, reduce the amount of information the PCB must gather and the amount of detailed control it must exercise, and so on. This, by any standard, is a tall order.

## Putting a Price on Pollution

And yet there is a very simple way to accomplish all this. *Put a price on pollution.* A price-based control mechanism would differ from an ordinary market transaction system only in that the PCB would set the prices, instead of their being set by demand-supply forces, and that the state would force payment. Under such a system, anyone could emit any amount of pollution so long as he pays the price which the PCB sets to approximate the marginal social cost of pollution. Under this circumstance, private decisions based on self-interest are efficient. If pollution consists of many components, each with its own social cost, there should be different prices for each component. Thus, extremely dangerous materials must have an extremely high price, perhaps stated in terms of "years in jail" rather than "dollars," although a sufficiently high dollar price is essentially the same thing. In principle, the prices should vary with geographical location, season of the year, direction of the wind, and even day of the week, although the cost of too many variations may preclude such fine distinctions.

Once the prices are set, polluters can adjust to them any way they choose. Because they act on self-interest they will reduce their pollution by every means possible up to the point where further reduction would cost more than the price. Because all face the same price for the same type of pollution, the marginal cost of abatement is the same everywhere. If there are economies of scale in pollution control, as in some types of liquid waste treatment, plants can cooperate in establishing joint treatment facilities. In fact, some enterprising individual could buy these wastes from various plants (at negative prices—i.e., they would get paid for carting them off), treat them, and then sell them at a higher price, making a profit in the process. (After all, this is what rubbish removal firms do now.) If economies of scale are so substantial that the provider of such a service becomes a monopolist, then the PCB can operate the facilities itself.

Obviously, such a scheme does not eliminate the need for the PCB. The board must measure the output of pollution from all sources, collect the fees, and so on. But it does not need to know anything about any plant except its total emission of pollution. It does not control, negotiate, threaten, or grant favors. It does not destroy incentive because development of new control methods will reduce pollution payments.

As a test of this price system of control, let us consider how well it would work when applied to automobile pollution, a problem for which direct control is usually considered the only feasible approach. If the price system can work here, it can work anywhere.

Suppose, then, that a price is put on the emissions of automobiles. Obviously, continuous metering of such emissions is impossible. But it should be easy to determine the average output of pollution for cars of different makes, models, and years, having different types of control devices and using different types of fuel. Through graduated registration fees and fuel taxes, each car owner would be assessed roughly the social cost of his car's pollution, adjusted for whatever control devices he has chosen to install and for his driving habits. If the cost of installing a device, driving a different car, or finding alternative means of transportation is less than the price he must pay to continue his pollution, he will presumably take the necessary steps. But each individual remains free to find the best adjustment to his particular situation. It would be remarkable if everyone decided to install the same devices which some states currently require; and yet that is the effective assumption of such requirements.

Even in the difficult case of auto pollution, the price system has a number of advantages. Why should a person living in the Mojave desert, where pollution has little social cost, take the same pains to reduce air pollution as a person living in Pasadena? Present California law, for example, makes no distinction between such areas; the price system would. And what incentive is there for auto manufacturers to design a less polluting engine? The law says only that they must install a certain device in every car. If GM develops a more efficient engine, the law will eventually be changed to require this engine on all cars, raising costs and reducing sales. But will such development take place? No collusion is needed for manufacturers to decide unanimously that it would be foolish to devote funds to such development. But with a pollution fee paid by the consumer, there is a real advantage for any firm to be first with a better engine, and even a collusive agreement wouldn't last long in the face of such an incentive. The same is true of fuel manufacturers, who now have no real incentive to look for better fuels. Perhaps most important of all, the present situation provides no real way of determining whether it is cheaper to reduce pollution by muzzling cars or industrial plants. The experts say that most smog comes from cars; but *even if true, this does not imply that it is more efficient to control autos rather than other pollution sources*. How can we decide which is more efficient without mountains of information? The answer is, by making drivers and plants pay the same price for the same pollution, and letting self-interest do the job.

In situations where pollution outputs can be measured more or less directly (unlike the automobile pollution case), the price system is clearly superior to direct control. A study of possible control methods in the Delaware estuary, for example, estimated that, compared to a direct control scheme requiring each polluter to reduce his pollution by a fixed percentage, an effluent charge which would achieve the same level of pollution abatement would be only half as costly—a saving of about $150 million. Such a price system would also provide incentive for further improvements, a simple method of handling new plants, and revenue for the control authority.

In general, the price system allocates costs in a manner which is at least superficially fair: those who produce and consume goods which cause pollution, pay the costs. But the superior efficiency in control and apparent fairness are not the only advantages of the price mechanism. Equally important is the case with which it can be put into operation. It is not necessary to have detailed information about all the techniques of pollution reduction, or estimates of all costs and benefits. Nor is it necessary to determine whom to blame or who should pay. All that is needed is a mechanism for estimating, if only roughly at first, the pollution output of all polluters, together with a means of collecting fees. Then we can simply pick a price—any price—for each category of pollution, and we are in business. The initial price should be chosen on the basis of some estimate of its effects but need not be the optimal one. If the resulting reduction in pollution is not "enough," the price can be raised until there is sufficient reduction. A change in technology, number of plants, or whatever, can be accommodated by a change in the price, even without detailed knowledge of all the technological and economic data. Further, once the idea is explained, the price system is much more likely to be politically acceptable than some method of direct control. Paying for a service, such as garbage disposal, is a well-established tradition, and is much less objectionable than having a bureaucrat nosing around and giving arbitrary orders. When businessmen, consumers, and politicians understand the alternatives, the price system will seem very attractive indeed.

## Who Sets the Prices?

An important part of this method of control obviously is the mechanism that sets and changes the pollution price. Ideally, the PCB could choose this price on the basis of an estimate of the benefits and costs involved, in effect imitating the impersonal workings of ordinary market forces. But because many of the costs and benefits cannot be measured, a less "objective," more political procedure is needed. This political procedure could take the form of a referendum, in which the PCB would present to the voters alternative schedules of pollution prices, together with the estimated effects of each. There would be a massive propaganda campaign waged by the interested parties, of course. Slogans such as "Vote NO on 12 and Save Your Job," or "Proposition 12 Means Higher Prices," might be overstatements but would contain some truth, as the individual voter would realize when he considered the suggested increase in gasoline taxes and auto registration fees. But the other side, in true American fashion, would respond by overstating *their* case: "Smog Kills, Yes on 12," or "Stop *Them* From Ruining *Your* Water." It would be up to the PCB to inform the public about the true effects of the alternatives; but ultimately, the voters would make the decision.

It is fashionable in intellectual circles to object to such democratic procedures on the ground that the uncultured masses will not make correct decisions. If this view is based on the fact that the technical and economic arguments are likely to be too complex to be decided by direct referendum, it is certainly a reasonable position; one obvious solution is to set up an elective or appointive board to make the detailed decisions, with the expert board members being ultimately responsible to the voters. But often there is another aspect to the antidemocratic position—a feeling that it is impossible to convince the people of the desirability of some social policy, not because the issues are too complex but purely because their values are "different" and inferior. To put it bluntly: many ardent foes of pollution are not so certain that popular opinion is really behind them, and they therefore prefer a more bureaucratic and less political solution.

The question of who should make decisions for whom, or whose desires should count in a society, is essentially a noneconomic question that an economist cannot answer with authority, whatever his personal views on the matter. The political structures outlined here, when combined with the economic suggestions, can lead to a reasonably efficient solution of the pollution problem in a society where the tastes and values of all men are given some consideration. In such a society, when any nonrepresentative group is in a position to impose its particular evaluation of the costs and benefits, an inefficient situation will result. The swimmer or tidepool enthusiast who wants Los Angeles Harbor converted into a crystal-clear swimming pool, at the expense of all the workers, consumers, and businessmen who use the harbor for commerce and industry, is indistinguishable from the stockholder in Union Oil who wants maximum output from offshore wells, at the expense of everyone in the Santa Barbara area. Both are urging an inefficient use of society's resources; both are trying to get others to subsidize their particular thing—a perfectly normal, if not especially noble, endeavor.

If the democratic principle upon which the above political suggestions are based is rejected, the economist cannot object. He will still suggest the price system as a tool for controlling pollution. With any method of decision—whether popular vote, representative democracy, consultation with the nature gods, or a dictate of the intellectual elite

—the price system can simplify control and reduce the amount of information needed for decisions. It provides an efficient, comprehensive, easily understood, adaptable, and reasonably fair way of handling the problem. It is ultimately the only way the problem will be solved. Arbitrary, piecemeal, stop-and-go programs of direct control have not and will not accomplish the job.

## Some Objections Aren't an Answer

There are some objections that can be raised against the price system as a tool of pollution policy. Most are either illogical or apply with much greater force to any other method of control.

For example, one could object that what has been suggested here ignores the difficulties caused by fragmented political jurisdictions; but this is true for any method of control. The relevant question is: what method of control makes interjurisdictional cooperation easier and more likely? And the answer is: a price system, for several reasons. First, it is probably easier to get agreement on a simple schedule of pollution prices than on a complex set of detailed regulations. Second, a uniform price schedule would make it more difficult for any member of the "cooperative" group to attract industry from the other areas by promising a more lenient attitude toward pollution. Third, and most important, a price system generates revenues for the control board, which can be distributed to the various political entities. While the allocation of these revenues would involve some vigorous discussion, any alternative methods of control would require the various governments to raise taxes to pay the costs, a much less appealing prospect; in fact, there would be a danger that the pollution prices might be considered a device to generate revenue rather than to reduce pollution, which could lead to an overly-clean, inefficient situation.

Another objection is that the Pollution Control Board might be captured by those it is supposed to control. This danger can be countered by having the board members subject to election or by having the pollution prices set by referendum. With any other control method, the danger of the captive regulator is much greater. A uniform price is easy for the public to understand, unlike obscure technical arguments about boiler temperatures and the costs of electrostatic collectors versus low-sulfur oil from Indonesia; if pollution is too high, the public can demand higher prices, pure and simple. And the price is the same for all plants, with no excuses. With direct control, acceptable pollution levels are negotiated with each plant separately and in private, with approved delays and special permits and other nonsense. The opportunities for using political influence and simple graft are clearly much larger with direct control.

A different type of objection occasionally has been raised against the price system, based essentially on the fear that it will solve the problem. Pollution, after all, is a hot issue with which to assault The Establishment, Capitalism, Human Nature, and Them; any attempt to remove the issue by some minor change in institutions, well within The System, must be resisted by The Movement. From some points of view, of course, this is a perfectly valid objection. But one is hopeful that there still exists a majority more concerned with finding solutions than with creating issues.

There are other objections which could be raised and answered in a similar way. But the strongest argument for the price system is not found in idle speculation but in the real world, and in particular, in Germany. The Rhine River in Germany is a dirty stream, recently made notorious when an insecticide spilled into the river and killed millions of fish. One tributary of the Rhine, a river called the Ruhr, is the sewer for one of the world's most concentrated industrial areas. The Ruhr River valley contains 40 per cent of German industry, including 80 per cent of coal, iron, steel and heavy chemical capacity. The Ruhr is a small river, with a low flow of less than half the flow on the Potomac near Washington. The volume of wastes is extremely large—actually exceeding the flow of the river itself in the dry season! *Yet people and fish swim in the Ruhr River.*

This amazing situation is the result of over forty years of control of the Ruhr and its tributaries by a hierarchy of regional authorities. These authorities have as their goal the maintenance of the quality of the water in the area at minimum cost, and they have explicitly applied the equimarginal principle to accomplish this. Water quality is formally defined in a technological rather than an economic way; the objective is to "not kill the fish." Laboratory tests are conducted to determine what levels of various types of pollution are lethal to fish, and from these figures an index is constructed which measures the "amount of pollution" from each source in terms of its fish-killing capacity. This index is different for each source, because of differences in amount and composition of the waste, and geographical locale. Although this physical index is not really a very precise measure of the real economic *cost* of the waste, it has the advantage of being easily measured and widely understood. Attempts are made on an *ad hoc* basis to correct the index if necessary—if, for example, a non-lethal pollutant gives fish an unpleasant taste.

Once the index of pollution is constructed, a price is put on the pollution, and each source is free to adjust its operation any way it chooses. Geographical variation in prices, together with some direct advice from the authorities, encourage new plants to locate where pollution is less damaging. For example, one tributary of the Ruhr has been converted to an open sewer; it has been lined with concrete and landscaped, but otherwise no attempt is made to reduce pollution in the river itself. A treatment plant at the mouth of the river processes all these wastes at low cost. Therefore, the price of pollution on this river is set low. This arrangement, by the way, is a rational, if perhaps unconscious, recognition of marginal principles. The loss caused by destruction of *one* tributary is rather small, if the nearby rivers are maintained, while the benefit from having this inexpensive means of waste disposal is very large. However, if *another* river were lost, the cost would be higher and the benefits lower; one open sewer may be the optimal number.

The revenues from the pollution charges are used by the authorities to measure pollution, conduct tests and research, operate dams to regulate stream flow, and operate waste treatment facilities where economies of scale make this desirable. These facilities are located at the mouths of some tributaries, and at several dams in the Ruhr. If the authorities find pollution levels are getting too high, they simply raise the price, which causes polluters to try to reduce their wastes, and provides increased revenues to use on further treatment. Local governments influence the authorities, which helps to maintain recreation values, at least in certain stretches of the river.

This classic example of water management is obviously not exactly the price system method discussed earlier. There is considerable direct control, and the pollution authorities take a very active role. Price regulation is not used as much as it could be; for example, no attempt is made to vary the price over the season, even though high flow on the Ruhr is more than ten times larger than low flow. If the price of pollution were reduced during high flow periods, plants would have an incentive to regulate their production and/or store their wastes for release during periods when the river can more easily handle them. The difficulty of continuously monitoring wastes means this is not done; as automatic, continuous measurement techniques improve and are made less expensive, the use of variable prices will increase. Though this system is not entirely regulated by the price mechanism, prices are used more here than anywhere else, and the system is much more successful than any other.[1] So, both in theory and in practice, the price system is attractive, and ultimately must be the solution to pollution problems.

1. For a more complete discussion of the Ruhr Valley system, see Allen V. Kneese, *The Economics of Regional Water Quality Management* (Baltimore, Md.: Johns Hopkins Press, 1964).

## "If We Can Go to the Moon, Why . . . etc?"

"If we can go to the moon, why can't we eliminate pollution?" This new, and already trite, rhetorical question invites a rhetorical response: "If physical scientists and engineers approached their tasks with the same kind of wishful thinking and fuzzy moralizing which characterizes much of the pollution discussion, we would never have gotten off the ground." Solving the pollution problem is no easier than going to the moon, and therefore requires a comparable effort in terms of men and resources and the same sort of logical hard-headedness that made Apollo a success. Social scientists, politicians, and journalists who spend their time trying to find someone to blame, searching for a magic device or regulation, or complaining about human nature, will be as helpful in solving the pollution problem as they were in getting us to the moon. The price system outlined here is no magic formula, but it attacks the problem at its roots, and has a real chance of providing a long-term solution.

# 2

ALLEN V. KNEESE

# *Analysis of Environmental Pollution*

*Allen V. Kneese is Director of the Quality of the Environment Program at Resources for the Future.*

## Introduction

Economic theory has provided a conceptual structure indispensable for understanding contemporary environmental problems and for formulating effective and efficient policy approaches toward them. Concepts like external diseconomies and public goods provide enormously useful insights. But economic theorizing and research that take place without being well informed about the substantive character of the problems under study is in danger of being somewhat arid because of extreme abstraction or of expending scarce energy and talent in the pursuit of relatively unimportant matters. The objective of the present essay is to provide an introduction to the substantive aspects of one of the major environmental problems facing both developed and some developing economies—environmental pollution.

*"Analysis of Environmental Pollution," by Allen V. Kneese, from* The Economics of Environment, *Peter Bohm and Allen V. Kneese, eds. (Macmillan. 1971). Originally published in* The Swedish Journal of Economics, *March 1971. Reprinted by permission of Macmillan, London and Basingstoke and St. Martin's Press, New York.*

Environmental pollution has existed for many years in one form or another. It is an old phenomenon,[1] and yet in its contemporary forms it seems to have crept up on governments and even on pertinent professional disciplines such as biology, chemistry, most of engineering,[2] and, of course, economics. A few economists, such as Pigou, wrote intelligently and usefully on the matter a long time ago, but generally even that subset of economists especially interested in externalities seems to have regarded them as rather freakish anomalies in an otherwise smoothly functioning exchange system. Even the examples commonly used in the literature have a whimsical air about them. We have heard much of bees and apple orchards and a current favorite example is sparks from a steam locomotive—this being some eighty years after the introduction of the spark arrester and twenty years after the abandonment of the steam locomotive.

Moreover, air and water continued until very recently to serve the economist as examples of free goods. A whole new set of scarce environmental resources presenting unusually difficult allocation problems seems to have appeared on the scene with the profession having hardly noticed. Fortunately, this situation is changing fast and much good work is appearing in the current economics literature.

Substantial and thoughtful attention from economists is especially needed because the economic and institutional sources of the problem are either neglected or thoroughly misunderstood by most of those currently engaged in the rather frantic discussion of it.

1. Many accounts attest that severe environmental degradation has existed for a long time in the western countries. In fact, the immediate surroundings of most of mankind in this part of the world were much worse a century ago than they are now. The following account of statements from an address of Charles Dickens may be interesting in this connection, especially to those who know contemporary London: "He knew of many places in it [London] unsurpassed in the accumulated horrors of their long neglect by the dirtiest old spots in the dirtiest old towns, under the worst old governments of Europe." He also said that the surroundings and conditions of life were such that "infancy was made stunted, ugly and full of pain—maturity made old—and old age imbecile." These statements are from *The Public Health a Public Question: First Report of the Metropolitan Sanitary Association,* address of Charles Dickens, Esq., London, 1850. Great achievements in the elementary sanitation of the close-in environment have been made as well as impressive gains in public health. The distinguishing feature of contemporary environmental pollution seems to be the large-scale and subtle degradation of common property resources. This point is developed in the text below.

2. There has been a relatively small group of sanitary engineers that has given close attention to environmental problems for a long time. I am here referring to the mainstream of work in these professions.

## I. "Global" Problems

I will begin this discussion of the substantive aspects of pollution problems by concentrating first on those problems, or potential problems, which affect the entire planet. Thereafter I will focus on "regional" problems. By regional I mean all those other than global. One must use a word like regional rather than terms pertaining to political jurisdictions such as nations, states, or cities because the scale of pollution resulting from the emissions of materials and energy follows the patterns, pulses, and rhythms of meteorological and hydrological systems rather than the boundaries of political systems—and therein lies one of the main problems.

The global problems to be discussed here pertain largely to the atmosphere because the marks of man have already been seen on that entire thin film of life-sustaining substance. It seems to have come as something of a shock to the natural science community that man not only can, *but has,* changed the chemical composition of the whole atmosphere.[3] Other large-scale problems, or potential problems, particularly those related to the "biosphere," will be discussed more briefly.

Before proceeding to what may be real global problems, it will be desirable to dispose of one red herring. One of the spectres raised by the more alarmist school of ecologists is that man will deplete the world's oxygen supply by converting it into carbon dioxide in the process of burning fossil fuels for energy. This idea has now been thoroughly discredited by two separate pieces of evidence. The first is measurement of changes in the oxygen supply over a period of years. There is currently *one* monitoring station in the world whose objective it is to identify long-term changes in the atmosphere. The station is operated at a high elevation in Hawaii by the U.S. Weather Bureau. Observations there have shown the oxygen content of the atmosphere to be remarkably stable. The other piece of evidence—perhaps more persuasive—is in the form of a "gedankenexperiment." If one burns, on paper, the entire known world supply of fossil fuels and all the present plant biomass, the impact on the oxygen supply is to reduce it by about 3%. This is much too small to be noticed in most areas of the earth.

Potentially real effects on the atmosphere and climate are thought to be connected with changes in carbon dioxide and particulate matter (including aerosols) in the atmosphere, petroleum in the oceans, waste energy rejection to the atmosphere, and the widespread presence of toxic

3. The fullest discussion of the range of problems considered in this section will be found in *Man's Impact on the Global Environment: Assessment and Recommendations for Action*. M.I.T. Press, Cambridge (Mass.), 1970.

agents in the coastal waters and oceans. I will discuss each of these briefly in turn.

The production of carbon dioxide is an inevitable result of the combustion of fossil fuels. In contrast to $O_2$, the relative quantity of $CO_2$ in the atmosphere has increased measurably. The possible significance of this is that $CO_2$ absorbs infrared radiations and therefore an increasing concentration of it in the atmosphere would tend to cause the surface of the earth to rise in temperature.[4] Some estimates have put the possible increase in $CO_2$,[5] if present rates of increase in the combustion of fossil fuel continue, at about 50% by the end of the century. An increase of this amount could raise the world's mean surface temperature several degrees with attendant melting of ice caps, inundation of seacoast cities, and undesirable temperature increases in densely inhabited areas. Estimates made in the summer of 1970 suggest, however, that the $CO_2$ increase will only be about 20% by the end of the century with lesser potential effects on climate.[6] The difference in estimates is accounted for by the newly recognized fact that less of the $CO_2$ generated by combustion is staying in the atmosphere than was previously supposed. Apparently one or more of the "sinks" for $CO_2$ is responding to the increased concentration, or possibly even a third force is leading to greater absorption. The main sinks for $CO_2$, or more specifically carbon, are solution in the oceans and conversion by the flora of the earth. Perhaps carbon is limiting to growth in some of these plant populations, and they are responding to its increased availability from the atmosphere.[7] Another possibility is less reassuring. Somewhat anomalously the mean temperature of the earth's surface has been falling over the past couple of decades according to Weather Bureau observations. As is true of many gases, the solubility of $CO_2$ in water increases when water temperature falls. Maybe that's where some of the $CO_2$ went.

But this brings us to another possible effect of man's activity on world climate. Some meteorologists—especially Bryson [8]—believe that man's industrial and agricultural activities are causing the world to cool off. The suspected mechanism is an increase in particulates and aerosols which, they think, are increasing the earth's albedo (ability to reflect incoming solar radiation). Farming and other activities in arid areas and the combustion of fuels send immense amounts of particulates and fine

4. Most incoming radiation is in the form of visible light, while most outgoing energy is in the form of infrared radiation.

5. Conservation Foundation, *Implications of Rising Carbon Dioxide Content of the Atmosphere,* New York, 1963.

6. *Man's Impact on the Global Environment . . . etc., op. cit.*

7. With higher concentrations of $CO_2$ in the atmosphere, one would naturally expect some increase in absorption by the oceans, since $CO_2$ solubility is a function of the partial pressure of $CO_2$.

8. R. A. Bryson and J. T. Peterson: Atmospheric aerosols: Increased concentrations during the last decade. *Science 162,* 3849 (Oct. 1968), 120–21.

water vapors (aerosols) into the atmosphere each day. This is an undisputed fact although observations indicate that a worldwide increase in particulate matter cannot yet be identified.[9] What is in dispute is the effect of the man-generated increase. Some not only believe this effect is significant, but that it may be sending us rather rapidly toward an ice age—perhaps the final ice age resulting in a perpetually frozen planet. Other factors might lead in the same direction as we shall see subsequently. The freeze-up hypothesis is, however, disputed by those other meteorologists who regard it as important to recognize the difference between particulates of different types and at different elevations in the atmosphere. Mitchell has pointed out that there has been a large amount of volcanic activity in recent years which has deposited great quantities of particulates at high elevations in the atmosphere.[10] The net effect of these is fairly clearly to reflect more energy away from the earth, and this could well be responsible for the observed temperature decline. On the other hand, he points out that particulates deposited at relatively low altitudes, such as those generated by man, could well have the reverse effect because they reflect energy back toward earth as well as away from it. Mitchell's calculations tend to show that the former effect outweighs the latter.[11] Thus, when the effect of the volcanic particulates wears off over a period of years, the lower altitude particulates could begin to reinforce $CO_2$ as a factor leading to rising world temperatures.

An additional, and possibly reinforcing factor, is the release of energy to the atmosphere due to the energy conversion activities of man. A large proportion of the energy from fuels man uses is transferred directly to the atmosphere—as, for example, the energy converted in automobile engines. Another large proportion is initially transferred to water—as, for example, when condensers in electric power plants are cooled with water. But this too is rather quickly rejected to the atmosphere by induced evaporation in watercourses or wet-cooling towers. Essentially, all of the energy converted from fuels is transferred to the atmosphere as heat. Because this is so, it is possible to make a rather precise estimate of this transfer by calculating the energy value of the fuels used in the world. On this basis there is an average emission of about $5.7 \times 10^{12}$ watts of energy per year from human conversion.[12] What does one make of such a monstrous number? More understandable perhaps is the statement that this is about 1/15,000 of the absorbed

9. It should be noted that this is a somewhat controversial conclusion.
10. J. M. Mitchell, Jr.: A preliminary evaluation of atmospheric pollution as a cause of the global temperature fluctuation of the past century. In *Global Effects of Environmental Pollution* (ed. Singer), pp. 97–112. Reidel, Holland, 1970.
11. Personal communication to William Frisken.
12. W. R. Frisken, "Extended Industrial Revolution and Climate Change," unpublished report prepared while he was a visiting scholar at RFF, July, 1970.

solar flux. That doesn't seem like much, but another important element in the picture is the fact that energy conversion is a rapidly growing activity all over the world. The most spectacular example is conversion to electric power which in the United States has been proceeding at a doubling time of ten years and even faster in one or two other large economies. Worldwide energy conversion (by far the largest proportion of which is from fossil fuels) as a whole has been proceeding at a growth rate of about 4% a year. If we project this growth rate for 130 years, we will reach a rate of energy rejection of about 1% of the absorbed solar flux. This is enough, some meteorologists believe, to have a substantial effect on world climate. If we proceed at the 4% growth rate for another 120 years, we will have reached 100% of the absorbed solar flux. This would be a total disaster. The resulting mean increase in world temperature would be about 50°C—a condition totally unsuitable for human habitation. We will never reach such a situation, but the important question is what circumstances will prevent us from so doing.

If one is given to apocalyptic visions, he can readily imagine a situation in which $CO_2$, particulates, and energy conversion reinforce each other and will, after a short reprieve from the volcanoes, make the earth into a kind of minihell.

But other things may happen too. For one thing, we are annually spilling on the order of $1.5 \times 10^6$ tons of oil directly into the oceans with perhaps another $4 \times 10^6$ tons being delivered by terrestrial streams. This may be enough "oil on troubled waters", some scientists believe, to smooth the sea surface sufficiently to cause its reflectivity to be increased significantly.[13] Again, the associated albedo effect would tend to cause cooling. But at the same time the reduction in the atmosphere–ocean interface would tend to diminish the absorption of $CO_2$ and thus possibly tend toward a warming condition.

And then there is the matter of the SST. . . . Aside from the major question of sonic booms, the emissions from SSTs may have substantial effects on the upper atmosphere. SSTs would fly at 65,000–70,000 feet and the atmosphere is very different up there. It is extremely dry and the layers at that elevation do not seem to mix much with the lower atmosphere. Five hundred SSTs might be in operation by the mid-1980s. If these were the American type, their emissions might cause an increase of water vapor in the upper atmosphere of 10% globally and possibly 60% over the North Atlantic where most of the flights would occur.[14] This could give rise to large-scale formations of very persistent cirrus cloud. Furthermore, the emissions of soot, hydrocarbons, nitrogen

13. Oil also may affect phytoplankton and other species directly.

14. L. Machta, "Stratospheric Water Vapor," a working paper written for the 1970 Summer Study on Critical Environmental Problems, sponsored by the Massachusetts Institute of Technology and held at Williams College in Williamstown, Massachusetts, July 1970.

oxides, and sulfate particles could cause stratospheric smog. The effects of all this would be somewhat uncertain but presumably not unlike those produced by particulates deposited into the upper atmosphere by volcanoes—in other words, increased albedo and consequent cooling at the earth's surface.

A final category of substances of possibly global significance are persistent organic toxins. DDT is a good example of these and has been found in living creatures all over the world. How it got to remote places like the Antarctic is still somewhat mysterious, but apparently substantial amounts are transmitted through the atmosphere as well as through the oceans. Aside from possible large-scale effects on ecological systems, these persistent toxins could affect the $O_2$–$CO_2$ balance by poisoning the phytoplankton which are involved in one of the important $CO_2$–$O_2$ conversion processes. We do not know whether this is happening or not.

Clearly, we are operating in a context of great uncertainty. Equally clearly, man's activities now and in the relatively near-term future may affect the world's climatic and biological regimes in a substantial way. It seems beyond question that a serious effort to understand man's effects on the planet and to monitor those effects is indicated. Should we need to control such things as the production of energy and $CO_2$ in the world, we will face an economic and political resource allocation problem of unprecedented difficulty and complexity.

The discussion of global effects of pollution was necessarily somewhat speculative, but now we turn to problems on a less grand scale. These regional problems are clear and present. A discussion is first presented under the traditional categories of waterborne, airborne, and solid residuals.[15] In the final section, I point explicitly to the interdependencies among these residuals streams and the implications of this for economic analysis. Unfortunately, most of the numbers given are from the United States. This is because I am simply not familiar with the data from other countries. The relationships in the United States may, however, be reasonably representative of those found in other industrialized countries.

15. Due to limitations of space, I will concentrate on material residuals as sources of environmental pollution. There is some discussion of energy residuals —especially where they interact in important ways with material residuals. Noise, an important energy residual, is not treated at all. A good introductory discussion of noise can be found in chapter 1 of the *Handbook of Noise Control* (ed. Cyril M. Harris). McGraw-Hill, New York, 1957.

# II. Waterborne Residuals

## *Degradable Residuals*

A somewhat oversimplified but useful distinction for understanding what happens when residuals are discharged to watercourses is between *degradable* and *non-degradable* materials. The most widespread and best known degradable residual is domestic sewage, but, in the aggregate, industry produces greater amounts of degradable organic residuals almost all of which is generated by the food processing, meat packing, pulp and paper, petroleum refining, and chemicals industries. Some industrial plants are fantastic producers of degradable organic residuals: a single uncontrolled pulp mill, for example, can produce wastes equivalent to the sewage flow of a large city.

When an effluent bearing a substantial load of degradable organic residuals is expelled into an otherwise "clean" stream, a process known as "aerobic degradation" begins immediately. Stream biota, primarily bacteria, feed on the wastes and break them down into their inorganic forms of nitrogen, phosphorus, and carbon, which are basic plant nutrients. In the breaking down of degradable organic material, some of the oxygen which is dissolved in any "clean" water is utilized by the bacteria. But this depletion tends to be offset by reoxygenation which occurs through the air–water interface and also as a consequence of photosynthesis by the plants in the water. If the waste load is not too heavy, dissolved oxygen in the stream first will drop to a limited extent (say, to 4 or 5 parts per million from a saturation level of perhaps 8–10 ppm, depending upon temperature) and then rise again. This process can be described by a characteristically shaped curve or function known as the "oxygen sag". The differential equations which characterize this process were first intorduced by Streeter and Phelps in 1925 and are often called the Streeter-Phelps equations.

If the degradable organic residual discharged to a stream becomes great enough, the process of degradation may exhaust the dissolved oxygen. In such cases, degradation is still carried forward but it takes place anaerobically, that is, through the action of bacteria which do not use free oxygen but organically or inorganically bound oxygen, common sources of which are nitrates and sulphates. Gaseous by-products result, among them carbon dioxide, methane, and hydrogen sulfide.

Water in which wastes are being degraded anaerobically emits foul odors, looks black and bubbly, and aesthetically is altogether offensive. Indeed, the unbelievably foul odors from the River Thames in mid-nineteenth century London caused the halls of Parliament to be hung with sheets soaked in quicklime and even induced recess upon occasion when the reek became too suffocating. So extreme a condition is rarely encountered nowadays, although it is by no means unknown. For example, a large lake near São Paulo, Brazil, is largely anaerobic, and most of the streams in the Japanese papermaking city Fuji are likewise lacking in oxygen. Other instances could be mentioned. But levels of dissolved oxygen low enough to kill fish and cause other ecological changes are a much more frequent and widespread problem.

High temperatures accelerate degradation. They also decrease the saturation level of dissolved oxygen in a body of water. So a waste load which would not induce low levels of dissolved oxygen at one temperature may do so if the temperature of the water rises. In such circumstances, heat may be considered a pollutant. Moreover, excess heat itself can be destructive to aquatic life. Huge amounts of heat are put into streams by the cooling water effluents of electric power plants and industry.

There is, in fact, increased concern about the impacts of heat residuals, particularly from power generation, in the face of the incessantly increasing demand for electric power already described and the development of nuclear power, the present "generation" of which requires more heat disposal per kwh generated than fossil fuel plants. Increasing use of cooling towers has been one response to this situation. But the use of cooling towers represents basically a transfer of the medium into which to reject the residual heat energy, that is, to the air instead of temporarily to the water. One author in the United States has discussed some aspects of what would happen over the central region of the United States under the alternative procedure, i.e., use of once-through cooling with discharge of waterborne heat to the main streams of the area, the Missouri and the Mississippi. About 540 million kilowatts of fossil fuel burning capacity are assumed installed and operating in this region by the year 2000. He writes:

> Imposing the requirement of at least 10 miles separation between stations and noting that such a generating capacity will raise the water temperature by about 20 deg F, we find approximately 3,000 miles of river spreading over the central region of the United States with a temperature 20 deg F higher than normal.[16]

Of course the ecological effects which would accompany such a large-scale heat discharge to our streams can only be speculated about at this time. If there were a substantial discharge of degradable organic residuals to those streams at the same time, they would almost certainly become anaerobic in the summer time. The freshwater life forms we are accustomed to would be lost.

16. S. M. Greenfield: *Science and the Natural Environment of Man,* p. 3, RAND Corporation, Santa Monica, Calif., February 17, 1969.

A conventional sewage treatment plant processing degradable organic residuals uses the same biochemical processes which occur naturally in a stream, but by careful control they are greatly speeded up. Under most circumstances, standard biological sewage treatment plants are capable of reducing the BOD (biochemical oxygen demand) in waste effluent by perhaps 90%. As with degradation occurring in a watercourse, plant nutrients are the end-product of the process.

Stretches of streams which persistently carry less than 4 or 5 ppm of oxygen will not support the higher forms of fish life. Even where they are not lethal, reduced levels of oxygen increase the sensitivity of fish to toxins. Water in which the degradable organic residuals have not been completely stabilized is more costly to treat for public or industrial supplies. Finally, the plant nutrients produced by bacterial degradation of degradable organic residuals, either in the stream or in treatment plants, may cause algae blooms. Up to a certain level, algae growth in a stream is not harmful and may even increase fish food, but larger amounts can be toxic to fish, produce odors, reduce the river's aesthetic appeal, and increase water supply treatment problems. Difficulties with algae are likely to become serious only when waste loads have become large enough to require high levels of treatment. Then residual plant nutrient products become abundant relative to streamflow and induce excessive plant growth.

Problems of this kind are particularly important in comparatively quiet waters such as lakes and tidal estuaries. In recent years certain Swiss and American lakes have changed their character radically because of the buildup of plant nutrients. The most widely known example is Lake Erie, although the normal "eutrophication" or aging process has been accelerated in many other lakes. The possibility of excessive algae growth is one of the difficult problems in planning for pollution control—especially in lakes, bays, and estuaries, for effective treatment processes today carry a high price tag.

In the United States, currently, BOD discharges by industry are apparently about twice as large as by municipalities. How fast BOD discharges grow depends on how effectively industrial wastes are controlled and municipal wastes treated. If current rates continue, BOD may grow about 3½% per year with plant nutrient discharges growing even faster.

Bacteria might also be included among what we have called the degradable pollutants since the enteric, infectious types tend to die off in watercourses, and treatment with chlorine or ozone is highly effective against them. Because of water supply treatment, the traditional scourges of polluted water—typhoid, paratyphoid, dysentery, gastroenteritis—have become almost unknown in advanced countries. One might say that public concern with environmental pollution peaked early in this century with the rapid spread of these diseases. But public health engineers were so successful in devising effective water supply treatment that attention to water pollution lapsed until its recent upsurge.

* * *

## *Non-degradable Pollution*

BOD serves as a good indicator of pollution where one aspect is concerned—the degradable residuals. But many residuals are non-degradable. These are not attacked by stream biota and undergo no great change once they get into a stream. In other words, the stream does not "purify itself" of them. This category includes inorganic substances—such materials as inorganic colloidal matter, ordinary salt, and the salts of numerous heavy metals. When these substances are present in fairly large quantities, they result in toxicity, unpleasant taste, hardness, and, especially when chlorides are present, in corrosion. These residuals can be a public health problem—usually when they enter into food chains. Two particularly vicious instances of poisoning by heavy metals have stirred the population of Japan. These are mercury poisoning through eating contaminated fish (Minimata disease) and cadmium poisoning through eating contaminated rice (Itai Itai disease). Several hundred people have been affected and more than a hundred have died. At the present time the Canadian government has forbidden the consumption of fish from both Lake Erie and Lake St. Clair because of feared mercury poisoning, and mercury has been discovered in many rivers in the United States.

## *Persistent Pollutants*

There is a third group of pollutants, mostly of relatively recent origin, which does not fit comfortably into either the degradable or non-degradable categories. These "persistent" or "exotic" pollutants are best exemplified by the synthetic organic chemicals produced in profusion by modern chemical industry. They enter watercourses as effluents from industrial operations and also as waste residuals from many household and agricultural uses. These substances are termed "persistent" because stream biota cannot effectively attack their complex molecular chains. Some degradation does take place, but usually so slowly that the persistents travel long stream distances, and in groundwater, in virtually unchanged form. Detergents (e.g., ABS), pesticides (e.g., DDT), and phenols (resulting from the distillation of petroleum and coal products) are among the most common of these pollutants. Fortunately the recent development and successful manufacture of "soft" or degradable detergents has opened the way toward reduction or elimination of the problems associated with them, especially that of foaming. However, another problem associated with dry detergents has not been dealt with. These detergents contain phosphate "fillers" which may aggravate the nutrients problem.

Some of the persistent synthetic organics, like phenols and hard detergents, present primarily aesthetic problems. The phenols, for example, can cause an unpleasant taste in waters, especially when they are treated with chlorine to kill bacteria. Others are under suspicion as possible public health problems and are associated with periodic fish kills in streams. Some of the chemical insecticides are unbelievably toxic. The material endrin, which until recently was commonly used as an insecticide and rodenticide, is toxic to fish in minute concentrations. It has been calculated, for example, that 0.005 of a pound of endrin in three acres of water one foot deep is acutely toxic to fish.

Concentrations of the persistent organic substances have seldom if ever risen to levels in public water supplies high enough to present an *acute* danger to public health. The public health problem centers around the possible *chronic* effects of prolonged exposure to very low concentrations. Similarly, even in concentrations too low to be acutely poisonous to fish, these pollutants may have profound effects on stream ecology, especially through biological magnification in the food chain; higher creatures of other kinds—especially birds of prey—are now being seriously affected because persistent pesticides have entered their food chains. No solid evidence implicates present concentrations of organic chemicals in water supplies as a cause of health problems, but many experts are suspicious of them.

The long-lived radio-nuclides might also be included in the category of persistent pollutants. They are subject to degradation but at very low rates. Atomic power plants may be an increasingly important source of such pollutants. Generation of power by nuclear fission produces fission waste products which are contained in the fuel rods of reactors. In the course of time these fuels are separated by chemical processes to recover plutonium or to prevent waste products from "poisoning" the reactor and reducing its efficiency. Such atomic waste can impose huge external costs unless disposed of safely. A large volume of low-level waste resulting from the day-to-day operation of reactors can for the time being be diluted and discharged into streams, although the permissible standards for such discharge have recently been severely questioned in the United States, both outside and inside the Atomic Energy Commission.

"Hot" waste, containing long-lived substances such as radioactive strontium, cesium, and carbon, is in a different category from any other pollutant. So far, the only practical disposal method for high-level wastes is permanent storage. The "ultimate" solution to this contamination problem may be fusion energy which leaves no residuals except energy. But while some promising developments have occurred recently

—especially in the Soviet Union—its development (if even possible) is, at least, decades away.

### *The Range of Alternatives*

One of the most important features of the waterborne residuals problem, from the point of view of economic analysis, is the wide range of technical options which exist both for reducing the generation and discharge of wastes and for improving the assimilative capacity of watercourses. In industry, in addition to treatment, changes in the quality and type of inputs and outputs, the processes used, and by-product recovery are important ways of reducing residuals discharge. The capability of watercourses to assimilate residuals can often be increased by using releases from reservoirs to regulate low river flows and by the direct introduction of air or oxygen into them by mechanical means.[17]

## III. Airborne Residuals [18]

### *Types, Sources, and Management Alternatives*

There is virtually an infinity of airborne residuals that may be discharged to the atmosphere, but the ones of central interest and most commonly measured are carbon monoxide, hydrocarbons, sulfur dioxide, oxides of nitrogen, and particulates. The quantities and main sources of these in the United States are shown in the following table.

TABLE 1. Summary of Gaseous Residuals from Energy Conversion, 1965 (Million Tons)

| Energy user | Carbon monoxide (CO) | Hydro-carbons (HC) | Sulfur dioxide ($SO_2$) | Oxides of nitrogen ($NO_x$) | Partic-ulates |
|---|---|---|---|---|---|
| Utility power | 1 | neg. | 13.6 | 3.7 | 2.4 |
| Industry and households | 5 | neg. | 8.4 | 7.0 | 7.0 |
| Transportation | 66 | 12 | 0.4 | 6.0 | 0.2 |
| Total | 72 | 12 | 22.4 | 16.7 | 9.6 |

neg.: Negligible.
Source: U.S. Public Health Service.

17. A fairly extensive discussion of technical options can be found in A. V. Kneese and B. T. Bower, *Managing Water Quality: Economics, Technology, Institutions*. The Johns Hopkins Press, Baltimore, 1968.
18. In preparing the section on air pollution, I have benefitted from an unpublished memorandum by Blair Bower and Derrick Sewell, 1969.

In the United States, by far the greatest tonnage of airborne residuals comes from the transportation sector, and virtually all of this is from internal combustion engines. They are especially important sources of carbon monoxide, hydrocarbons, and oxides of nitrogen. There are a number of ways in which emissions can be reduced from internal combustion engines, and some of these have been implemented. Carbon monoxide and hydrocarbons can be controlled to some extent by various means of achieving more complete combustion of the fuel delivered to the fuel tank. Oxides of nitrogen are much harder to deal with because they are not a result of incomplete combustion but are synthesized from atmospheric gases when combustion takes place under high temperatures and pressures. It is now thought that the best way to control these would be through catalytic afterburners which, however, could add substantially to the cost and complexity of engines. Many people outside the automobile industry believe that large reductions in emissions can be effectively and economically achieved by abandoning the internal combustion engine in favor of other engine types (such as steam and electric) and heavier reliance on mass transit.[19]

Stationary sources (utility power, industry and households) are the main sources of sulfur oxides, particulates, and oxides of nitrogen. Control of emissions from these sources is a large and complex subject, but the main possibilities can be grouped into four categories: (1) fuel preparation (such as removing sulfur-bearing pyrites from coal before combustion), (2) fuel substitutions (such as substituting natural gas and low-sulfur oil and coal for high-sulfur coal), (3) redesigning burners (for example, in oil-burning furnaces two-stage combustion can reduce oxides of nitrogen), and (4) the treatment of stack gases (for example, stack gases can be scrubbed with water or dry removal processes can extract sulfur and particulates).[20]

Of course, all of these control technologies are likely to involve net costs even when they result in usable recovered materials. Furthermore, none of these processes inherently results in a reduction of $CO_2$. The possible significance of this was discussed in the opening section.

## *Assimilative Capacity of the Atmosphere*

The capacity of the atmosphere to assimilate discharges of residuals varies with time, space, and the nature of the materials being discharged. From a resources management point of view it is necessary to

19. R. U. Ayres and R. P. McKenna: *Alternatives to the Internal Combustion Engine: Impacts on Environmental Quality*. The Johns Hopkins Press, Baltimore, 1971.
20. A good source on the technology of pollution control from stationary sources is Arthur B. Stern (ed)., *Air Pollution,* vols. I, II, III. Academic Press, New York, 1968.

be able to translate a specified time and location pattern of discharges of gaseous residuals into the resulting time and spatial pattern of ambient (environmental) concentrations, because in most cases there are multiple sources of discharge. With variations in type, quantity, and time pattern of discharge, the problem is compounded in complexity. However, imaginative applications of atmospheric diffusion models analogous to the water diffusion models described earlier have been used to help define "air sheds" for analysis of air quality management strategies.[21]

Another complication in environmental modeling results from the interactions between gaseous residuals and water quality. Such interactions can involve large geographic areas. For example, atmospheric scavenging—particularly washout by precipitation—appears to be becoming an increasing problem. High stacks are often used to reduce the local impact of air pollution, but they result in spreading the residual more thinly over larger areas. Thus, high stacks on power plants in England are said to be causing "acid rain" in Scandinavia.

Precipitation is the primary cleansing mechanism for airborne gases and fine particles. Since sulfur dioxide is highly soluble in water, the washout process involves the absorption of the gas by drops of rain (or flakes of snow) as they fall through the gaseous discharge from a stack. Where the washout occurs over a body of water, adverse effects on water quality can occur. For example, atmospheric scavenging is believed to be contributing to deterioration of water quality in the Great Lakes Basin, at least with respect to the presence of trace elements in the lakes.

The areal extent of the atmospheric scavenging phenomenon is illustrated by data from the atmospheric chemical network stations in Europe relating to the acidity and sulfur contents of precipitation and the consequences on soils, surface waters, and biological systems. In 1958, pH values (pH is a measure of acidity—the lower the pH, the higher the acidity) below 5 were found only in limited areas over the Netherlands. In 1966, values below 5 were found in an area that spreads over Central Europe, and pH values in the Netherlands were less than 4.

## *Impacts of Gaseous Residuals on Receptors*

Perhaps of most immediate concern are direct effects on people, ranging in severity from the lethal to the merely annoying. Except for extreme air pollution episodes, fatalities are not, as a rule, traceable in-

21. A. A. Teller: The Use of Linear Programming to Estimate the Cost of Some Alternative Air Pollution Abatement Policies", *Proceedings* of the IBM Scientific Computing Symposium on Water and Air Resource Management, held on Oct. 23–25, 1967, at the Thomas J. Watson Research Center, Yorktown Heights, N.Y.

dividually to the impact of air pollution, primarily because most of the effects are synergistic. Thus, air pollution is an environmental stress which, in conjunction with a number of other environmental stresses, tends to increase the incidence and seriousness of a variety of pulmonary diseases, including lung cancer, emphysema, tuberculosis, pneumonia, bronchitis, asthma, and even the common cold. Clearly, however, acute air pollution episodes have raised death rates. Such occurrences have been observed in Belgium, Britain, Mexico, and the United States, among others. But the more important health effects appear to be associated with persistent exposure to the degraded air which exists in most cities.

The preponderance of evidence suggests that the relationship between such pollutants as $SO_2$, CO, particulates, and heavy metals and disease is real and large.[22] But one should not underrate the difficulties of establishing such relationships in an absolutely firm manner.

Direct effects on humans have parallels in the animal and plant worlds. Animals of commercial importance (livestock) are not located to any appreciable extent within cities, so effects on them are usually minor. Effects on pets (dogs, cats, and birds) almost certainly exist, although they have not been much documented.

As far as plants are concerned, much the same situation holds. Crops are mostly some distance away from cities, and hazards are likely to be rather special in nature (e.g., fluorides from superphosphate plants, or sulfur oxides from copper smelters). However, there are some districts where truck crops—mostly fruits and vegetables—are grown in close juxtaposition to major cities and are substantially affected by air pollution. In suburban gardens and city parks, there are deleterious effects on shrubs, flowers, shade trees, and even on forests in the air sheds of cities.

## *Damage to Property*

A third category of effects comprises damage to property. Here again, sulfur oxides and oxidants are perhaps equally potent. Sulfur oxides combine with water to form sulfurous acid ($H_2SO_3$) and the much more corrosive sulfuric acid ($H_2SO_4$). These acids will damage virtually any exposed metal surface and will react especially strongly with limestone or marble (calcium carbonate). Thus many historic buildings and objects (like "Cleopatra's Needle" in New York) have suffered extremely rapid deterioration in modern times.

Sulfur oxides will also cause discoloration, hardening and embrittlement of rubber, plastic, paper, and other materials. Oxidants such as

22. L. B. Lave and E. P. Seskin: Air pollution and human health. *Science 169*, no. 3947 (Aug. 1970) [p. 356 in this volume].

ozone will also produce the latter type of effect. Of course, the most widespread and noticeable of all forms of property damage is simple dirt (soot). Airborne dirt affects clothing, furniture, carpets, drapes, exterior paintwork, and automobiles. It leads to extra washing, vacuum cleaning, dry-cleaning, and painting; and, of course, all of these activities do not entirely eliminate the dirt, so that people also must live in darker and dirtier surroundings.

## *A Few Comments Comparing Air and Water Pollution Problems*

There are important parallels and contrasts between the effects and possible modes of management of water and air pollution.

1. In the United States and abroad, air pollution is heavily implicated as a factor affecting public health. Water pollution may be more costly in terms of non-human resources, but the current link of water pollution to public health problems on any large scale in advanced countries is a matter of suspicion concerning chronic effects rather than of firm evidence. Much stronger evidence links air pollution to public health problems.

2. As in the case of water pollution, a great many of the external costs imposed by air pollutants would appear to be measurable, but very little systematic measurement has yet been undertaken. The more straightforward effects are, for example, soiling, corrosion, reduction in property values, and agricultural losses.[23]

3. Current technology apparently provides fewer classes of means of dealing with air pollution than with water pollution. In part this is because it is easier for man to control hydrological events than meteorological events. The assimilative capacity of the air mantle cannot be effectively augmented. In part it is because air is not delivered to users in pipes as water frequently is, so that it is only to a limited extent that polluted air is treatable before it is consumed. Therefore, we are in somewhat the same position in regard to polluted air as the fish are with polluted water. We live in it. Furthermore, it is also more difficult and costly to collect gaseous residuals for central treatment. Accordingly, control of air pollution is largely a matter of preventing pollutants from escaping from their source, eliminating the source, or of shifting location of the source or the recipient. Water pollution, on the other hand, is in present intricate problems of devising optimal control systems.

4. To the extent that air sheds are definable, air shed authorities or compacts of districts are conceivable and may be useful administrative devices. In the United States the current federal policy approach points strongly in this direction and in this respect (but not others) is more advanced than the water pollution control programs.

23. Some efforts to provide economically useful estimates of damages are discussed elsewhere in this issue.

## IV. Solid Residuals

Just about every type of object made and used by man can and does eventually become a solid residual. Some of the main categories of importance are organic material, which includes garbage, and industrial solid wastes, such as from the canning industry, for example. Newspapers, wrappings, containers, and a great variety of other objects are found in household, commercial, office, industrial solid wastes. A very important source of solids in the United States is automobiles which will be discussed separately, further on. In the United States, about 5 lb. per day per capita of solid wastes are collected of which about three are household and commercial. Industrial, demolition, and agricultural wastes constitute most of the other. Altogether the United States generates (exclusive of agricultural wastes) each year approximately 3–5 billion tons of solid residuals from household, commercial, animal, industrial, and mining activities and spends about 4–5 billion dollars to handle and dispose of them.[24] In addition, there is a large amount of uncollected solids which litter the countryside.

The disposal of solid wastes can have a number of deleterious effects on society. Littering, dumps, and landfills produce visual disamenities. The disposal of solid wastes can cause adverse effects on air and water quality. Incineration of solid wastes is an obvious source of air pollution in many areas, as are burning dumps. Dumps tend to catch fire by spontaneous combustion unless they are relatively carefully controlled. Furthermore, drainage from disposal sites can reduce water quality in watercourses, and the sites may also provide a habitat for rodents and insect vectors. The disposal of collected solid wastes (which excludes much industrial waste, automobiles, and all of agricultural waste) in the United States is roughly in the following proportions: about 90% goes into landfill operations of one kind or another; another 8% is incinerated; and a small amount, about 2%, goes into hog-feeding and miscellaneous categories. Landfill can be an effective and low external cost way of disposing of wastes. However, many landfills are poorly operated and impose external costs via effects on the air, water, and landscapes.

As previously mentioned, automobiles are a special problem. Of the 10–20 million junk cars in existence at any one time in the United States, about 73% are in the hands of wreckers, in other words, in junk

24. R. Black, A. Muhich, A. Klee, H. Hickman and R. Vaughn: *The National Solid Waste Survey*. U.S. Department of Health, Welfare, and Education, 24 Oct. 1968. Also W. O. Spofford, Jr. Solid waste management: Some economic considerations. *Natural Resources Journal*, vol. II (1971).

yards; about 6% in the hands of scrap processers; and about 21% abandoned and littering the countryside. Recovery could be made much more economical by slight design changes, but presently there are no incentives to do so. Furthermore, unless it is managed to avoid them, the recycle or "secondary materials" industry itself can cause substantial external costs. For example, automobiles are usually burned prior to being prepared for scrap metal, and this can be an important source of local air pollution. Some of the processes involved in recycling automobiles have very high noise levels. While it seems clear that recycle of materials is underused under present circumstances, suffering as it does from tax, labelling, and other disadvantages with respect to new materials, it is also true that it is not a total panacea as one might gather from some of its more ecstatic adherents. For example, the paint on automobiles could not be recycled except at the expense of immense quantities of energy and other resources. Moreover, some materials such as paints, thinners, solvents, cleaners, fuels, etc., cannot perform their functions without being dissipated to the environment.

## V. The Flow of Materials [25]

To tie together some of the points made in previous sections, it is useful to view environmental pollution and its control as a materials balance problem for the entire economy. Energy residuals could be treated in an entirely parallel fashion, but I will not discuss this here.[26]

The inputs to the system are fuels, foods, and raw materials which are partly converted into final goods and partly become residuals. Except for increases in inventory, final goods also ultimately enter the residuals stream. Thus goods which are "consumed" really only render services temporarily. Their material substance remains in existence and must either be reused or discharged to the environment.

In an economy which is closed (no imports or exports) and where there is no net accumulation of stocks (plant, equipment, inventories, consumer durables, or residential buildings), the amount of residuals inserted into the natural environment must be approximately equal to the weight of basic fuels, food, and raw materials entering the processing and production system, plus gases taken from the atmosphere. This re-

25. This section is based heavily on R. U. Ayres and A. V. Kneese: "Production, consumption, and externalities." *American Economic Review 59,* no. 3 (June 1969).

26. While very little direct exchange between material and energy occurs, it is important to note that there are significant tradeoffs between these residuals streams. For example, an effort to achieve complete recycle with present levels of materials flow would require monstrous amounts of energy to overcome entropy.

sult, while obvious upon reflection, leads to the at first rather surprising corollary that residuals disposal involves a greater tonnage of material than basic materials processing, although many of the residuals, being gaseous, require no physical "handling."

Fig. 1 shows a materials flow of the type I have in mind in a little greater detail and relates it to a broad classification of economic sectors. In an open (regional or national) economy, it would be necessary to add flows representing imports and exports. In an economy undergoing stock or capital accumulation, the production of residuals in any given year would be less by that amount than the basic inputs. In the entire U.S. economy, accumulation accounts for about 10–15% of basic annual inputs, mostly in the form of construction materials, and there is some net importation of raw and partially processed materials amounting to 4 or 5% of domestic production. Table 2 shows estimates of the weight of raw materials produced in the United States in several recent years, plus net imports of raw and partially processed materials.

Of the active inputs,[27] perhaps three-quarters of the overall weight is eventually discharged to the atmosphere as carbon (combined with atmospheric oxygen in the form of CO or $CO_2$) and hydrogen (combined with atmospheric oxygen as $H_2O$) under current conditions. This results from combustion of fossil fuels and from animal respiration. Discharge of carbon dioxide can be considered harmless in the short run, as we have seen, but may produce adverse climatic effects in the long run.

The remaining residuals are either gases (like carbon monoxide, nitrogen dioxide, and sulfur dioxide—all potentially harmful even in the short run), dry solids (like rubbish and scrap), or wet solids (like garbage, sewage, and industrial wastes suspended or dissolved in water). In a sense, the dry solids are an irreducible, limiting form of waste. By the application of appropriate equipment and energy, most undesirable substances can, in principle, be removed from water and air streams [28]—but what is left must be disposed of in solid form, transformed, or reused. Looking at the matter in this way clearly reveals a primary interdependence among the various residuals streams which casts into doubt the traditional classification, which I have used earlier in this article, of air, water, and land pollution as individual categories for purposes of planning and control policy.

Residuals do not necessarily have to be discharged to the environment. In many instances, it is possible to recycle them back into the productive system. The materials balance view underlines the fact that the throughput of new materials necessary to maintain a given level of production and consumption decreases as the technical efficiency of energy conversion and materials utilization and reutilization increases.

27. See footnote to Table 2.
28. Except $CO_2$, which may be harmful in the long run, as noted.

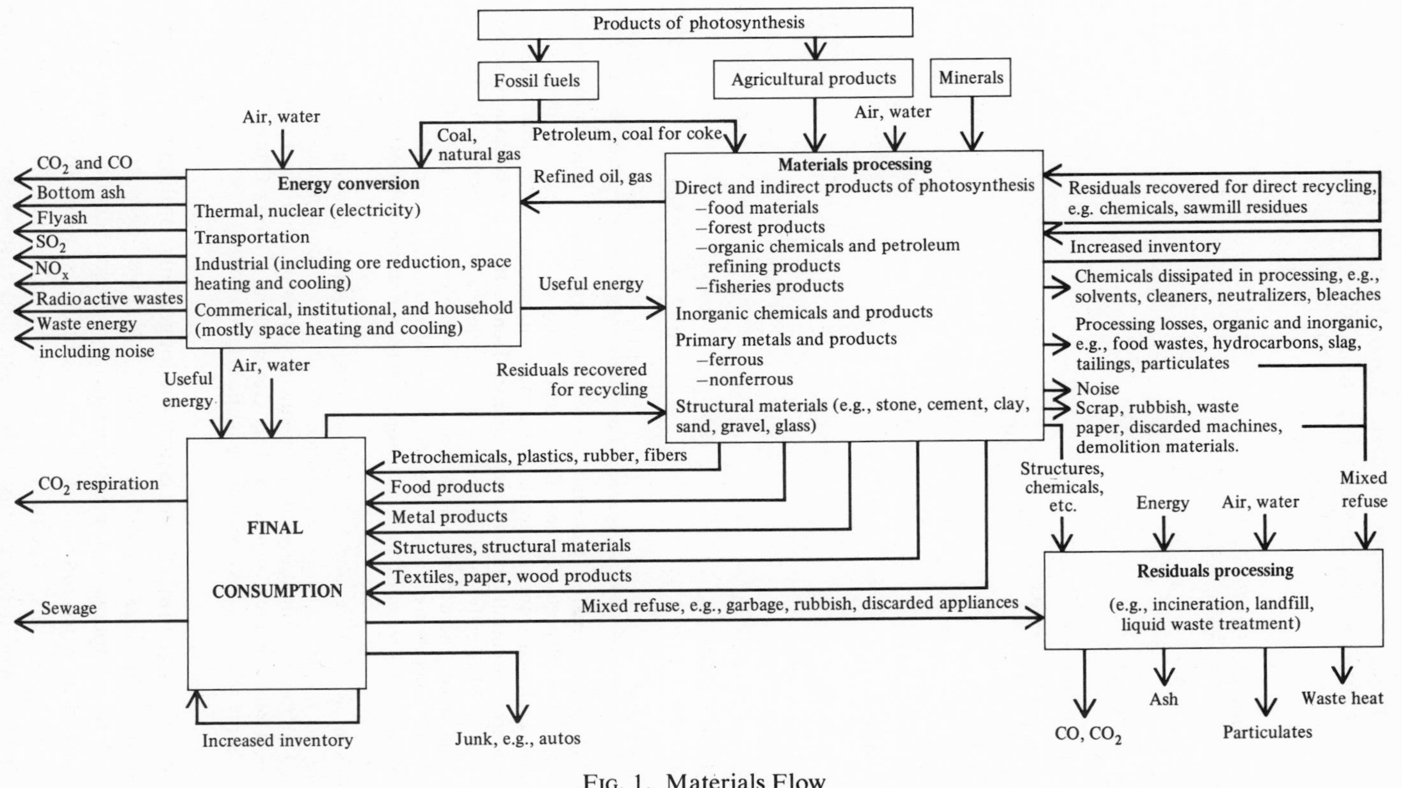


FIG. 1. Materials Flow

TABLE 2. Weight of Basic Materials Production in the United States, plus Net Imports, 1963–65 (Million Tons)

| | 1963 | | 1964 | | 1965 | |
|---|---|---|---|---|---|---|
| *Agricultural (incl. fishery and wildlife and forest) products* | | | | | | |
| Food and Fiber | | | | | | |
| Crops | 350 | | 358 | | 364 | |
| Livestock and dairy | 23 | | 24 | | 23.5 | |
| Fishery | 2 | | 2 | | 2 | |
| Forestry Products (85% dry wt. basis) | | | | | | |
| Sawlogs | 107 | | 116 | | 120 | |
| Pulpwood | 53 | | 55 | | 56 | |
| Other | 41 | | 41 | | 42 | |
| Total | | 576 | | 596 | | 607.5 |
| *Mineral Fuels* | | | | | | |
| Total | | 1 337 | | 1 399 | | 1 448 |
| *Other Minerals* | | | | | | |
| Iron ore | 204 | | 237 | | 245 | |
| Other metal ores | 161 | | 171 | | 191 | |
| Other non-metals | 125 | | 133 | | 149 | |
| Total | | 490 | | 541 | | 585 |
| Grand Total[a] | | 2 403 | | 2 536 | | 2 640.5 |

[a] Excluding construction materials, stone, sand, gravel, and other minerals used for structural purposes, ballast, fillers, insulation, etc. Gangue and mine tailings are also excluded from this total. These materials account for enormous tonnages but undergo essentially no chemical change. Hence, their use is more or less tantamount to physically moving them from one location to another. If these were to be included, there is no logical reason to exclude material shifted in highway cut-and-fill operations, harbor dredging, landfill plowing, and even silt moved by rivers. Since a line must be drawn somewhere, we chose to draw it as indicated above.

SOURCE: R. U. Ayres and A. V. Kneese: Environmental Pollution. In *Federal Programs for the Development of Human Resources,* a Compendium of Papers submitted to the Subcommittee on Economic Progress of the Joint Economic Committee, United States Congress, Vol. 2. Government Printing Office, Washington, 1968. Some revisions have been made in the original table.

Similarly, other things being equal, the longer cars, buildings, machinery, and other durables remain in service, the fewer new materials are required to compensate for loss, wear, and obsolescence—although the use of old or worn machinery (e.g., automobiles) tends to increase other residuals problems. Technically efficient combustion of (desulfurized) fossil fuels would leave only water, ash, and carbon dioxide as residuals, while nuclear energy conversion need leave only negligible quantities of material residuals (although thermal pollution and radiation hazards cannot be dismissed by any means).

Given the population, industrial production, and transport service in an economy (a regional rather than a national economy would normally be the relevant unit), it is possible to visualize combinations of social policy which could lead to quite different relative burdens placed on the various residuals-receiving environmental media; or, given the possibilities for recycle and less residual-generating production processes, the overall burden to be placed upon the environment as a whole. To take one extreme, a region which went in heavily for electric space heating, electric transportation systems, and wet-scrubbing of stack gases (from steam plants and industries), which ground up its garbage and delivered it to the sewers and then discharged the raw sewage to watercourses, would protect its air resources to an exceptional degree. But this would come at the sacrifice of placing a heavy residuals load upon water resources. On the other hand, a region which treated municipal and industrial waste water streams to a high level and relied heavily on the incineration of sludges and solid wastes would protect its water and land resources at the expense of discharging waste residuals predominantly to the air. Finally, a region which practiced high-level recovery and recycle of waste materials and fostered low residual production processes to a far-reaching extent in each of the economic sectors might discharge very little residual waste to any of the environmental media.

Further complexities are added by the fact that sometimes it is, as we have seen, possible to modify an environmental medium through investment in control facilities so as to improve its assimilative capacity. The easiest to see but far from only example is with respect to watercourses where reservoir storage can be used to augment low river flows that ordinarily are associated with critical pollution (high external cost situations). Thus, internalization of external costs associated with particular discharges, by means of taxes or other restrictions, even if done perfectly, cannot guarantee Pareto optimality. Collective investments involving public good aspects must enter into an optimal solution.

To recapitulate the main points these considerations raise for economic analysis briefly: (1) Technological external diseconomies are not freakish anomalies in the processes of production and consumption but an inherent and normal part of them. Residuals generation is inherent in virtually all production and consumption activities, and there are only two ways of handling them—recycle, or discharge into environmental media without or with modifications. (2) These external diseconomies are apt to be quantitatively negligible in a low-population or economically undeveloped setting, but they become progressively (nonlinearly) more important as the population rises and the level of output increases (i.e., as the natural reservoirs providing dilution and other assimilative properties become exhausted). (3) They cannot be properly dealt with by considering environmental media, such as air and water, in isolation. (4) Isolated and ad hoc taxes and other restrictions are not sufficient for their optimum control, although taxes and restrictions are essential elements in a more systematic and coherent program of environmental quality management. (5) Public investment programs, particularly including transportation systems, sewage disposal systems, and river flow regulation, are intimately related to the amounts and effects of residuals and must be planned in light of them. (6) There is a wide range of (technological) alternatives for coping with the environmental pollution problems stemming from liquid, gaseous, and solid residuals. Economic tools need to be selected and developed which can be used to approximate optimal combinations of these alternatives.

# 3

CLIFFORD S. RUSSELL AND
HANS H. LANDSBERG

# *International Environmental Problems—A Taxonomy*

*Clifford S. Russell is a Research Associate, and Hans H. Landsberg is Director of the Appraisals Program at Resources for the Future.*

The last few years have seen an explosion of interest in environmental problems among citizens of the developed countries, both East and West. Most of this interest has focused on domestic situations and on possible changes in domestic policies designed to provide remedies. Increasingly, however, the focus has widened to embrace environmental concerns that transcend national borders. A high point may be reached in June 1972, when the United Nations Conference on the Human Environment will be held in Stockholm. Wide-ranging discussions and the signing of international treaties on specific international environmental issues are on the agenda. Even though it will not be the first conference on these subjects,[1] both the auspices of the United Na-



1. Notable examples are the Study on Critical Environmental Problems held at Williams College during 1970, the results of which have quickly become available in an initial report [*Man's Impact on the Global Environment: Assessments and Recommendations for Action* (M.I.T. Press, Cambridge, Mass., 1970)], the

tions and the publicity that it is bound to receive will give it special importance.

The growth of interest and enthusiasm is not matched by accomplishments. That little action has been taken is perhaps easily explained, since sovereign states are involved in these issues, which are old as a class but essentially novel in degree. So far, it has even proved difficult for concerned parties to discuss the problems.[2]

A major reason for the lack of communication has been the general failure to look beyond the label, "international environmental problems," to the disparate elements it covers and to limit, in advance, the number of such elements that can be discussed at any one time and in any given group. A second reason may be that environmentalists have sometimes couched their arguments in terms that impugn the morality and intelligence of the parties concerned, thus guaranteeing defensive, hostile reactions.[3]

A third reason may be that management of international environmental problems is most often thought of in terms of "police actions" and regulatory authorities rather than as a component of growth and development. It should be realized that this component of growth and development, neglected by the now developed countries (and being paid for

---

Prague Conference in 1971, sponsored by the United Nations Economic Commission for Europe, and the United Nations Conference on the Human Environment in Stockholm, June 1972.

2. The exception has been the increasing activity by international scholarly bodies. For example, the Scientific Committee on Problems of the Environment was formally established late in 1970 by the International Council of Scientific Unions and so far has set up at least three study groups. These have begun to conduct research on matters ranging from the effect of chlorinated aromatic compounds on human tissue to the scope and methods of worldwide environmental monitoring systems and the structure and functioning of ecosystems as influenced by man. Specialized agencies like the World Meteorological Organization (with its World Weather Watch) or the planned Global Atmospheric Research Program (to be undertaken as part of the World Weather Program) are similar ambitious programs. The International Biological Programme has been functioning for some years now. Others, such as UNESCO's Man and Biosphere, have barely begun to function. Below the global level, study programs are under way in the Organization for Economic Cooperation and Development, at NATO, at the Council of Europe, and others, but none of these are at this time action-oriented.

3. Thus, some scientists appear to feel that the setting of environmental standards is a "scientific" question, like that of understanding environmental mechanisms. One set of standards then is "correct"; others are "wrong." In fact, however, scientific inquiry only furnishes society with the understanding and the data on which to base decisions. Given identical facts, the differing tastes, preferences, and prejudices of different members of society will lead them to advocate different standards. Because of the impossibility of trading the results of environmental standards in a private market, the problem of combining individual preferences to arrive at a social decision becomes a political problem. Politicians make these decisions not by default, but because such decisions are the very substance of democratic government.

dearly by them), can still be built into the development of emerging countries, probably with long-lasting benefits. Finally, political problems, in terms of a lack of new institutions and mechanisms, have played a role. For example, the growing pollution of the Baltic Sea involves eight countries, three of which are in the Soviet orbit, and one of which (East Germany) has a sufficiently undefined international status to make any international agreement difficult, at best, to achieve.

Many environmental problems involve citizens of two or more countries and hence are "international." Confusion and controversy arise easily: an individual or a government is usually concerned with (or even aware of) only one or two specific problems and incorrectly assumes that other individuals or governments are talking about the same problem when they use the same general label. Consider, for example, the prospects for agreement when one group's mind is on the long-term buildup of carbon dioxide in the atmosphere and particulates in the lower stratosphere; another worries over the dangers associated with increasing storage of radioactive wastes; a third focuses on the ecological implications of large-scale hydroelectric developments in the tropics; a fourth is concerned with the effect of domestic air pollution controls on export prices and hence trade patterns; yet another is concerned about a specific regional problem in which one nation's pollution, or attempt at protection against pollution, imposes costs on another nation; and, finally, a group of developing nations views matters through the prism of its overwhelming interest in increasing per capita income.

Some of these situations affect all the world's people, though significant contributors may number only a handful. Others are problems of a particular region and do not concern nations outside that region. To developing nations, all environmental problems may appear to be potential threats to their domestic development. At the least, they seem to be concerns of those nations that have incomes sufficiently high to permit concern with esthetics and that have health standards high enough to permit detection of the effects on morbidity and mortality of concentrations of sulfur dioxide. To lay the basis for more successful discussion, this article suggests a first cut at a taxonomy of international environmental problems and solutions, as well as areas in which further research can contribute to this discussion. We take the point of view of the social scientist, since that view is most likely to speak directly to the concerns of those who must ultimately do the discussing and deciding, but the categories we suggest are based on characteristics of the physical world.

## A Taxonomy of Problems

International environmental problems may profitably be divided into two broad categories, depending on the nature and scope of the international linkages involved: physical-linkage effects and social-linkage effects. The first may be divided again into global and regional effects, and the second into pecuniary and nonpecuniary effects.

Global problems are those problems that physically involve all or nearly all nations of the world, either as contributing parties (emitters) or damaged parties (receptors) or both. Some of the most widely discussed environmental issues fall into this category. For example, since World War II, persistent pesticides have been used all over the world in programs to control disease vectors and agricultural pests. The residues directly affect animal life and potentially affect human life, not only in the country in which a specific application is made, but also through the actions of wind, water, and living carriers, even in regions remote from the point of origin. Notice that it is the combination of persistence and mobility that makes the pesticide problem a global one. If any significant user remains outside a control agreement and continues application of pesticides, the impact on everyone may still be felt.[4] Even if they were applied by virtually every nation, highly toxic pesticides that were used in small quantities and that broke down quickly to inert residual chemicals might not constitute an international problem. Each nation could manage its own environmental quality problem by enacting its own laws: It would not be dependent for effectiveness on simultaneous action by all other user nations.[5]

Two familiar examples of global problems, the balance of carbon dioxide in the atmosphere and the particulate content of the stratosphere, are closely related to man's burning of fossil fuels, and both tend to affect the earth's temperature.[6] Here again, the essential elements of the problem are persistence in the atmosphere (for carbon

4. Persistence alone is not sufficient to cause a global problem. A discarded tire is extremely persistent and creates a visual blight that is surely an environmental quality problem. But the tire is not subject to global transportation by the natural system into which it is discharged. Similarly, mobility without persistence would not create the problem discussed.

5. How subtle are the distinctions involved is evidenced by the possibility that even nonmobile, nonpersistent pesticides could have affects beyond national borders if (i) they affected migratory birds, for example, and thus broke the food chain, or (ii) they were used near international boundaries. But control of these cases would require agreement among a much smaller set of nations.

6. For a recent comprehensive survey of the subject, see H. E. Landsberg [*Science* 170, 1265 (1970)].

dioxide, the extended time scale of the carbon cycle and its components) and the global span of the physical systems involved. Carbon dioxide is relatively stable, and the molecules are not removed from the atmosphere very rapidly. The very small particulates do not settle out rapidly, but tend to remain in suspension in the stratosphere over long periods. Thus global agreement will eventually be needed to assure a long-term solution. However, since the sources of fossil-fuel emissions are highly concentrated now in the developed countries (North America, Western Europe, the Soviet Union, and Japan), an agreement on limitations among this group would probably result in a solution good for several decades at least.[7]

Carbon monoxide and larger particulates are also produced in great quantities in some forms of fossil-fuel combustion and cause specific environmental problems. Carbon monoxide, however, is not stable; it does not survive long in that form in the air and is not a problem at some distance from the source. Similarly, larger particulates tend to settle out relatively quickly after emission. Thus, even though these products of combustion are as ubiquitous as carbon dioxide and the very small particulates, they create almost entirely intranational or local difficulties.

A fourth example of a global environmental problem is the dumping or spillage of oil on the high seas. Here, a look at the registry of the world tanker fleet makes it clear that relatively few nations are emitters. Through its effects on marine life, however, the dumping affects the much larger group of nations that engage in ocean fishing (or that, by a slightly more remote linkage, depend on another nation's catch). Ultimately, it could affect every nation directly: if, for example, the as yet controversial possibility that continued large-scale dumping could affect the photosynthetic activity of the seas and hence could threaten one of our largest sources of oxygen were to be demonstrated; or if changes in the surface reflectivity of the water affected the earth's heat balance. Once again, the global nature of the problem results from the discharge of a persistent residual into a natural system that spreads the effects of the residual over long distances. For example, Thor Heyerdahl sighted numerous, large globs of oil in a recent Atlantic crossing. The list of global environmental problems, which, prior to the nuclear test ban treaty, also included worldwide transport of radionuclides, may become longer as our ability to measure trace elements and track their movements increases and we are alerted to interdependencies as yet unseen. But for the moment, the issues listed appear to constitute the major, truly global phenomena.[8]

7. See W. O. Spofford, "Decision-making under uncertainty: The case of carbon dioxide build-up in the atmosphere," paper prepared for the Study of Critical Environmental Problems, Williams College, July 1970.

8. Dealing with global issues is, in one sense, not a totally new experience. Control of infectious diseases affecting human, animal, and plant life has long been

## Regional Problems

Regional problems result from physical, including biological, linkages between two or more nations, with little or no spillover to the world at large because of the particular combinations of relatively low persistency of pollutants and relatively limited scope of the natural systems involved in transporting them. Regional problems often resemble the domestic environmental quality issues now facing most developed nations. That is, because geographic proximity frequently permits identification of "upstream" and "downstream" countries, the *assignment* of costs of control and the benefits from damages avoided presents no difficulty, even if the *estimation* of damages avoided may, in principle, be impossible in many cases. Many of the same analytical techniques developed for dealing with domestic quality issues are directly applicable to regional situations. Examples of such situations are most common in the highly industrialized parts of the world: their natural systems are subject to the most stress, both in quantity and variety of pollution discharges; agriculture tends to be more intensive, with the attendant use of fertilizers and pesticides; [9] "conventional" pollutants associated with lack of sanitary facilities (toilets, sewers, waste disposal, and so on) are at a low level, thus giving greater visibility to the pollutants associated with high technology (detergents, scrapped automobiles, carbon monoxide, and so on); and incomes are high enough to permit people to concern themselves with damages to esthetic values and recreational opportunities.

There is no lack of examples of regional problems of the upstream-downstream variety. The Rhine serves France and Germany as a sewer, but it serves the Netherlands as a part of its water supply. Acid rainfall over western Sweden and eastern Norway has been attributed by some scientists to sulfur oxide emissions originating in industrial operations in Germany's Ruhr and England's Midlands. As a result, trout fishing in southern Norway is threatened, and there is a suspicion that the growth of trees is being slowed down. The Environ-

---

practiced on a world-wide basis with considerable success, since in many instances carriers could be identified and thus isolated. Note also that global "concern," of course, extends far beyond this list, to include population growth, the nuclear threat, and so on.

9. Even this situation is not without its fuzzy edges. Agriculture in the developed countries is conducted with more care for soil erosion; thus, dispersion of nutrients into the environment is lessened. On the other hand, poor farming practices in the less developed countries may be the cause of the major portion of particulates in the atmosphere, according to the Williams College report (see footnote 1, above). If so, poor agricultural practices would have to be considered a true global environmental issue.

mental Committee of the Organization for Economic Cooperation and Development has been asked to investigate the problem, and management of the resulting project has been entrusted to Norway. The seaborne flow of pollutants from Italy to France along the Riviera is another instance of the upstream-downstream syndrome. The use of Arctic waters by U.S. oil tankers, should this development occur, would be yet another, involving Canada, the United States, and, depending on the implication of ocean currents and the like, possibly other countries.

In other cases, because of the natural system involved, all parties become both emitters and receptors. This is true, for example, of the Baltic Sea and of Lake Erie. Countries around the Baltic are concerned, above all, about oil transportation and mercury pollution from pulp mills. The narrow link with the North Sea makes much of the Baltic practically an inland sea body, and the results of a major oil spill could bring great harm to any or all of the eight countries involved.

Other cases involve despoliation rather than pollution in the conventional sense, but still present the upstream-downstream pattern of damage. Thus, European conservationists are concerned about the effects that the Italian practice of netting will have on migratory bird populations. Large numbers of birds that winter in Africa and summer in the north of Europe are trapped each year as they migrate up the Italian peninsula. Finally, there are the major environmental alterations that involve neither pollution nor despoliation. A case in point is the Aswan Dam. By cutting the flow of silt and organic debris in the Nile, it appears to have adversely affected the eastern Mediterranean sardine fishery (apart from other consequences that are purely domestic at this time).

Although regional and global problems have many similarities, it is useful to distinguish between them in order to emphasize that not every international environmental problem need—or should—be grist for the mill of the United Nations. It may, in fact, be helpful in seeking a solution to involve only the smallest possible group of nations—generally those directly interested.[10] This is not, however, to suggest that the distinction between the two classes will always be clear. Realistically, one must expect that some large-scale regional problems will be most conveniently dealt with as global issues, while the interests of a very few powerful nations may so dominate a global problem that its solution rests, at least initially, entirely with them.[11]

10. Recent work in political theory on legislatures, individual preferences, and social decisions suggests that it is preferable not to have a decision made on an issue by a legislature in which few individual legislators have a direct interest in that issue [see E. Haefele, *Amer. Econ. Rev.* 61, 217 (1971)]. An analogy between legislators and national government representatives in international organizations does not seem far-fetched.

11. This is true of nuclear disarmament, the extreme case of a threat to life on earth. On a less extreme level, G. F. Kennan has suggested that global problems

## Social-Linkage Effect

Social-linkage effect is the term we use to refer to a second class of international environmental problems in which no physical linkages exist but in which, nonetheless, the policies of one national government impinge directly on the well-being of citizens of one or more other nations. This may occur through established economic relationships between nations (that is, trade and investment, including foreign aid), or in a way that is not, in the first instance, pecuniary. We deal with the second class first and call it, for want of a better term, nonpecuniary linkage.

A classic case of nonpecuniary linkage is that of one country's possessing unique natural or historical gifts that citizens of other nations value as part of the human cultural and natural heritage.[12] Thus, for example, Uganda's plan to develop a hydroelectric scheme that would involve cutting by 90 percent the flow through Murchison Falls (a very narrow gorge on the Upper White Nile) has aroused considerable ire among conservationists, particularly those in Europe.[13] The filling of Lake Nasser, behind the Aswan High Dam, stimulated an international effort to rescue a number of tombs, temples, and statues erected along that stretch of the river by early Egyptian civilizations. Similarly, the Italian government's care of remnants of antiquity has postponed the completion of Rome's subway, but has probably prevented it from becoming an international issue.

Many countries, particularly the developing nations, have been criticized for failing to take effective action to protect animal species that are valued by conservationists around the world. Perhaps the most widely known examples are the African cats, especially the leopard, which are endangered by poachers, and the Ceylonese elephant, threatened by the large-scale clearing of forests for agriculture.

These situations are local phenomena, but they become international problems when citizens of other countries protest and attempt to obtain actions that accord with their own values rather than with those of the country concerned. Aroused citizens implicitly or explicitly involve their own governments in their efforts. Thus, the opponents of the Mur-

are best handled by the developed nations in any event—not only because they are the principal polluters but because it will be too much to expect the rest of the world to take an interest in a problem that does not loom large at that stage of economic development [*Foreign Aff.* 48, 401 (April 1970)].

12. J. V. Krutilla, C. J. Cicchetti, A. M. Freeman. C. S. Russell, in *Environmental Quality Analysis; Theory and Method in the Social Sciences,* V. Kneese and B. T. Bower, Eds. (Johns Hopkins Press, Baltimore, 1972).

13. C. S. Russell, *ibid.*

chison Falls Dam would persuade the British government to renege on its promise of financing the initial construction phase. Thus the nonpecuniary interaction can become pecuniary.

It is worth noting that motives are of the highest character in both camps; and, as is always the case when neither side is villainous, the problem appears in its purest and most difficult to solve form.

This is equally true of yet another variety of nonpecuniary interaction—international altruism, in which citizens of one nation endeavor to help citizens of another nation avoid mistakes in dealing with the environment. For example, agricultural experts in the West may be anxious to help African nations avoid exhaustion and erosion of laterite soils. This type of interaction will only lead to problems if the country to be assisted does not agree that the proposal is in its best interests, or if the outside altruists become too insistent or paternalistic.

This situation is apt to arise, above all, in large-scale, agriculture-oriented engineering works. As an example, consider the controversy over the eventual benefits of the Aswan High Dam. Leaving aside the question—often hard to judge—of whether or not some of the effects were or could reasonably have been anticipated, it is useful to distinguish between adverse effects that diminish the chances of success of the primary project objectives, and those that adversely affect some other environmental facet.

In the first instance, decreased soil fertility or increased salinity, for example, if indeed resulting from the changed characteristics of the river, would directly diminish the project's objective—that is, higher agricultural production. Provided there was no dispute on the scientific findings, it would be a straightforward computation to evaluate the size of the loss in dollars and cents. There would be no room for dispute over the consequences. By contrast, the spread of schistosomiasis, while also damaging to the country, could not, in the same sense, be calculated as a direct offset to agricultural production. Outsiders might view it as part of an ecological horror story and consider it a serious offset to the value of the project as a whole, but the national government, given the already very wide diffusion of the disease in rural Egypt, might view it less severely. Indeed, it might even consider that increased output and income might, in the long run, provide a better basis for a successful battle against the disease *anywhere*.

Weighing of environmental effects is, then, unlikely to lead to controversy in the first instance (direct relation to project objective) but apt to do so in the second (adverse side effects related only tenuously, if at all, to project objective).

Divergences of judgment, as described above, sometimes leading to the attempt to impose some kind of sanctions on another country, are not new. Boycotts, embargoes, and other measures have been used in the past to express disapproval of a country's behavior and force it to comply with more acceptable standards. What is new here is the issue that gives rise to such conflict and pressures: and as that issue gains increased status among the aspirations of mankind, the opportunities for intervention, as well as the felt justification, are bound to rise.

## Pecuniary Effects

The issue becomes at once more pedestrian and more pointed when we turn to another class of social linkage—namely, the pecuniary effects upon country B of specific environmental policies followed in country A. Here the cases shade into well-known phenomena in foreign trade, even though the impetus lies in a newly prominent field, the environment.

Thus, when the United States adopts strict automobile emission standards, it raises the costs of European and Japanese auto manufacturers who wish to export to this country. Even if the sales price of U.S. automobiles should rise in proportion, foreign manufacturers will need to make special provisions for cars sold in the U.S. market. This fact, together with the likelihood of a proportionately greater financial burden on smaller cars, will directly affect the income of the owners and employees of foreign firms, and will indirectly affect the income of other citizens of those nations. Similarly, limits on the permissible sulfur content of fuels burned in U.S. and European cities imply gains for those nations that own low-sulfur fuel reserves, and lost markets or decreased profits (because of the costs of desulfurization) for others. These examples are only extensions of long-standing regulations in food imports, for example, where tolerances for specific ingredients or impurities, or observation of stated sanitary procedures, are prerequisites for admission to the U.S. market. They bring us squarely up against trade effects as international environmental problems; and for many people trade effects are the most immediate and important of such problems. Thus, it will be valuable to pause and consider how trade effects fit into the taxonomy we are proposing.

There are generally two kinds of trade effects: (i) loss of export markets as a consequence of the increased costs of maintaining high environmental quality in the exporting country, or (ii) the erection of barriers to imports in line with the importing country's policies on environmental quality. In the first case, by forcing domestic manufacturers to absorb the costs of disposing of production residuals, environmental quality legislation will tend to diminish the competitiveness of domestic products in the world market. This, in turn, will lead to a decline in domestic income and employment and to losses in the value of invested capital.[14] The second effect is exemplified by the standards on auto emissions and sulfur content of fuels. Here the major losses will arise through action in the importer's country, which raises the additional specter of retaliation.

There is no doubt that both situations can create friction between nations and be the subject of negotiation, unilateral action, and so on. Hence, both are international problems. This is obvious with barriers to imports, but the implications of the loss of export markets are no less disconcerting. As put by Germany's Minister of the Interior Genscher:

> We must . . . avoid a situation in which individual countries exclude themselves from making investments for environmental protection, thereby securing competitive advantages for their own economy vis-à-vis those countries who do meet their responsibilities.

What is noteworthy here is the use of the term "responsibility," placing environmental effects in a context beyond voluntary action.

It is important to realize, however, that trade effects are quite different from the direct, physical-linkage effects. The loss of export markets due to action in the exporting country is simply an international facet of the classical adjustment process necessary within an economy when tastes or ground rules shift markedly, causing capital and labor to move out of some industries and into others. Some owners of capital suffer unanticipated losses, and, in the short run, lower incomes and some unemployment will result. But the real cost of producing the same goods (or of obtaining the goods internationally) has increased by what it costs to achieve the higher level of environmental quality. Hence, given government policies designed to maintain aggregate demand, there is no reason that full employment cannot again be obtained.[15] In the long run, and in the absence of similar environmental policies by other governments, the nation will tend to import those goods that involve the greatest environmental costs and to export those involving the least. It will, in effect, be exporting pollution.

The emergence of so-called "pollution havens" is not a theoretical

14. We emphasize that income, as measured by such market-based indices as gross national product, will decline relative to what it would have been in the absence of the domestic policies. The fact that the policies were adopted, however, shows that the society judged it would be better off with them than without them. Nonmarket income has increased enough to offset the decline in market income.

15. Even in the short run, wise government policies may ease the transition considerably. For example, worker retraining allowances and expanded unemployment coverage, as well as subsidized loans for machinery conversions, can cut the frictional costs of the required employment shifts. Such policies to offset trade effects are possible under the current tariff law but have almost never been used. There are signs that this situation is changing. For a discussion of assistance being given the Massachusetts shoe industry, see *New York Times,* 21 February 1971, section F, p. 2.

consideration. For example, air pollution standards have led to reductions in copper smelting operations in Arizona, Texas, Montana, and Washington, and to an increase in shipments of ore to smelters in West Germany, Canada, and Japan. Japan is reported to be shipping ore to Indonesia for smelting, though on a very small scale. A movement in the opposite direction is our export of coal and lumber to Japan. To the extent that those exports are stimulated by lower prices, made possible by a lack of strip-mining regulation and of control of timber overcutting, the United States is functioning as a pollution haven, suffering land erosion, acid mine drainage, reduction of wildlife stocks, and disruption of natural vistas in order to support domestic employment in export industries.

Environmental policies that act as direct obstacles to trade affect the outside world as do other nontariff barriers such as product quotas or import prohibitions. For the domestic economy of the importing countries, the higher real cost of imports is, again, simply one facet of the overall cost of attaining the desired level of environmental quality.

Because international trade problems have been dealt with extensively elsewhere, little need be said about them here, even though they arise in a novel context. However, the novelty must be qualified. Nontariff barriers that confer large gains on specific industries are frequently justified in terms of noble objectives—to further health or some other aspect of the well-being of some or all citizens. When environmental objectives act as trade barriers, it is not, therefore, surprising that the motives of those who are responsible for them will be suspected.

It is difficult to predict the effects on international trade of domestic environmental quality policies that are imposed unilaterally. The one attempt with which we are familiar is limited to five major developed countries (France, Japan, West Germany, the United States, and the United Kingdom). The reported results must be interpreted with caution because of the rather strict assumptions and limited data on which they are based.[16] Nevertheless, this study does indicate the order of magnitude of the short-term impact on national income and the balance of payments of domestic environmental control policies, if it is assumed that each nation, in turn, acts alone while others continue present policies and that the government of the nation imposing the controls does not pursue policies designed to reduce the resulting economic dislocations. In this situation, the predicted effect on the balance of payments and national income of the United States is quite small; on those of West Germany, moderate; and on those of the other three nations, substantial. These results suggest that U.S. environmental policy need not

16. R. C. d'Arge, Appendix F to A. V. Kneese, "The Economics of Environmental Pollution in the United States," a paper prepared for a meeting of the Atlantic Council, Washington, D.C., 1970.

be constrained by fears of serious national income and balance of payments implications, but that other major developed nations will probably be extremely cautious about acting, except in concert with their major trading partners, in such a way as to force domestic industry to take account of environmental damages. There is the additional complication that the degree to which adjustments are borne by society as a whole (for example, through subsidies paid out of general revenue), as against being reflected in the price of the affected commodity, will impinge upon relative prices and thus competitiveness in international markets.

Environmental considerations are equally likely to affect the allied field of investment. We have pointed out the tendency to import, rather than produce at home, goods that have a highly adverse effect on the environment. In addition, some countries do not impose strict regulations, either because conditions (whether natural or otherwise) do not yet require them, or because they wish to attract capital. As a result of these two factors, investments may be shifted to those countries. It is easy to see how such moves can lead to international friction.

These investment effects also appear in the field of foreign aid, whether bilateral or multilateral, except that here the reverse situation obtains. By U.S. law, development projects financed by the U.S. government must now be evaluated for their impact on the environment. In a parallel development, the World Bank has recently established a program designed to look into adverse effects of foreign aid on the environment. From such evaluations are likely to come actions to prevent or remedy adverse environmental consequences. Hence, the cost of a given project is likely to be higher than it would otherwise have been, and the host country will be concerned over the competitive status of the goods and services that will result from the investment. Recipients of aid do not look kindly on the need for additional foreign exchange, perhaps foreign technicians, and further delay in achieving economic independence—all for benefits often little understood or valued.

## Shared Experiences

Before abandoning the taxonomy of problems and proceeding to that of solutions, a comment is needed on a range of matters that are not in any real sense international environmental problems but that do relate to them. These are the domestic environmental experiences common to most countries at specific stages in their growth.

Problems of human settlement, especially those of large urban areas, come to mind at once. These are not new problems. Most of them have merely been given a new label. Thus Calcutta, Rio de Janeiro, Lima,

Tokyo, and New York all suffer from problems related to large concentrations of population. Similar problems were noted in London 150 years ago. The international aspect lies in the commonality of such problems, not in any interaction. And so it is with matters like soil erosion, poor drainage and resulting salinity (encountered many thousands of years ago in the plains of the Tigris and Euphrates), deforestation to meet the needs of shipbuilders (Rome), or settlers (United States), and a host of other environmental problems. Here the opportunities for international cooperation are greatest: in the exchange of information, technology, and so forth. In short, here is the possibility of progress without conflict. But, by the same token, it is not here that truly perplexing international issues are found.

A final comment is in order on the above taxonomy. A problem in any of the three categories may be—and usually is—complicated by considerations of the income distribution among the nations (and their citizens) involved. These will be particularly obvious and important when one or more of the interested parties are developing nations, but they will also be present when only developed nations are concerned. Any particular solution to an international environmental problem will involve transfers of real income from nation to nation. These can generally be identified and at least partially quantified at a technical level. But the desirable direction and size of such income transfers become two variables for consideration in the political process of choosing between alternative outcomes.

## A Taxonomy of Solutions

Solutions to international environmental problems may be either negotiated or imposed. If negotiated, the appropriate group of interested parties will, as we have suggested, be defined by the scale of the natural system involved, although considerable improvement may be obtained over a fairly long period through agreements among the smaller group of nations responsible for most of the problem.

Solutions may be imposed by a single nation or by a group of nations that has the required economic—or, in extreme cases, military—power. The imposition may be directly by force: for example, if one nation invades another to destroy a dam that has changed the flow of a river.[17] More likely are impositions based on the terms of foreign aid (as in the

17. It is hard to visualize anything less drastic, such as water pollution, as a casus belli, even though the consequences in the long run may be just as damaging. But, of course, the invocation of an environmental insult may merely mask a more traditional objective of foreign policy, economic or otherwise.

Murchison Falls example mentioned above), trade restrictions (as in auto emission standards for imported vehicles and the prohibition against importing certain furs), or by internal law operating as a trade restriction (restrictions on the sulfur content of fuels or prohibitions against the landing of SST's at domestic airports).

An imposed solution generally implies that the costs and benefits have been assessed by the imposing nation from its own point of view. But if negotiation is to be attempted, the problem of evaluating alternative solutions becomes extremely difficult. There are the usual problems of making cost comparisons among nations with different internal factor–cost structures and correcting the nominal rate of exchange to reflect at least the most serious distortions. In addition, the task of getting any real notion of the benefits will be all but impossible. Nations' preferences for the changes in environmental quality being sought will vary in accordance with their stage of economic development, cultural matrix, political structure, and so on. Moreover, these changes are associated not with private goods, where the market provides a test of preferences, but with public goods, which are consumed willy-nilly in equal amounts by all. Therefore, there exists an opportunity, if not an incentive, to conceal true preferences (for a nation as much as for an individual) and report falsely on the evaluation of benefits. Solutions may be implemented through the setting of standards or through levying charges on contributing nations, although there are tremendous difficulties in achieving either on an international scale.

Standards, in turn, may be "ambient" (that is, applying to the quality of the environment of the receptor nations) or "discharge" (applying only to the contributing nations).[18] If the mechanisms of the natural world are sufficiently understood, a set of ambient standards, if attainable at all, can be translated into a set of discharge standards.[19]

Because demands for "minimum standards" crop up so frequently in proposals for safeguarding the environment, it is well to stress that such minimums are unlikely to be either unambiguously defined or easily agreed upon. For example, is the minimum standard for oil tanker de-

18. We place the words "ambient" and "discharge" in quotes to indicate that the same notions may be used more broadly. Thus, for example, in the trapping of migrating birds, a discharge standard would require that the trapping nation allow some number or percentage of the flock to escape. An "ambient" standard would require the maintenance of some population in each of the host countries.

19. In addition, of course, the ground rules chosen in translating "ambient" to "discharge" standards will make a difference to economic efficiency. If all individual dischargers may be subject to different discharge constraints, an ambient standard may be attained most efficiently. On the other hand, if notions of "equity" seem to require a uniform standard, then some dischargers will be cutting back too far when the ambient standard is just attained. In any event, however, we are a long way from approaching such sophistication in domestic affairs, let alone international conduct.

sign simply to require hull thickness and tank sizes such that at least tankers won't break up in storms for their first 10 years? Or is it to require some maximum of oil spillage resulting from a design collision or grounding incident? Who chooses the design incident and the maximum acceptable spillage? Or, consider minimum standards for mercury contamination of foods. For any particular food, the level of contamination that a nation will be willing to tolerate will vary with the importance of that food in the diet of its citizens, *and* with the incidence of poisoning it finds acceptable. Since individual susceptibility to mercury undoubtedly varies, any given minimum standard is likely to result in some (albeit very few) cases of poisoning. The only minimum standard that can, with certainty, prevent poisoning is a zero level. Any nonzero level implies choices of probabilities of incidence; and any level high enough to be a minimum (in the sense that no nation would insist that a higher tolerance be agreed on) would undoubtedly lead to significant incidence in one or more nations.

Notice also that *any* standard which changes the status quo will create costs for some nations and benefits for others. Thus, in the tanker example, nations whose citizens own tanker fleets or ship oil will absorb costs if tankers are made more expensive. Nations with coastlines near busy international shipping lanes will reap obvious benefits from stronger construction, smaller tank size, and so on. Similarly for the mercury example, nations producing and consuming a particular food are likely to differ on the desirability of any standard, even a very low one.

The point is that minimum standards are no more objectively determined than would be optimum standards, and attempting to find and agree on a set of global minimum standards will not make the negotiating problems appreciably easier. Any standard that finds immediate and nearly unanimous international support is likely to be quite meaningless.

The term "monitoring" also tends to give rise to much confusion, even though it is a prominent activity apt to draw nations together rather than push them apart. One kind of monitoring is directed toward exploring basic processes and flows in natural systems, setting baselines, discovering what needs to be measured, and assuring compatibility of measurements carried on by different nations. Some of this activity is underway. A second sense of the term refers to compliance with set standards and is used in the context of regulation. In terms of sequence, the second sense follows the first. A clear distinction between the two meanings is helpful for avoiding unnecessary conflict and suspicion.

## Other Dimensions of Variation

International environmental problems differ in a number of dimensions other than the one we have chosen for our basic taxonomy. Thus, problems may involve different time scales between cause and effect and, hence, a different level of immediacy for the present population of the world. For example, the buildup of carbon dioxide is a long-term problem, with the possibility of any detrimental global effects many decades in the future, if they occur at all. This is apart from the fact that the environmental effects of fuel combustion and energy conversion and use generally are as yet poorly understood. Thus, any attempts at timing are highly speculative.

The persistent pesticide problem, on the other hand, is much more immediate, with consequences of past applications observable today and with every indication that the situation will worsen unless action is taken now. Related to this dimension of timing is the degree of certainty with which an event will or will not occur. Generally, the further in advance effects are predicted, the more uncertain the outcome; on the other hand, the continuation of presently observable effects is far more certain.

Two other important dimensions of environmental problems are magnitude and degree of irreversibility of effects. Because the effects will be of different types, occur in different places, and affect different facets of human life, the relative magnitude of particular problems is not easily determined, except in the infrequent cases where some estimates of monetary damages can be obtained.

The scope of effects, however, can roughly be compared: For example, a global warming trend is a "greater" effect than is the extinction of a species. However, this approach is too intuitive, and it becomes progressively less useful as one moves away from extremes.

Reversibility refers to the possibility of returning the world, or one of its subsystems, to the state it was in before some effect occurred. Thus, sulfur dioxide pollution in the atmosphere, because of the speed with which it is scrubbed out, is highly reversible. The construction of a dam, on the other hand, is generally considered to be irreversible. As a matter of fact, many effects are reversible *at some cost;* irreversibility, in everyday parlance, generally means that the cost of returning to an earlier state is very high. True irreversibility can be seen in species extinction, destruction of scenic areas, and changes in global climate and weather (except sub specie aeternitatis).

The dimensions of timing, certainty, magnitude, and reversibility of effects all contribute to the broader dimension of urgency. A situation that produces immediate (thus certain), serious, and irreversible effects is perceived as more urgent than a situation that produces long-deferred (hence uncertain), minor, or reversible effects. There is, of course, no one scale on which all environmental problems may be ranked according to urgency, but individuals and governments make subjective evaluations of this sort all the time, and these evaluations help to determine willingness to negotiate solutions.

The ability to proceed from some mutual willingness to negotiate to an acceptable settlement will depend, in general, on the number of interested parties, the degree of diversity in their development and needs, and the nature of the issues involved. For example, if "national pride" somehow becomes an issue in its own right, negotiation will be far more difficult than if only economic or esthetic questions were involved (though in practice it will be hard to make distinctions). More important, objectives differ. Thus, developing countries will generally desire to exploit their natural resources more rapidly and process more of them at home, in order to earn foreign exchange and raise per capita income. "Some of us would rather see smoke coming out of a factory and men employed than no factory at all. It is, after all, a matter of priority," commented the president of the Consumer Association of Malaysia at the recent meeting of the International Organization of Consumers Unions. At the same conference, the director of the Consumer Council of India is quoted with this remark: "The wealthy countries worry about car fumes. We worry about starvation." Presence or creation of appropriate institutions will also make a difference. The pooling of the Scandinavian countries' research efforts in "Nordforsk" is likely to promote internationally advantageous action, just as failure to do so in the context of the Baltic will retard action. Thus, building institutions is an important element of progress in negotiated settlements.

## Some Research Needs

We have attempted simply to outline and categorize the range of issues subsumed in the broad heading "international environmental problems." Solutions to these problems must rest on difficult international negotiations, but social scientists can play a useful role in setting the stage for such negotiations by undertaking research to answer questions that are currently being answered, without sound data or analysis, by advocates of one or another solution. The following are examples of such questions:

1. What are the pitfalls of various types of international agreements as revealed by previous experience, for example, in the field of marine fisheries or international communications via the radio spectrum? What

forms and safeguards have proven to be successful in, for instance, international control of uses of nuclear energy? in control of the movement of toxic materials, such as drugs?

2. What are the costs of altering the behavior currently giving rise to the problems? For example, what would be the costs (economic, political, and in terms of additional radiation hazard) of the widespread substitution of atomic energy for fossil fuel in generating electricity, in order to cut down on global emissions of carbon dioxide and particulates? What would be the cost of abandoning the use of DDT in specified areas and for specified purposes, and of replacing it with a range of alternatives? What would be the costs of undertaking alternatives to the schemes giving rise to the direct nonpecuniary interactions defined above? As a specific example, what would be the additional costs to Uganda of obtaining the generating capacity of the Murchison Falls project by some other method? Or how valuable is the generating capacity that would be lost in some compromise scheme to preserve a flow through Murchison Falls?

3. What policies could be devised to assure that the cost differences calculated above could be made up by nations voicing concern? As a subproblem, how could one effectively assure the country making the substitution—as one would have to—that the cost difference would not come out of development funds that had already been committed? And how would it be possible to erect safeguards against "ecological blackmail" (that is, the threat of an adverse undertaking as a means of securing financial indemnity for acceptable modifications, or abstention from action)?

4. What are the costs of meeting various objective sets of ambient environmental quality standards in such classic cases as the Rhine and the Baltic? Who would bear these costs in the first instance? And what mechanism could encourage payments from nations deriving benefits to those bearing costs?

This list is not meant to be comprehensive, but to outline a set of specific problems that can be tackled by social scientists with, for the most part, techniques that have already been developed in the context of domestic problems. It is purposely confined to the cost side of the problems and is designed to modify arguments about these problems, at least to the extent that proponents of change know what costs they are imposing on other nations. The task of quantifying benefits is probably beyond the competence of any scientist and must remain a matter for political judgment as exercised in the process of negotiation.

# II

# *FORMAL ANALYSIS*

The first group of papers presented the environmental problem along with a foretaste of the economist's approach to it. The second group delves more deeply into the economic analysis. The next group will pursue some of the policy implications of this analysis.

Above all, the economist sees the environmental problem as the leading instance of resource misallocation caused by "externalities"—consequences or side effects of one's acts which are borne by others. Externalities have always existed; the reasons they have recently come to be of such critical importance will be explored in Section IV. But policy recommendations and causal inferences must rest on a firm grasp of the economic forces at work. This sets the objective of this second group of papers.

The basic conceptual tools are set forth by Otto A. Davis and Morton I. Kamien. The key concept is "Pareto-optimality." When resources are allocated according to this criterion, any change in the way in which they are used will necessarily reduce the welfare of some members of the society. By the same token, when they are not so used, there will be room for a universal increase in well-being. A decentralized, market-guided competitive economy will tend toward a Pareto-optimal use of resources under certain conditions, of which the most pertinent is that there be no significant externalities, for market-guided decisions ignore external effects. When such effects are present, the market, if left to itself, will use resources, and particularly the environment, in ways that leave room for improvement, in principle at least, from everyone's point of view. Since everyone could gain from a shift to Pareto-optimal resource use, the object of environmental policy should be to see that resources are used according to this criterion. The theoretical framework of this analysis is set forth in same detail in the Introduction to this volume. But Pareto-optimal use of the environment is not easy to achieve. The latter part of the Davis and Kamien paper surveys the main strategies for moving in this direction.

The trouble with the environment is that it is a "common-property resource"—one that we all share without owning. This is a basic theme in J. H. Dales's thinking (represented later in this collection). It has been said that "everyone's property is no one's property." The economic theory of the use and misuse of common property has nowhere been better explained than in H. Scott Gordon's analysis of the overuse of a typical common-property resource, an ocean fishery. The analysis applies, of course, to all resources to which potential users have unrestricted access. But in examining the use of a common-property resource that serves only a single purpose—fish production—and a commercial one at that, Gordon has been able to abstract from two of the more cumbersome problems that plague efforts to determine both what is optimal in the use of the environment and how to achieve it. He avoids having to contend with either the problem of conflicting demands upon the same environmental resource—as occurs when a watercourse is wanted for both swimming and waste disposal—or the difficulty of placing a value on environmental services—such as personal amenities—for which there are no market prices. Thus he manages to focus on the single issue of how the common-property characteristic of the environment leads to its overexploitation.

Undoubtedly the most widely read selection in this volume is Ronald Coase's "The Problem of Social Cost." It provides the basis for much of the subsequent analysis. Coase, like many of the contributors to this volume, is concerned with the single question of what sort of social arrangement can be relied on, in the presence of externalities, to bring about an optimal allocation of resources. At the outset he demonstrates the principle—now generally accepted but by no means intuitively obvious—that, even where externalities do exist, no misallocation of resources will occur if the parties involved are in a position to negotiate to their mutual advantage at no cost. Negotiation will lead to an agreement either for the party imposing the damage to compensate the sufferer or for the potential sufferer to pay the potential damager to refrain partially or altogether from his harmful acts. In either case external costs will be "internalized."

But where common-property resources are concerned, the cost of negotiation, far from being zero, is likely to be prohibitive because of requirements for getting large numbers of people together and for avoiding freeloaders. The question, then, is: What sort of social mechanism can bring about the same results as negotiation? Coase's major contribution here is to recognize that responsibility for damages is a reciprocal one, falling no less on the afflicted party to take steps to avoid them, than on the perpetrator. This insight has some profound implications for environmental policy. It suggests that, if optimal use of environmental resources is the sole object of policy, the responsibility for limiting the cost of environmental damage should be made to fall wherever it can be shouldered most cheaply rather than entirely on the initiator of the damages. In no event, for example, should a polluter be charged more than what it would cost the sufferers to take steps to avoid injury. A similar theme is developed in Section III by Guido Calabresi.

Ralph Turvey's paper synthesizes the major ideas in several influential papers on externalities, including Coase's. He concludes, along with Coase, that no single device such as a tax will eliminate the divergence between social and private costs caused by externalities in all instances. He compounds the difficulties by pointing out that in some circumstances an effluent charge or tax may itself induce behavior that misallocates resources. In many cases, he argues, the costs of trying to achieve an optimal allocation of resources may outweigh the gain.

The final selection, by Allen V. Kneese and Blair T. Bower, is a careful analysis of the consequences of effluent charges and other measures that might be used to control the pollution of public waters. It formalizes some of Coase's arguments and reemphasizes the genuine intricacy of the analysis required to assure that policies intended to improve the quality of the environment actually have that result. The general principles developed here are, of course, applicable as well to the analysis of environmental resources other than water.

The papers in this group all share one fundamental shortcoming: they disregard distributional implications and consequences. This is most pointedly true of Coase, whose principal theme is that in the interests of economic efficiency the costs of avoiding external damages should be borne by the party that can most cheaply avoid them. But it is equally true not only of all the other papers in the group but of most of the formal literature on the subject. The article by Dorfman and Jacoby in the next section is one of the few that deal directly with distributional issues. The neglect can no doubt be traced to the absence of a satisfactory criterion for assessing the desirability of different distributions of income, for in these matters there is no analog to the Pareto-optimality standard for measuring economic efficiency. Yet distributional consequences are too important to be neglected in practical policy formation.

# 4

OTTO A. DAVIS AND
MORTON I. KAMIEN

# *Externalities, Information, and Alternative Collective Action*

*Otto A. Davis is Professor of Managerial Economics at Northwestern University, and Morton I. Kamien is Professor of Economics at Carnegie-Mellon University.*

## Introduction

Awareness that an action often entails subsidiary as well as direct consequences is commonplace. In choosing an occupation we consider not only the direct monetary remunerations involved, but also the associated security, power, and prestige. When purchasing apparel we take into account its attractiveness as well as the protection and comfort which it affords. In using drugs, we should be acutely conscious of their possibly harmful side effects as well as of their direct curative powers. When

*This selection contains most of a paper that appeared in Joint Economic Committee, United States Congress,* The Analysis and Evaluation of Public Expenditures: The PPB System, *Vol. I, (Washington, D.C.; U.S. Government Printing Office, 1969), pp. 67–86.*

purchasing a house, we are likely to take into account not only the size and age but also the quality of the neighborhood in which it is located, its proximity to good schools, and the possible availability of public transportation.

In everyday parlance we refer to these secondary attributes of products or actions as "side-effects," "fringe-benefits," or "occupational disease." Our concern with these matters is not wasted on the advertising industry which promotes many products by stressing their desirable side-benefits. Witness the number of advertisements that allude to the masculinity, femininity, youthfulness, and glamour that are to be derived from the use of this or that product. Indeed, some products and occupations have become better known by their side effects than by their direct benefits.

Of course, concern with secondary consequences is not solely confined to the advertising industry. Regard for side-effects finds expression in the selection of products or occupations and the amounts we are willing to pay or sacrifice to avoid or incur them. For example, a desire to live in a "better" neighborhood manifests itself by a willingness to pay more for a house in the preferred neighborhood than for a comparable house elsewhere. Likewise, a strong preference for a relatively secure occupation is satisfied by a willingness to sacrifice potentially higher monetary gains in other occupations. Businessmen find it profitable to be responsive to these desires of consumers. Though the primary function of an automobile is transportation, manufacturers provide a wide variety of models to satisfy the secondary features desired by purchasers. Drug producers attempt to develop new drugs that possess the same beneficial properties as existing ones while reducing undesirable side-effects. Moreover, the responsiveness of producers is spurred by the knowledge that competitors will cater to the preferences of customers. For this reason, competition among producers is thought to be desirable. Similarly, competition among buyers assures that goods and services will be allocated in conformity with the relative desires and abilities of the participants to pay.

From the above account, one might be tempted to conclude that a freely competitive economy should provide the goods and services desired by consumers in such a way as to preclude the possibility that another allocative mechanism (such as government) might be judged to be more appropriate for given situations. Given certain conditions and a plausible criterion, one of the major contributions of modern economic theory is the confirmation of this conclusion. Yet, even the most casual observation of the real world discloses that our society often takes recourse to collective governmental action for the provision of certain goods and services. It might be supposed, therefore, that modern economic theory must be either wrong or irrelevant for the real world. While there probably is considerable sympathy for such a conclusion in some circles, it is taken here to be an obviously incorrect deduction. An alternative explanation of this apparent divergence between theory and reality is that collective decisions are necessarily bad and that governments act in a nonsensical manner. That conclusion is also rejected here. Of course, the denial of this alternative does not imply that governments always make the wisest or best decisions since it is clearly possible that governmental decision processes can be improved and that we should do what we can to improve them. Yet another explanation of the apparent divergence between theory and reality (accepted here), is that the conditions or assumptions which are the basis for the conclusion regarding the efficiency of a market system are not always satisfied in reality. According to this viewpoint, it is advantageous to study certain aspects of economic theory because it produces insight into situations where market systems cannot be expected to work very well. The existence of such situations raises the problem of selecting the proper institutional arrangements for the performance of specific activities. Although solutions to these problems that would elicit a consensus are not currently available, there is nonetheless advantage in knowing where markets might, and where they might not, work tolerably well.

The purpose of this essay is to present and amplify the major conclusion of modern welfare economics which may be viewed as a theorem concerning the allocation of resources by a market mechanism. Particular attention will be paid to certain of the assumptions upon which this theorem is based. Examples will be used to help make the major points clear. The question of selecting proper criteria for institutional choice will be addressed. Perhaps the outcome of the discussion will be a better understanding of some of the issues and difficulties involved in selecting institutional mechanisms capable of attaining an acceptable allocation of limited resources.

## The Criterion, the Market, and Optimality

To talk meaningfully about selection among alternatives, one must have in mind a method of ordering or weighing the various possibilities. Economists employ the notion of efficiency or (to use the more technical term), the concept of Pareto optimality, for weighing different allocative mechanisms. An allocation of resources is inefficient or non-optimal if it is possible to make at least one member of society better off without making any other member worse off via a reallocation. An allocation is Pareto optimal if it is impossible to better someone's condition without simultaneously worsening another's, via reallocation. It is worth noting that this criterion of efficiency or Pareto optimality need not lead to an unambiguous ordering of alternative allocations since (theoretically) there exists an infinity of positions which are Pareto optimal. Still, there is good reason to insist (within the limits of practicality) that all solutions be efficient since, by definition, a non-Pareto optimal solution permits bettering someone's condition without worsening things for anyone else. The qualifying phrase "within the limits of practicality" here denotes that, although the theoretical possibility of improving at least one person's position without inflicting harm on anyone else must be admitted whenever the situation is not Pareto optimal, the practical means of actually accomplishing such an improvement need not be at all obvious to frail human minds. An allocative mechanism is deemed desirable if it leads to Pareto optimal allocations of resources.

The notion of Pareto optimality probably would be neither interesting nor useful were it not for some of the developments of modern welfare economics. The most important of these can be regarded as one of the fundamental theorems of economics. It can be stated informally as follows: Given certain assumptions about the technology, the availability of information, the characteristics of goods and services, and the absence of monopoly power, there exists a set of market prices which, responded to by profit-maximizing firms and utility-maximizing consumers, will result in the attainment of a Pareto optimum position for the economic system. The theorem is a powerful argument for the organization of our society so that resource allocation proceeds via the mechanism of competitive markets. If the assumptions of the theorem were universally satisfied, then government could limit itself largely to programs aimed at the attainment of desirable income distribution and be quite certain that competition would cause the system to be efficient.

There is little need to review in detail the entire set of assumptions upon which the above theorem rests. Indeed, economists have long been searching for a minimal set of assumptions which are sufficient for markets to attain Pareto optimality, and it is doubtful that the end of the search is anywhere in sight. Accordingly, it is appropriate to focus here only on those which seem to cause the great difficulties in the real world. Fortunately or unfortunately, it will be seen that these difficulties appear to be interrelated in ways that are not always obvious.

Consider the technology first. A strong assumption here is that all firms have convex production possibility sets. This supposition means that there cannot be increasing returns to scale. Or, it must *not* be true that ever larger firms can produce the same product at a lower per unit cost than relatively smaller ones.

A second consideration is the availability of information. Producers are assumed to have complete knowledge about the available technology. Consumers are supposed to know whether particular goods and services are available, and their characteristics. Finally, both producers and consumers are presumed to know the relevant set of prices.

The third condition concerns the characteristics of goods and services that the economic system is supposed to produce. Not only are there supposed to be no "public" goods—e.g., goods such as radio waves or television signals which can be "used" by one listener or viewer via reception, without diminishing their availability to others—but the consumption of other goods and services (called "private" goods) by any one decision unit is not supposed to directly affect other decision units. Thus, although the "side effects" mentioned in the introduction are allowed, there is presumed an absence of what economists call non-pecuniary externalities. One should note that the presence of externalities is not sufficient cause for the market to avoid optimality, but such a presence means that attainment of optimality by the market mechanism cannot be guaranteed.

A fourth condition worth noting is the absence of monopoly power. It is the *competitive* market that, under certain circumstances, is capable of attaining Pareto optimality. Since there will be little further discussion of monopoly in this essay, we mention at this point that monopoly is often related to the other conditions under consideration. For example, it is acknowledged that one of the difficulties associated with increasing returns is the emergence of monopoly. Likewise, initial monopoly power can sometimes be maintained because technological knowledge is not always available to all and, even when it can be made available, there may be barriers to its transmission and assimilation.

All this assumes that markets either do or can be made to exist for all goods. Unfortunately, this does not always seem to be the case.

In the remainder of this essay, we concentrate upon the consideration of externality. The discussion will be testimony to the fact that all of these matters are interwoven and cannot always be separated.*

. . .

## Problems Associated with Externalities

. . . Effects upon [persons] not associated with specified purchases or activities, are called externalities. Alternative terms are spillovers, external effects, or social effects. While the literature distinguishes many kinds of externalities, it is necessary for the purposes of this essay to identify only two types: technological (or nonpecuniary) and pecuniary externalities.

Let us first deal with the concept of the pecuniary externality. When deciding whether or not to purchase an item, an individual will ordinarily take into account his own desire for the item, its price, and his budgetary situation. Rarely, and generally only in the case of a monopsony,

* A section entitled "Problems of Information" has been omitted [Editor].

does the individual consider that his decision to purchase can contribute to—maybe even increase—the demand for that product and thereby cause its price to rise. Usually, this disregard is justified, for the individual's purchase of a commodity is such a small fraction of the total amount sold, that his decision has a negligible impact upon price. Whenever an individual's decision does effect price, not only he, but also all other purchasers bear the resulting increase or decrease. This change in price, caused by individual decisions, is termed a pecuniary externality. If the individual decision causes the price to rise—the usual case associated with an increase in demand—then the phenomenon is a pecuniary external diseconomy to other consumers. Whenever the decision causes the price to fall—as illustrated by the decision to join a group travel arrangement which is not yet at capacity—the phenomenon is termed a pecuniary external economy to other consumers. Of course, by symmetry, a pecuniary external diseconomy to consumers is a pecuniary external economy to sellers, and a pecuniary external economy to consumers is a diseconomy to sellers.

The important point to note is that pecuniary externalities, be they economies or diseconomies, pose no problem for the market economy. Indeed, they are the central ingredient of the market place. Changing demands cause prices to rise and fall; these fluctuations provide the essential signals for the market place to ration the available goods and services efficiently.

Technological externalities are quite another matter. These refer to more or less direct effects, other than price changes, that one decision unit might impose on another. Technological externalities can, and in many instances do, prevent the market mechanism from functioning efficiently i.e. giving rise to a Pareto optimal allocation. In these instances, there exists a possibility for action to improve society's well-being, i.e. bettering one citizen's lot but not at another's expense. Some examples may serve to illustrate the issue here.

Since both of the authors resided in Pittsburgh, it may be appropriate to begin with an example relating to steel manufacture. For expository purpose, imagine that Pittsburgh does not have a smoke control ordinance. Then, according to the production process employed, more or less smoke may be discharged into the atmosphere as a by-product of steel manufacture. Insofar as the producer is interested in profits (most manufacturers are), there is an incentive to choose that method of production which is most profitable without regard for the associated level of smoke discharge. The manufacturer may regard smoke disposal as another resource which contributes to the production of steel. The justification for conceiving of disposal as a resource is that reduction in smoke discharge can only be achieved by adopting an alternative (and more expensive), production method which emits less smoke, or by using the same process with the addition of smoke control devices. Either alternative involves the use of additional resources such as labor and capital. These additional resources are not free; since there is no charge for the emission of smoke into the atmosphere, there is little if any motivation to limit the use of the resource which might be called smoke disposal.

Although the discharge of smoke into the atmosphere might be viewed as a free resource by the firm, it is certainly not without consequence to those residing within adjoining communities. Not only does smoke cause more rapid deterioration of the exteriors of buildings and certain kinds of equipment—which will certainly mean that compensatory resources will have to be spent in more intensive cleaning, maintenance and repair—it also contributes to smog creation—which probably has a direct, though not yet fully documented, effect upon the health of some residents in the community. In other words, to the community at large the discharge of smoke into the atmosphere is not a free resource. Economists call the situation, in which the firm does not bear the full costs of its actions, an instance where private costs diverge from social costs. The essential point to notice about the situation as outlined here, is that—without some kind of institutional change—the steel producer has nothing more than possible humanitarian concerns as a counterbalance to his profit interests, to make him recognize that smoke discharge imposes costs upon his neighbors.

Of course, industrial smoke is not the only cause of smog. One of the most oft-mentioned contributors today is the automobile. To comprehend the nature of the relevant motivations, imagine the situation prior to the regulation requiring installation of smog control devices in all new cars. It is obvious that if consumers demand and are willing to pay for smog control devices, the automobile industry would develop and sell these devices in much the same way that it supplies other optional equipment. Moreover, competition among manufacturers, foreign and domestic, would assure improvement of these devices over time. Thus, all that would be required (technical difficulties aside), to call forth a supply of smog control devices is the potential of profits. Unfortunately, there may not be any foreseeable profits.

Imagine, for the sake of argument, that the auto industry had developed an effective smog control device that it offers as optional equipment for all new cars. A person deciding whether or not to order this optional equipment might reason as follows: Suppose I purchase the smog control devices for my new car, and everyone else also purchases them, then we will have less smog in the city. But, since my car adds only a negligible amount to the smog problem, it follows that if everyone else purchases a device and I do not, smog will be diminished by almost exactly the same amount and I will have saved the cost of the device. Hence, if everyone else purchases a device, I will be better off if I do not get one installed on my car. Suppose now that no one else purchases a device. Obviously, there will be a smog problem. However, if I do purchase a device, the problem will not be noticeably better and I will be out the money which I paid for the smog control device. Thus, if no one else purchases, I should not purchase either. Obviously, the analysis is the same if some of the other people purchase and some do not, for my contribution to smog is negligible. Conclusion: I will be better off, no matter what other people do, if I do not purchase a smog control device.

Since all potential new car buyers will reason roughly as the representative individual does, there will be a zero demand for smog control devices. In the absence of some kind of regulation or collective decision, the automobile manufacturers will have no motivation to develop and market smog control devices. This conclusion holds even if—and it is an if—everyone would be better off when all cars were equipped with smog control devices. For each prospective purchaser of a device, the benefits from his purchase are widely dispersed while the costs accrue to him. The technological externality associated with the exhaust of a car can prevent the unregulated market from leading the system to a Pareto optimum.

For the final example of this section, consider the problem of the pollution of Lake Erie. Biologists tell us that Lake Erie is dying and that it has "aged" 15,000 years in the past half century. It was long believed that Lake Erie's pollution was caused largely by the dumping of raw sewage and industrial wastes into the lake. A major source of the raw sewage is antiquated sewage systems, some of which are combined sanitary and storm sewers so that the overflow runs directly into the lake during periods of rain. For the moment, imagine that the entire pollution problem is caused by the raw sewage so that treatment—designed to remove organic material which otherwise is digested in the lake via biological process that depletes its dissolved oxygen content—could solve the problem. The now familiar dilemma would act to frustrate a pure market solution. Each municipality or sewage district would reap only a fraction of the benefits of its own efforts at treating sewage, but would bear the full costs of that treatment. Employing the same reasoning as the customer considering the purchase of a smog control device, each would come to the rational decision to continue to allow raw sewage to flow into the lake even though all municipalities together might be better off if each installed treating devices. Thus, the technological externality reflected in reverse by failure to receive the full benefit of one's expenditure for treatment—the fact that the decision-making entity does not bear the full costs of its decision to forego treatment and allow a flow of raw sewage—results in failure of the pure market solution unless a financial incentive to come to the opposite decision is offered by a higher level of government.

In actuality, the pollution of Lake Erie is a much more complex phenomenon than indicated by the above discussion. Even after treatment to remove indigestible solids and to break down organic material so that the sewage is discharged as mostly inorganic products, the residual inorganic matter contains large amounts of nitrate and phosphate which, instead of being swept harmlessly to the sea, tend to remain in the lake long enough to fertilize monstrous growths of algae which use up to an estimated eighteen times as much oxygen as the present flow of organic matter from inadequate sewage plants. Thus the standard treatment of sewage, primarily aimed at organic matter, is not likely to solve the problem even if such treatment were undertaken.

It might be suggested that one of the "essential" nutrients such as the phosphate should be removed from the waste so that the algae would not grow. Unfortunately some two-thirds of the phosphorus in municipal waste, which is roughly three-quarters of the total wastes, stems from detergents. If the housewife or commercial laundry knew that the detergents they use contribute importantly to the pollution of the lake, which they probably do not know, would there be any incentive to economize on the use of detergents or to demand a new kind that contained less phosphorus? Again, the familiar dilemma appears. Even if they were aware of contributing to pollution, each could rationalize that their own contribution was negligible, so that the rational decision would be to ignore the entire situation. Thus the manufacturers of detergents would have no incentive to try to develop products which contain less phosphorus, the municipal sewage systems would have no more incentive than in the previous instance to remove the material, with the result that the pure market solution would be to continue pollution of the lake. Again the existence of this technological externality—the fact that those causing the pollution do not bear the full costs of their actions—can cause the market mechanism to generate an allocation of resources which is not Pareto optimal.

## The Possibility of a Solution

The preceding illustrations indicate that the existence of technological externalities can necessitate modifications to claims of the unregulated market mechanism's efficiency. It must be admitted that the problems created by technological externalities are most perplexing. Only a few years ago, economists may have thought that an adequate solution was available, but the consensus has now vanished. Instead, one finds that a variety of solutions have been suggested by various persons—many are not economists—and a goodly number of these have even been tried or implemented in certain situations.

The belief that a universal solution might be found is present in many discussions and analyses of technological externalities in the literature. Thus, proposals are often treated as if they are supposed to be "the" solution to the problem of technological externalities. Unfortunately, this belief has not yet been justified and it may be that there is no simple and universally acceptable solution to the problem. Perhaps, at least for the foreseeable future, our society has no alternative but to seek pragmatic solutions to the problem.

Accordingly, it is necessary to examine many of the suggested solutions to the problem. The remainder of this essay is devoted to the description of some proposed solutions. It is desirable, of course, to make the point of view adopted in this discussion, explicit. Specifically, we view the situation as one which can be conceptually accommodated to cost-benefit analysis. All of the proposals have associated costs and benefits. The trick is to identify which is most suitable for a given situation. It is hoped that the following discussion will be useful for this purpose.

## Solution by Prohibition

When one is convinced that collective action is necessary to correct the abuses caused by a technological externality, the first impulse is to simply prohibit the action giving rise to it. After all, if creation of the externality is prohibited, will not the market system then bring the economy to a Pareto optimal position?

Although this course of action may seem appealing at first, it takes little thought to realize that simple prohibition of activities that create technological externalities is a poor approach. Obviously, one could not seriously propose that car owners stop driving, that steel manufacturers stop producing, or that municipalities stop disposing of their sewage. Some might conclude, however, that we should have perfectly clean water or perfectly clean air, so that full treatment of effluents is needed. That conclusion, however, misses the fundamental point. Optimality does not call for the complete elimination of externalities. Instead, optimality requires that externalities be present in the "right amount." Some examples may help clarify this point.

Consider the case of water pollution. Natural biological processes in both lakes and streams give them a certain capability of cleansing themselves. If absolutely no untreated wastes are allowed to flow into these waters, this natural capability will not be used. From an economic standpoint, however, this natural capability is a resource available for use in the production of desired goods and services. In establishing the desired level of water quality, the benefits of cleaner water must be balanced against the costs of attaining it. Moreover, the assimilative capacity of a body of water can always be used up to the point at which objectionable side effects occur.

Air pollution affords a similar example. It would be prohibitively expensive to prevent any contaminants whatsoever from escaping into the air. Further, there is no reason not to use the natural absorbic capacity of the atmosphere. Air pollution is said to occur when this capacity is exceeded.

The above examples should illustrate that strict prohibition of whatever causes a technological externality is almost certain to prevent attainment of Pareto optimality. An appropriate level of externality, not necessarily equal to zero, is needed. So, in the case of water pollution a Pareto optimal solution may in fact mean some deterioration of water quality in certain streams and possibly complete deterioration of water quality in other streams.

## Solution by Directive

Having seen that the crux of the problem is to get just the right amount of technological externality, it is tempting to suggest that the government decide how much of it should be produced. This procedure would involve, for example, governmental determination of the extent to which municipalities bordering Lake Erie should treat their sewage in terms of, say, the percentage of organic matter removed, phosphorous content, and setting an absolute quantity limit above which sewage would have to be completely purified. Similarly, in the air pollution example the government would have to specify just how much smoke a factory could emit.

There are several difficulties with this procedure. First, as already noted there is the problem of determining just how much of the externality is desirable. This question is related to the problem of determining the overall quality standard. In principle this could be done by careful weighing of costs and benefits. To be specific, consider again the example of the pollution of Lake Erie. Suppose that it would cost 50 billion dollars per year in operating costs alone to process the sewage to an extent that pollution in the lake would diminish from present levels. Although the benefits to be derived from an unpolluted Lake Erie may be substantial, it is rather doubtful that they would be valued at anything near 50 billion dollars per year. The costs in this case would simply outweigh the benefits. Hence, the rational decision would be to tolerate an even higher level of pollution and not increase the level of sewage processing. On the other hand, imagine that all the sewage could be processed with an increase in annual expenditures of only five dollars. Clearly, the yearly benefits from an unpolluted lake would exceed this figure, so that the quality standard should be set to cause total sewage treatment. In between these extremes, however, the computations become very difficult. The difficulties of determining the benefits associated with various degrees of pollution are almost insurmountable. Consequently, there must be a great deal of arbitrariness in setting the overall standard. Another problem, overlooked in the above discussion, is the absence of complete understanding of lake ecology; there is still some uncertainty about the effects of treatment upon pollution.

Even if the overall standard could be determined despite the stated difficulties, other difficulties would remain. The overall standard must be translated into directives for each of the entities that emit pollutants. In principle, the directives should cause the marginal effectiveness of the last dollar spent upon the processing of wastes to be equalized for each of the polluters. In practice, the marginal effectiveness of dollars spent for treating wastes cannot be determined accurately for any given polluter because it depends upon the policies pursued by other polluters. Thus, a degree of arbitrariness results at this level too.

In the case of air pollution, implementation of the procedure is even more difficult, although the same principles are involved. The overall standard should still be determined by weighing and comparing the benefits and the costs of different alternatives. However, the degree of arbitrariness is greater because even less is known about the relationship between the level of air pollution and the health of the residents, than about stream ecology. Also, there is the complication that in most urban areas the amount of pollutants which can be released into the atmosphere for any given standard depends upon weather conditions and prevailing winds.

The above comments are not intended to convey that the control of externalities by directives is to be dismissed as being obviously inappropriate. Our intention is to point out some of the difficulties associated with the procedure. We might add that this procedure also involves administrative costs of policing the directives, which cannot be ignored in its evaluation.

## Solution by Voluntary Action

Some argue that collective action is not needed to correct the market solution when there are technological externalities. It has often been pointed out that there is motivation for private parties to act to correct the situation by a variety of methods. Two methods frequently discussed are bribes and merger.

Consider again the example of a steel producer, unchecked by a smoke control ordinance, discharging smoke into the atmosphere. The previous discussion indicated that this situation potentially gives rise to a divergence between private and social cost or between the private and social benefit of steel production. To avoid the adverse effects of smoke discharge, the community might resort to bribing the steel producer to decrease or discontinue altogether the discharge of smoke. The rationale for this behavior is that as long as the amount of the bribe needed to induce the steel manufacturer to reduce smoke discharge is less than the damage inflicted on the community, then the community will be better off by paying the bribe. Of course, the community acting rationally in its own interest would never proffer a bribe that exceeded the value of the damage inflicted via smoke discharge. The steel producer should (in turn) accept or reject the bribe in accordance with his best interests. Therefore, if the bribe exceeded the amount he would have to spend on means to reduce smoke discharge, he should accept the bribe and effect the desired reduction; if the costs were too great, he should not. In any case, a quantitative measure of the damage suffered by the community from smoke would have been presented to the manufacturer in such a way as to make him cognizant of this figure when deciding how much smoke to discharge. Moreover, whatever the final level of smoke discharge, that level would be Pareto optimal if there were such a thing as perfect bargaining. One can reason as follows: Acceptance of the bribe by the manufacturer indicates that he is at least as well off as before, while payment of the bribe by the community indicates that it is at least as well off as before. Consequently, the situation is improved. If the bargaining were perfect, any departure from the agreed-upon position would only improve the position of one of the parties at the expense of the other. It is also true that rejection of the bribe by the manufacturer, under perfect bargaining, leads to a Pareto optimal solution. By rejecting the bribe, the manufacturer would disclose that the value of this resource (release of smoke into the air) is of greater value to him than to the community.

The method of avoiding a divergence between private and social cost described above is purely voluntary and leads, when bargaining is perfect, to a Pareto optimal allocation of resources. In the absence of distributional considerations, it would, therefore, appear to be the ideal way in which to resolve problems. Unfortunately, bargaining is not perfect and there are several impediments to its widespread use. The first difficulty is associated with the valuation of smoke damage suffered by the community. The most direct way of estimating the damage is to ask each member of the community how much he would be willing to contribute to the bribe to be offered the manufacturer. In principle, each individual should be willing to contribute the amount he would have to expend to avoid the damage from smoke by other means. Unfortunately, the individual would realize that if he contributes nothing towards the bribe but others contribute positive amounts and smoke abatement is effected, he will reap the benefits of smoke abatement at no cost to himself. If all members of the community adopt this attitude, no bribe will be offered and the scheme will fail. In other words, the fact that the benefit from smoke abatement is distributed to many individuals impedes the realization of the necessary collective action by the community. The second difficulty with the bribe procedure is that it requires that the community know all the available methods for manufacturing steel, as these are related to smoke control, and the associated costs so that they might prevent the manufacturer from cheating. Suppose (to illustrate the point) that after the bribe has been accepted by the manufacturer, the demand for steel rises and output increases. The producer can now argue legitimately that a larger bribe is required for him to maintain the previously agreed upon level of smoke discharge. Unless the community is completely knowledgeable about steel-making technology, it cannot be sure that the manufacturer is not expanding his output more than would be optimal for him in the absence of the bribe. Thus, a seemingly ideal scheme for avoidance of a divergence between private and social cost is marred by difficulties in implementation.

Another voluntary scheme for internalizing non-pecuniary externalities, free of some of the implementation difficulties mentioned, is the merger of the involved entities whenever merger is a feasible possibility. To illustrate how this procedure might work, consider the following situation: A firm discharges wastes which are harmful to fish life into a stream. Assume, further, that a fishery operates downstream from the firm. In the absence of any governmental regulation, the upstream firm will discharge waste into the stream without regard to the damage—smaller catches or tainted fish—inflicted on the fishery. Were the firm and the fishery to merge under a single ownership, it would be in the consolidated firm's best interest to take account of the losses incurred by its downstream subsidiary as a consequence of the actions of its upstream plant. The consolidated firm should balance the cost of disposing of waste at the upstream plant by other means than discharge into the stream, against the costs incurred by the downstream fishery as a result of waste discharge into the stream, if it is to maximize the combined profit from the two operations. Since, in this example, Pareto optimality corresponds to joint profit maximization by the two entities, merger will assure a Pareto optimal allocation of the resource in question—viz. the stream. It might also be noted that the profit of the consolidated firm will always be at least as great as the combined profits of the two firms operating in isolation. The reason for this is that the merged firm always has the option of adopting the operating policies of the two firms working independently. The difference between the profits of the consolidated firm and the combined profits of the individual firms reflects the loss to society from the presence of a non-pecuniary externality.

Two difficulties with the merger solution can be pointed out. The first is the practical consideration that the entities have to be firms. The second difficulty is that merger is feasible only when the number of entities involved is small. As the number of decisionmaking units entering the consolidation increases, the chances of effecting the merger decline. This is because it becomes increasingly more difficult to persuade potential participants that it is in their best interest to join the coalition as the number of participants increases. Individual units may find it profitable to postpone entry into the coalition in order to extract a larger portion of the joint profits from the merged entity. A more practical consideration involves the increase in the difficulties of coordination and computation with the number of participants; it becomes increasingly difficult for the merged entity to actually realize the potential gains. Yet another difficulty is that of a merged entity becoming big enough to cause a distortion in the allocation of resources via monopoly or monopsony power. In this case, the losses from the presence of non-pecuniary externalities have to be weighed against the losses to society from the resource allocation distortions engendered by imperfectly competitive markets.

## Solution by Taxes and Subsidies

If voluntary arrangements among the entities effected by non-pecuniary externalities are impractical or not forthcoming, collective governmental action might be justified. In economics literature, the classic form of government intervention in this situation is the payment of a subsidy to units whose actions confer external economies upon other units, and the levying of taxes upon those entities whose actions confer external diseconomies upon other units. In essence, the idea is to encourage those activities that contribute to the "common good" and discourage those that detract from the "common good."

To illustrate how this scheme might work, consider again the example of an upstream firm and a downstream fishery. Suppose that the waste discharged into the stream by the firm provides food for fish in the stream and is therefore *beneficial* to the fishery. Since, by hypothesis, voluntary negotiation by the parties involved is ruled out here, the fishery has no way of communicating the magnitude of the benefits it derives from the firm's discharge of waste into the stream. Consequently, the amount of food provided to the fish may not be ideal. In this case, a government subsidy to the firm for the discharge of waste can in principle be devised to achieve the desired result. Likewise, if (as in our earlier description of the situation) the waste discharged is harmful to fish life, then in principle a tax can be imposed on the firm that reflects the damage imposed on the fishery.

We have stressed the "in principle" nature of these conclusions because of the immense informational requirements necessary for the implementation of this scheme. A little reflection will make it apparent that the governmental agency imposing a tax or offering a subsidy will need to know the production technologies of all the entities involved. In effect, the governmental agency will have to solve the same problem that the directors of the merged firm solve. They would have to know the effect upon fish life which the discharge of waste caused so that a tax or subsidy could be imposed that would result in just the right amount of waste discharge. Instead of issuing orders regarding the quantities of each product that every subsidiary should produce to maximize joint profits (a practice that the executives of the merged firm might follow), the agency would attempt to achieve the same results under this scheme by determining the "proper" tax or subsidy. Suffice it to say the information available to the directors of the merged firm is rarely available to an outside governmental agency. Bits of this information may of course be available and it might be possible—at a cost—to obtain additional information.

The amount of information required also depends on the nature of the productive technologies involved. Less information is required by the agency for the successful implementation of a tax-subsidy scheme if the underlying productive technologies are separable or additive, than if they are not. For example, if the cost of producing the upstream firm's product and the cost of waste treatment are additive, then the tax on waste discharge simply depends on the amount of waste discharge. However, if these costs are not additive, then the tax must vary not only in accordance with the level of waste discharge, but also with the amount of the firm's primary product. Of course, the informational requirements mount enormously as the number of involved economic entities increases. Despite all these difficulties, an attempt to achieve optimal resource allocation by taxes and subsidies might be justified if the losses to society from the presence of non-pecuniary externalities is large enough. In essence, what has to be balanced in this situation is the cost of acquiring the needed information against the losses to society if nothing is done or another imperfect policy is followed.

## Solution by Regulation

Another collective action which is often suggested is governmental regulation. For example, the official governmental response to the fact that cars contribute to the air pollution in our cities has been to reduce the range of consumer choice by simply requiring that all new cars be equipped with devices designed to reduce the level of pollutants in the exhaust gases. This regulation obviously permits an escape from the dilemma described earlier where a rational calculation would cause the consumer to refrain from purchasing a control device.

Regulation also has implementation difficulties associated with it. In the case of automobiles, for example, there is uncertainty as to whether the devices will be effective in reducing the discharge of pollutants, especially as the cars grow old. There are also problems of enforcement. For instance, there is speculation that if the devices are effective and pollutants are kept in the engine instead of being spewed into the atmosphere, then the life of the engine might be shortened and repairs will have to be made more frequently. These are costs of the regulation which, along with the costs of the devices, have to be weighed against possible benefits that are rather difficult to compute. If the devices do have the anticipated effect upon engines, then each owner has an incentive to take action that will render the devices ineffective and thereby increase the life of the engine and reduce his repair bills. Certainly, the owner cannot be expected to keep the device in good working order, or to repair it when it breaks, since this action would not be in his own self-interest. Consequently, the regulation cannot be expected to be successful in reducing air pollution, even if the devices work, unless it is accompanied by the practice of periodically inspecting all cars and requiring that the devices be maintained in good working order.

This discussion should indicate that solution by regulation is not as simple as it first seems. The administrative costs of enforcing the regulations are relevant and cannot be overlooked. Neither should the fact that a regulatory solution is necessarily inflexible be ignored. This point can also be better understood by recourse to the prior example. Many of our nation's motor vehicles are operated for a considerable portion of their life in non-urban areas where air pollution is not considered a problem. Ideally, vehicles operating in these areas should not be required to have smog control devices so that the natural ability of the atmosphere to accommodate a certain level of pollutants could be utilized. Obviously, it is impossible to design regulations which would accomplish this ideal due to the very mobility of motor vehicles and the population. Solution by regulation is inherently incapable of bringing the system to a Pareto optimal solution for many externalities, because regulations are inherently inflexible.

## Solution by Payment

One of the obvious ways of trying to accommodate the market system to the presence of technological externalities is to provide a financial incentive for the desired actions to be taken. In the pollution of Lake Erie, for example, one of the problems supposedly is caused by Cleveland's archaic sewage system which combines storm and sanitary sewers. The limited capacity of the treatment facilities for anything beyond modest rainfall causes raw sewage to flow directly into the lake. Although it certainly contributes to the pollution of Lake Erie, the citizens of Cleveland do not bear the full cost of their archaic sewage system since persons residing outside of Cleveland desire use of the lake too. Other people, therefore, bear part of the costs of pollution, including that portion of the pollution which stems from Cleveland's sewage system. Of course, what is true for Cleveland is also true for many other cities and towns in the lake area. Because each entity does not bear the full costs of its own contribution to pollution, none have enough incentive to remedy the situation. A possible remedial policy is for a higher level of government—e.g., the federal government—to provide the proper incentive. A federal subsidy for the capital costs of improving the sewage facilities might do the trick.

The major drawback of this policy is its crudeness. It does not easily provide proper coordination for all of the relevant units in the relevant system. Another limitation of this policy is that it is suited only to those kinds of externalities where capital costs are the only really significant block to the improvement of the situation.

## Solution by Action

Sometimes, there are simple and direct actions which can be taken to ameliorate the effects of an externality. Probably the clearest example involves fishing. Consider a lake or stream where many people come to fish. Beyond some level of activity, the future fish population is endangered. Thus, when a fisherman makes a catch, he can affect the future population of fish and thus lower the pleasure and profits of fishermen in the future. The individual fisherman, of course, has no incentive to consider the effect his own activity has upon others. In extreme cases, of course, the fish population could be exhausted.

An immediate remedy for this situation is for the government to continually stock the lake or river so that the fish population is never diminished to the danger point. The externality is then more or less eliminated by this direct action. Needless to say, there are drawbacks to this policy—not the least is its limited applicability.

## Concluding Remarks

It should be obvious by now that there is a whole menu of policies which can be formed to deal with problems caused by technological externalities. None of these policies, at our present level of knowledge, appears to be perfect. Nor do we believe that any one clearly dominates all others as the best of the imperfect lot for dealing with each and every technological externality. Thus, it is argued here that policies must be designed with particular situations in mind and that what is best for dealing with one externality may be inappropriate for another. Accordingly, it is appropriate to conclude this essay with the seemingly suitable procedure for policy selection.

The tools of cost-benefit analysis appear to provide the proper perspective. In a given situation, the policymaker should consider the problem and imagine the application of the alternative approaches to it. The principle of selection is simple. Each measure of policy (including that of doing nothing) will have costs and benefits associated with it. The policymaker should select that measure for implementation which produces the greatest net benefits.

# 5

H. SCOTT GORDON

# *The Economic Theory of a Common-Property Resource: The Fishery*

*H. Scott Gordon is Professor of Economics at Indiana University.*

## I. Introduction

The chief aim of this paper is to examine the economic theory of natural resource utilization as it pertains to the fishing industry. It will appear, I hope, that most of the problems associated with the words "conservation" or "depletion" or "overexploitation" in the fishery are, in reality, manifestations of the fact that the natural resources of the sea yield no economic rent. Fishery resources are unusual in the fact of their common-property nature; but they are not unique, and similar problems are encountered in other cases of common-property resource industries, such as petroleum production, hunting and trapping, etc. Although the theory presented in the following pages is worked out in terms of the fishing industry, it is, I believe, applicable generally to all

*"The Economic Theory of a Common-Property Resource: The Fishery," by H. Scott Gordon, from* The Journal of Political Economy, *April 1954, Sections I, II, III.* 

cases where natural resources are owned in common and exploited under conditions of individualistic competition.

## II. Biological Factors and Theories

The great bulk of the research that has been done on the primary production phase of the fishing industry has so far been in the field of biology. Owing to the lack of theoretical economic research,[1] biologists have been forced to extend the scope of their own thought into the economic sphere and in some cases have penetrated quite deeply, despite the lack of the analytical tools of economic theory.[2] Many others, who have paid no specific attention to the economic aspects of the problem, have nevertheless recognized that the ultimate question is not the ecology of life in the sea as such, but man's use of these resources for his own (economic) purposes. Dr. Martin D. Burkenroad, for example, began a recent article on fishery management with a section on "Fishery Management as Political Economy," saying that "the Management of fisheries is intended for the benefit of man, not fish; therefore effect of management upon fishstocks cannot be regarded as beneficial *per se*." [3] The great Russian marine biology theorist, T. I. Baranoff, referred to his work as "bionomics" or "bio-economics," although he made little explicit reference to economic factors.[4] In the same way, A. G. Huntsman, reporting in 1944 on the work of the Fisheries Research Board of Canada, defined the problem of fisheries depletion in economic terms: "Where the take in proportion to the effort fails to yield a satisfactory living to the fisherman"; [5] and a later paper by the same author con-

1. The single exception that I know is G. M. Gerhardsen, "Production Economics in Fisheries," *Revista de economía* (Lisbon), March, 1952.
2. Especially remarkable efforts in this sense are Robert A. Nesbit, "Fishery Management" ("U.S. Fish and Wildlife Service, Special Scientific Reports," No. 18 [Chicago, 1943]) (mimeographed), and Harden F. Taylor, *Survey of Marine Fisheries of North Carolina* (Chapel Hill, 1951); also R. J. H. Beverton, "Some Observations on the Principles of Fishery Regulation," *Journal du conseil permanent international pour l'exploration de la mer* (Copenhagen), Vol. XIX, No. 1 (May, 1953); and M. D. Burkenroad, "Some Principles of Marine Fishery Biology," *Publications of the Institute of Marine Science* (University of Texas), Vol. II, No. 1 (September, 1951).
3. "Theory and Practice of Marine Fishery Management," *Journal du conseil permanent international pour l'exploration de la mer,* Vol. XVIII, No. 3 (January, 1953).
4. Two of Baranoff's most important papers—"On the Question of the Biological Basis of Fisheries" (1918) and "On the Question of the Dynamics of the Fishing Industry" (1925)—have been translated by W. E. Ricker, now of the Fisheries Research Board of Canada (Nanaimo, B.C.), and issued in mimeographed form.
5. "Fishery Depletion," *Science,* XCIX (1944), 534.

tains, as an incidental statement, the essence of the economic optimum solution without, apparently, any recognition of its significance.[6] Upon the occasion of its fiftieth anniversary in 1952, the International Council for the Exploration of the Sea published a *Rapport Jubilaire,* consisting of a series of papers summarizing progress in various fields of fisheries research. The paper by Michael Graham on "Overfishing and Optimum Fishing," by its emphatic recognition of the economic criterion, would lead one to think that the economic aspects of the question had been extensively examined during the last half-century. But such is not the case. Virtually no specific research into the economics of fishery resource utilization has been undertaken. The present state of knowledge is that a great deal is known about the biology of the various commercial species but little about the economic characteristics of the fishing industry.

The most vivid thread that runs through the biological literature is the effort to determine the effect of fishing on the stock of fish in the sea. This discussion has had a very distinct practical orientation, being part of the effort to design regulative policies of a "conservation" nature.

* * *

The term "fisheries management" has been much in vogue in recent years, being taken to express a more subtle approach to the fisheries problem than the older terms "depletion" and "conservation." Briefly, it focuses attention on the quantity of fish caught, taking as the human objective of commercial fishing the derivation of the largest sustainable catch. This approach is often hailed in the biological literature as the "new theory" or the "modern formulation" of the fisheries problem.[7] Its limitations, however, are very serious, and, indeed, the new approach comes very little closer to treating the fisheries problem as one of human utilization of natural resources than did the older, more primitive, theories. Focusing attention on the maximization of the catch neglects entirely the inputs of other factors of production which are used up in fishing and must be accounted for as costs. There are many references to such ultimate economic considerations in the biological literature but no analytical integration of the economic factors. In fact, the very conception of a *net economic yield* has scarcely made any appear-

6. "The highest take is not necessarily the best. The take should be increased only as long as the extra cost is offset by the added revenue from sales" (A. G. Huntsman, "Research on Use and Increase of Fish Stocks," *Proceedings of the United Nations Scientific Conference on the Conservation and Utilization of Resources* [Lake Success, 1949]).

7. See, e.g., R. E. Foerster, "Prospects for Managing Our Fisheries," *Bulletin of the Bingham Oceanographic Collection* (New Haven). May, 1948; E. S. Russell, "Some Theoretical Considerations on the Overfishing Problem," *Journal du conseil permanent international pour l'exploration de la mer,* 1931, and *The Over; fishing Problem,* Lecture IV.

ance at all. On the whole, biologists tend to treat the fisherman as an exogenous element in their analytical model, and the behavior of fishermen is not made into an integrated element of a general and systematic "bionomic" theory. In the case of the fishing industry the large numbers of fishermen permit valid behavioristic generalization of their activities along the lines of the standard economic theory of production. The following section attempts to apply that theory to the fishing industry and to demonstrate that the "overfishing problem" has its roots in the economic organization of the industry.

## III. Economic Theory of the Fishery

In the analysis which follows, the theory of optimum utilization of fishery resources and the reasons for its frustration in practice are developed for a typical demersal fish. Demersal, or bottom-dwelling fishes, such as cod, haddock, and similar species, and the various flat-fishes, are relatively nonmigratory in character. They live and feed on shallow continental shelves where the continual mixing of cold water maintains the availability of those nutrient salts which form the fundamental basis of marine-food chains. The various feeding grounds are separated by deep-water channels which constitute barriers to the movement of these species; and in some cases the fish of different banks can be differentiated morphologically, having varying numbers of vertebrae or some such distinguishing characteristic. The significance of this fact is that each fishing ground can be treated as unique, in the same sense as can a piece of land, possessing, at the very least, one characteristic not shared by any other piece: that is, location.

(Other species, such as herring, mackerel, and similar pelagic or surface dwellers, migrate over very large distances, and it is necessary to treat the resource of an entire geographic region as one. The conclusions arrived at below are applicable to such fisheries, but the method of analysis employed is not formally applicable. The same is true of species that migrate to and from fresh water and the lake fishes proper.)

We can define the optimum degree of utilization of any particular fishing ground as that which maximizes the net economic yield, the difference between total cost, on the one hand, and total receipts (or total value production), on the other.[8] Total cost and total production can each be expressed as a function of the degree of fishing intensity or, as the biologists put it, "fishing effort," so that a simple maximization solution is possible. Total cost will be a linear function of fishing effort, if

8. Expressed in these terms, this appears to be the monopoly maximum, but it coincides with the social optimum under the conditions employed in the analysis, as will be indicated below.

we assume no fishing-induced effects on factor prices, which is reasonable for any particular regional fishery.

The production function—the relationship between fishing effort and total value produced—requires some special attention. If we were to follow the usual presentation of economic theory, we should argue that this function would be positive but, after a point, would rise at a diminishing rate because of the law of diminishing returns. This would not mean that the fish population has been reduced, for the law refers only to the *proportions* of factors to one another, and a fixed fish population, together with an increasing intensity of effort, would be assumed to show the typical sigmoid pattern of yield. However, in what follows it will be assumed that the law of diminishing returns in this pure sense is inoperative in the fishing industry. (The reasons will be advanced at a later point in this paper.) We shall assume that, as fishing effort expands, the catch of fish increases at a diminishing rate but that it does so because of the effect of catch upon the fish population.[9] So far as the argument of the next few pages is concerned, all that is formally necessary is to assume that, as fishing intensity increases, catch will grow at a diminishing rate. Whether this reflects the pure law of diminishing returns or the reduction of population by fishing, or both, is of no particular importance.

Our analysis can be simplified if we retain the ordinary production function instead of converting it to cost curves, as is usually done in the theory of the firm. Let us further assume that the functional relationship between average production (production-per-unit-of-fishing-effort) and the quantity of fishing effort is uniformly linear. This does not distort the results unduly, and it permits the analysis to be presented more simply and in graphic terms that are already quite familiar.

In Figure 1 the optimum intensity of utilization of a particular fishing ground is shown. The curves *AP* and *MP* represent, respectively, the average productivity and marginal productivity of fishing effort. The relationship between them is the same as that between average revenue and marginal revenue in imperfect competition theory, and *MP* bisects any horizontal between the ordinate and *AP*. Since the costs of fishing supplies, etc., are assumed to be unaffected by the amount of fishing effort, marginal cost and average cost are identical and constant, as shown by the curve *MC, AC*.[10] These costs are assumed to include an oppor-

9. Throughout this paper the conception of fish population that is employed is one of *weight* rather than *numbers*. A good deal of the biological theory has been an effort to combine growth factors and numbers factors into weight sums. The following analysis will neglect the fact that, for some species, fish of different sizes bring different unit prices.

10. Throughout this analysis, fixed costs are neglected. The general conclusions reached would not be appreciably altered, I think, by their inclusion, though the

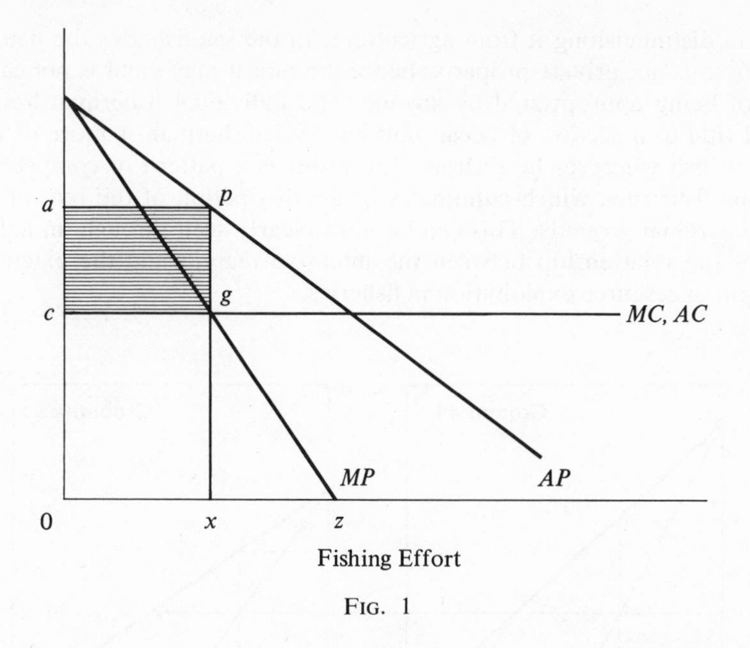


FIG. 1

tunity income for the fishermen, the income that could be earned in other comparable employments. Then *Ox* is the optimum intensity of effort on this fishing ground, and the resource will, at this level of exploitation, provide the maximum net economic yield indicated by the shaded area *apgc*. The maximum sustained physical yield that the biologists speak of will be attained when marginal productivity of fishing effort is zero, at *Oz* of fishing intensity in the chart shown. Thus, as one might expect, the optimum economic fishing intensity is less than that which would produce the maximum sustained physical yield.

The area *apgc* in Figure 1 can be regarded as the rent yielded by the fishery resource. Under the given conditions, *Ox* is the best rate of exploitation for the fishing ground in question, and the rent reflects the productivity of that ground, not any artificial market limitation. The rent here corresponds to the extra productivity yielded in agriculture by soils of better quality or location than those on the margin of cultivation, which may produce an opportunity income but no more. In short, Figure 1 shows the determination of the intensive margin of utilization on an intramarginal fishing ground.

We now come to the point that is of greatest theoretical importance in understanding the primary production phase of the fishing industry

presentation would be greatly complicated. Moreover, in the fishing industry the most substantial portion of fixed cost—wharves, harbors, etc.—is borne by government and does not enter into the cost calculations of the operators.

and in distinguishing it from agriculture. In the sea fisheries the natural resource is not private property; hence the rent it may yield is not capable of being appropriated by anyone. The individual fisherman has no legal title to a section of ocean bottom. Each fisherman is more or less free to fish wherever he pleases. The result is a pattern of competition among fishermen which culminates in the dissipation of the rent of the intramarginal grounds. This can be most clearly seen through an analysis of the relationship between the intensive margin and the extensive margin of resource exploitation in fisheries.

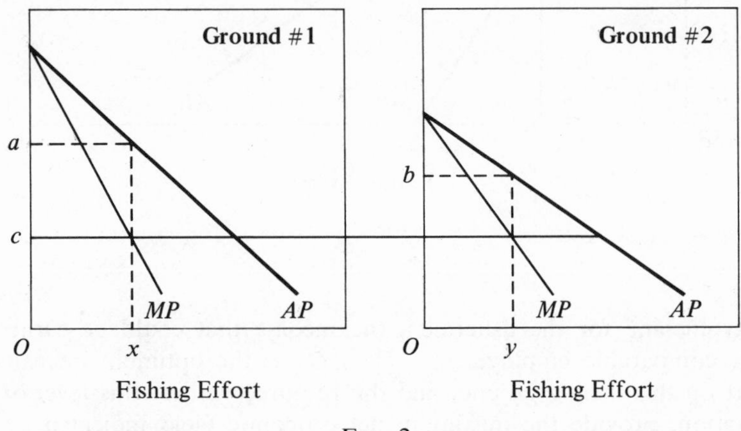


FIG. 2

In Figure 2, two fishing grounds of different fertility (or location) are shown. Any given amount of fishing effort devoted to ground *2* will yield a smaller total (and therefore average) product than if devoted to *1*. The maximization problem is now a question of the allocation of fishing effort between grounds *1* and *2*. The optimum is, of course, where the marginal productivities are equal on both grounds. In Figure 2, fishing effort of $Ox$ on *1* and $Oy$ on *2* would maximize the total net yield of $Ox + Oy$ effort if marginal cost were equal to $Oc$. But if under such circumstances the individual fishermen are free to fish on whichever ground they please, it is clear that this is not an equilibrium allocation of fishing effort in the sense of connoting stability. A fisherman starting from port and deciding whether to go to ground *1* or *2* does not care for *marginal* productivity but for *average* productivity, for it is the latter that indicates where the greater total yield may be obtained. If fishing effort were allocated in the optimum fashion, as shown in Figure 2, with $Ox$ on *1,* and $Oy$ on *2,* this would be a disequilibrium situation. Each fisherman could expect to get an average catch of $Oa$ on *1* but only $Ob$ on *2*. Therefore, fishermen would shift from *2* to *1*. Stable equilibrium would not be reached until the average productivity of both grounds was equal. If we now imagine a continuous gradation of fishing grounds, the extensive margin would be on that ground which yielded nothing more than outlaid costs plus opportunity income—in short, the one on which average productivity and average cost were equal. But, since average cost is the same for all grounds and the average productivity of all grounds is also brought to equality by the free and competitive nature of fishing, this means that the intramarginal grounds also yield no rent. It is entirely possible that some grounds would be exploited at a level of *negative* marginal productivity. What happens is that the rent which the intramarginal grounds are capable of yielding is dissipated through misallocation of fishing effort.

This is why fishermen are not wealthy, despite the fact that the fishery resources of the sea are the richest and most indestructible available to man. By and large, the only fisherman who becomes rich is one who makes a lucky catch or one who participates in a fishery that is put under a form of social control that turns the open resource into property rights.

Up to this point, the remuneration of fishermen has been accounted for as an opportunity-cost income comparable to earnings attainable in other industries. In point of fact, fishermen typically earn less than most others, even in much less hazardous occupations or in those requiring less skill. There is no effective reason why the competition among fishermen described above must stop at the point where opportunity incomes are yielded. It may be and is in many cases carried much further. Two factors prevent an equilibration of fishermen's incomes with those of other members of society. The first is the great immobility of fishermen. Living often in isolated communities, with little knowledge of conditions or opportunities elsewhere; educationally and often romantically tied to the sea; and lacking the savings necessary to provide a "stake," the fisherman is one of the least mobile of occupational groups. But, second, there is in the spirit of every fisherman the hope of the "lucky catch." As those who know fishermen well have often testified, they are gamblers and incurably optimistic. As a consequence, they will work for less than the going wage.[11]

The theory advanced above is substantiated by important developments in the fishing industry. For example, practically all control measures have, in the past, been designed by biologists, with sole attention paid to the production side of the problem and none to the cost side. The result has been a wide-open door for the frustration of the purposes of such measures. The Pacific halibut fishery, for example, is often hailed as a great achievement in modern fisheries management. Under

11. "The gambling instinct of the men makes many of them work for less remuneration than they would accept as a weekly wage, because there is always the possibility of a good catch and a financial windfall" (Graham, *op. cit.*, p. 86).

international agreement between the United States and Canada, a fixed-catch limit was established during the early thirties. Since then, catch-per-unit-effort indexes, as usually interpreted, show a significant rise in the fish population. W. F. Thompson, the pioneer of the Pacific halibut management program, noted recently that "it has often been said that the halibut regulation presents the only definite case of sustained improvement of an overfished deep-sea fishery. This, I believe, is true and the fact should lend special importance to the principles which have been deliberately used to obtain this improvement." [12] Actually, careful study of the statistics indicates that the estimated recovery of halibut stocks could not have been due principally to the control measures, for the average catch was, in fact, greater during the recovery years than during the years of decline. The total amount of fish taken was only a small fraction of the estimated population reduction for the years prior to regulation.[13] Natural factors seem to be mainly responsible for the observed change in population, and the institution of control regulations almost a coincidence. Such coincidences are not uncommon in the history of fisheries policy, but they may be easily explained. If a long-term cyclical fluctuation is taking place in a commercially valuable species, controls will likely be instituted when fishing yields have fallen very low and the clamor of fishermen is great; but it is then, of course, that stocks are about due to recover in any case. The "success" of conservation measures may be due fully as much to the sociological foundations of public policy as to the policy's effect on the fish. Indeed, Burkenroad argues that biological statistics in general may be called into question on these grounds. Governments sponsor biological research when the catches are disappointing. If there are long-term cyclical fluctuations in fish populations, as some think, it is hardly to be wondered why biologists frequently discover that the sea is being depleted, only to change their collective opinion a decade or so later.

Quite aside from the *biological* argument on the Pacific halibut case, there is no clear-cut evidence that halibut fishermen were made relatively more prosperous by the control measures. Whether or not the recovery of the halibut stocks was due to natural factors or to the catch limit, the potential net yield this could have meant has been dissipated through a rise in fishing costs. Since the method of control was to halt fishing when the limit had been reached, this created a great incentive on the part of each fisherman to get the fish before his competitors. During the last twenty years, fishermen have invested in more, larger, and faster boats in a competitive race for fish. In 1933 the fishing sea-

12. W. F. Thompson, "Condition of Stocks of Halibut in the Pacific," *Journal du conseil permanent international pour l'exploration de la mer,* Vol. XVIII, No. 2 (August, 1952).

13. See M. D. Burkenroad, "Fluctuations in Abundance of Pacific Halibut," *Bulletin of the Bingham Oceanographic Collection,* May, 1948.

son was more than six months long. In 1952 it took just twenty-six days to catch the legal limit in the area from Willapa Harbor to Cape Spencer, and sixty days in the Alaska region. What has been happening is a rise in the average cost of fishing effort, allowing no gap between average production and average cost to appear, and hence no rent.[14]

Essentially the same phenomenon is observable in the Canadian Atlantic Coast lobster-conservation program. The method of control here is by seasonal closure. The result has been a steady growth in the number of lobster traps set by each fisherman. Virtually all available lobsters are now caught each year within the season, but at much greater cost in gear and supplies. At a fairly conservative estimate, the same quantity of lobsters could be caught with half the present number of traps. In a few places the fishermen have banded together into a local monopoly, preventing entry and controlling their own operations. By this means, the amount of fishing gear has been greatly reduced and incomes considerably improved.

That the plight of fishermen and the inefficiency of fisheries production stems from the common-property nature of the resources of the sea is further corroborated by the fact that one finds similar patterns of exploitation and similar problems in other cases of open resources. Perhaps the most obvious is hunting and trapping. Unlike fishes, the biotic potential of land animals is low enough for the species to be destroyed. Uncontrolled hunting means that animals will be killed for any short-range human reason, great or small: for food or simply for fun. Thus the buffalo of the western plains was destroyed to satisfy the most trivial desires of the white man, against which the long-term food needs of the aboriginal population counted as nothing. Even in the most civi-

14. The economic significance of the reduction in season length which followed upon the catch limitation imposed in the Pacific halibut fishery has not been fully appreciated. E.g., Michael Graham said in summary of the program in 1943: "The result has been that it now takes only five months to catch the quantity of halibut that formerly needed nine. This, *of course,* has meant profit, where there was none before" (*op. cit.,* p. 156; my italics). Yet, even when biologists have grasped the economic import of the halibut program and its results, they appear reluctant to declare against it. E.g., W. E. Ricker: "This method of regulation does not necessarily make for more profitable fishing and certainly puts no effective brake on waste of effort, since an unlimited number of boats is free to join the fleet and compete during the short period that fishing is open. However, the stock is protected, and yield approximates to a maximum if quotas are wisely set; as biologists, perhaps we are not required to think any further. Some claim that any mixing into the economics of the matter might prejudice the desirable biological consequences of regulation by quotas" ("Production and Utilization of Fish Population," in a Symposium on Dynamics of Production in Aquatic Populations, Ecological Society of America, *Ecological Monographs,* XVI [October, 1946], 385). What such "desirable biological consequences" might be, is hard to conceive. Since the regulatory policies are made by man, surely it is necessary they be evaluated in terms of human, not piscatorial, objectives.

lized communities, conservation authorities have discovered that a bag-limit *per man* is necessary if complete destruction is to be avoided.

The results of anthropological investigation of modes of land tenure among primitive peoples render some further support to this thesis. In accordance with an evolutionary concept of cultural comparison, the older anthropological study was prone to regard resource tenure in common, with unrestricted exploitation, as a "lower" stage of development comparative with private and group property rights. However, more complete annals of primitive cultures reveal common tenure to be quite rare, even in hunting and gathering societies. Property rights in some form predominate by far, and, most important, their existence may be easily explained in terms of the necessity for orderly exploitation and conservation of the resource. Environmental conditions make necessary some vehicle which will prevent the resources of the community at large from being destroyed by excessive exploitation. Private or group land tenure accomplishes this end in an easily understandable fashion.[15] Significantly, land tenure is found to be "common" only in those cases where the hunting resource is migratory over such large areas that it cannot be regarded as husbandable by the society. In cases of group tenure where the numbers of the group are large, there is still the necessity of co-ordinating the practices of exploitation, in agricultural, as well as in hunting or gathering, economies. Thus, for example, Malinowski reported that among the Trobriand Islanders one of the fundamental principles of land tenure is the co-ordination of the productive activities of the gardeners by the person possessing magical leadership in the group.[16] Speaking generally, we may say that stable primitive cultures appear to have discovered the dangers of common-property tenure and to have developed measures to protect their resources. Or, if a more Darwinian explanation be preferred, we may say that only those primitive cultures have survived which succeeded in developing such institutions.

Another case, from a very different industry, is that of petroleum production. Although the individual petroleum producer may acquire undisputed lease or ownership of the particular plot of land upon which his well is drilled, he shares, in most cases, a common pool of oil with other drillers. There is, consequently, set up the same kind of competi-

15. See Frank G. Speck, "Land Ownership among Hunting Peoples in Primitive America and the World's Marginal Areas," *Proceedings of the 22nd International Congress of Americanists* (Rome, 1926), II, 323–32.

16. B. Malinowski, *Coral Gardens and Their Magic,* Vol. I, chaps. xi and xii. Malinowski sees this as further evidence of the importance of magic in the culture rather than as a means of co-ordinating productive activity; but his discussion of the practice makes it clear that the latter is, to use Malinowski's own concept, the "function" of the institution of magical leadership, at least in this connection.

tive race as is found in the fishing industry, with attending overexpansion of productive facilities and gross wastage of the resource. In the United States, efforts to regulate a chaotic situation in oil production began as early as 1915. Production practices, number of wells, and even output quotas were set by governmental authority; but it was not until the federal "Hot Oil" Act of 1935 and the development of interstate agreements that the final loophole (bootlegging) was closed through regulation of interstate commerce in oil.

Perhaps the most interesting similar case is the use of common pasture in the medieval manorial economy. Where the ownership of animals was private but the resource on which they fed was common (and limited), it was necessary to regulate the use of common pasture in order to prevent each man from competing and conflicting with his neighbors in an effort to utilize more of the pasture for his own animals. Thus the manor developed its elaborate rules regulating the use of the common pasture, or "stinting" the common: limitations on the number of animals, hours of pasturing, etc., designed to prevent the abuses of excessive individualistic competition.[17]

There appears, then, to be some truth in the conservative dictum that everybody's property is nobody's property. Wealth that is free for all is valued by none because he who is foolhardy enough to wait for its proper time of use will only find that it has been taken by another. The blade of grass that the manorial cowherd leaves behind is valueless to him, for tomorrow it may be eaten by another's animal; the oil left under the earth is valueless to the driller, for another may legally take it; the fish in the sea are valueless to the fisherman, because there is no assurance that they will be there for him tomorrow if they are left behind today. A factor of production that is valued at nothing in the business calculations of its users will yield nothing in income. Common-property natural resources are free goods for the individual and scarce goods for society. Under unregulated private exploitation, they can yield no rent; that can be accomplished only by methods which make them private property or public (government) property, in either case subject to a unified directing power.

17. See P. Vinogradoff, *The Growth of the Manor* [London, 1905], chap. iv; E. Lipson, *The Economic History of England* [London, 1949], I, 72.

# 6

RONALD COASE

# The Problem of Social Cost[1]

*Ronald Coase is Professor of Economics at the University of Chicago.*

## I. The Problem to Be Examined

This paper is concerned with those actions of business firms which have harmful effects on others. The standard example is that of a factory the smoke from which has harmful effects on those occupying neighbouring properties. The economic analysis of such a situation has usually proceeded in terms of a divergence between the private and social product of the factory, in which economists have largely followed the treatment of Pigou in *The Economics of Welfare*. The conclusion to which this kind of analysis seems to have led most economists is that it would be desirable to make the owner of the factory liable for the damage caused to those injured by the smoke, or alternatively, to place a tax on the factory owner varying with the amount of smoke produced and equiva-

*"The Problem of Social Cost," by Ronald H. Coase, from* The Journal of Law and Economics, *October 1960. (Several passages devoted to extended discussions of legal decisions have been omitted.)*

1. This article, although concerned with a technical problem of economic analysis, arose out of the study of the Political Economy of Broadcasting which I am now conducting. The argument of the present article was implicit in a previous article dealing with the problem of allocating radio and television frequencies ("The Federal Communications Commission," 2 *J. Law & Econ.* [1959]) but comments which I have received seemed to suggest that it would be desirable to deal with the question in a more explicit way and without reference to the original problem for the solution of which the analysis was developed.

lent in money terms to the damage it would cause, or finally, to exclude the factory from residential districts (and presumably from other areas in which the emission of smoke would have harmful effects on others). It is my contention that the suggested courses of action are inappropriate, in that they lead to results which are not necessarily, or even usually, desirable.

## II. The Reciprocal Nature of the Problem

The traditional approach has tended to obscure the nature of the choice that has to be made. The question is commonly thought of as one in which A inflicts harm on B and what has to be decided is: how should we restrain A? But this is wrong. We are dealing with a problem of a reciprocal nature. To avoid the harm to B would inflict harm on A. The real question that has to be decided is: should A be allowed to harm B or should B be allowed to harm A? The problem is to avoid the more serious harm. I instanced in my previous article [2] the case of a confectioner the noise and vibrations from whose machinery disturbed a doctor in his work. To avoid harming the doctor would inflict harm on the confectioner. The problem posed by this case was essentially whether it was worth while, as a result of restricting the methods of production which could be used by the confectioner, to secure more doctoring at the cost of a reduced supply of confectionery products. Another example is afforded by the problem of straying cattle which destroy crops on neighbouring land. If it is inevitable that some cattle will stray, an increase in the supply of meat can only be obtained at the expense of a decrease in the supply of crops. The nature of the choice is clear: meat or crops. What answer should be given is, of course, not clear unless we know the value of what is obtained as well as the value of what is sacrificed to obtain it. To give another example, Professor George J. Stigler instances the contamination of a stream.[3] If we assume that the harmful effect of the pollution is that it kills the fish, the question to be decided is: is the value of the fish lost greater or less than the value of the product which the contamination of the stream makes possible? It goes almost without saying that this problem has to be looked at in total *and* at the margin.

2. Coase, "The Federal Communications Commission," 2 *J. Law & Econ.* 26–27 (1959).
3. G. J. Stigler, *The Theory of Price,* 105 (1952).

## III. The Pricing System with Liability for Damage

I propose to start my analysis by examining a case in which most economists would presumably agree that the problem would be solved in a completely satisfactory manner: when the damaging business has to pay for all damage caused *and* the pricing system works smoothly (strictly this means that the operation of a pricing system is without cost).

A good example of the problem under discussion is afforded by the case of straying cattle which destroy crops growing on neighbouring land. Let us suppose that a farmer and cattle-raiser are operating on neighbouring properties. Let us further suppose that, without any fencing between the properties, an increase in the size of the cattle-raiser's herd increases the total damage to the farmer's crops. What happens to the marginal damage as the size of the herd increases is another matter. This depends on whether the cattle tend to follow one another or to roam side by side, on whether they tend to be more or less restless as the size of the herd increases and on other similar factors. For my immediate purpose, it is immaterial what assumption is made about marginal damage as the size of the herd increases.

To simplify the argument, I propose to use an arithmetical example. I shall assume that the annual cost of fencing the farmer's property is $9 and that the price of the crop is $1 per ton. Also, I assume that the relation between the number of cattle in the herd and the annual crop loss is as follows:

| Number in herd (steers) | Annual crop loss (tons) | Crop loss per additional steer (tons) |
|---|---|---|
| 1 | 1 | 1 |
| 2 | 3 | 2 |
| 3 | 6 | 3 |
| 4 | 10 | 4 |

Given that the cattle-raiser is liable for the damage caused, the additional annual cost imposed on the cattle-raiser if he increased his herd from, say, 2 to 3 steers is $3 and in deciding on the size of the herd, he will take this into account along with his other costs. That is, he will not increase the size of the herd unless the value of the additional meat produced (assuming that the cattle-raiser slaughters the cattle) is greater than the additional costs that this will entail, including the value of the additional crops destroyed. Of course, if, by the employment of dogs, herdsmen, aeroplanes, mobile radio and other means, the amount of damage can be reduced, these means will be adopted when their cost is less than the value of the crop which they prevent being lost. Given that the annual cost of fencing is $9, the cattle-raiser who wished to have a herd with 4 steers or more would pay for fencing to be erected and maintained, assuming that other means of attaining the same end would not do so more cheaply. When the fence is erected, the marginal cost due to the liability for damage becomes zero, except to the extent that an increase in the size of the herd necessitates a stronger and therefore more expensive fence because more steers are liable to lean against it at the same time. But, of course, it may be cheaper for the cattle-raiser not to fence and to pay for the damaged crops, as in my arithmetical example, with 3 or fewer steers.

It might be thought that the fact that the cattle-raiser would pay for all crops damaged would lead the farmer to increase his planting if a cattle-raiser came to occupy the neighbouring property. But this is not so. If the crop was previously sold in conditions of perfect competition, marginal cost was equal to price for the amount of planting undertaken and any expansion would have reduced the profits of the farmer. In the new situation, the existence of crop damage would mean that the farmer would sell less on the open market but his receipts for a given production would remain the same, since the cattle-raiser would pay the market price for any crop damaged. Of course, if cattle-raising commonly involved the destruction of crops, the coming into existence of a cattle-raising industry might raise the price of the crops involved and farmers would then extend their planting. But I wish to confine my attention to the individual farmer.

I have said that the occupation of a neighbouring property by a cattle-raiser would not cause the amount of production, or perhaps more exactly the amount of planting, by the farmer to increase. In fact, if the cattle-raising has any effect, it will be to decrease the amount of planting. The reason for this is that, for any given tract of land, if the value of the crop damaged is so great that the receipts from the sale of the undamaged crop are less than the total costs of cultivating that tract of land, it will be profitable for the farmer and the cattle-raiser to make a bargain whereby that tract of land is left uncultivated. This can be made clear by means of an arithmetical example. Assume initially that the value of the crop obtained from cultivating a given tract of land is $12 and that the cost incurred in cultivating this tract of land is $10, the net gain from cultivating the land being $2. I assume for purposes of simplicity that the farmer owns the land. Now assume that the cattle-raiser starts operations on the neighbouring property and that the value of the crops damaged is $1. In this case $11 is obtained by the farmer from sale on the market and $1 is obtained from the cattle-raiser for damage suffered and the net gain remains $2. Now suppose that the cattle-raiser finds it profitable to increase the size of his herd, even though the amount of damage rises to $3; which means that the value of the additional meat production is greater than the additional costs, including the additional $2 payment for damage. But the total payment for damage is now $3. The net gain to the farmer from cultivating the land is still $2. The cattle-raiser would be better off if the farmer would agree not to cultivate his land for any payment less than $3. The farmer would be agreeable to not cultivating the land for any payment greater than $2. There is clearly room for a mutually satisfactory bargain which would lead to the abandonment of cultivation.[4] But the same argument applies not only to the whole tract cultivated by the farmer but also to any subdivision of it. Suppose, for example, that the cattle have a well-defined route, say, to a brook or to a shady area. In these circumstances, the amount of damage to the crop along the route may well be great and if so, it could be that the farmer and the cattle-raiser would find it profitable to make a bargain whereby the farmer would agree not to cultivate this strip of land.

But this raises a further possibility. Suppose that there is such a well-defined route. Suppose further that the value of the crop that would be obtained by cultivating this strip of land is $10 but that the cost of cultivation is $11. In the absence of the cattle-raiser, the land would not be cultivated. However, given the presence of the cattle-raiser, it could well be that if the strip was cultivated, the whole crop would be destroyed by the cattle. In which case, the cattle-raiser would be forced to pay $10 to the farmer. It is true that the farmer would lose $1. But the cattle-raiser would lose $10. Clearly this is a situation which is not likely to last indefinitely since neither party would want this to happen. The aim of the farmer would be to induce the cattle-raiser to make a payment in return for an agreement to leave this land uncultivated. The farmer would not be able to obtain a payment greater than the cost of fencing off this piece of land nor so high as to lead the cattle-raiser to

4. The argument in the text has proceeded on the assumption that the alternative to cultivation of the crop is abandonment of cultivation altogether. But this need not be so. There may be crops which are less liable to damage by cattle but which would not be as profitable as the crop grown in the absence of damage. Thus, if the cultivation of a new crop would yield a return to the farmer of $1 instead of $2, and the size of the herd which would cause $3 damage with the old crop would cause $1 damage with the new crop, it would be profitable to the cattle-raiser to pay any sum less than $2 to induce the farmer to change his crop (since this would reduce damage liability from $3 to $1) and it would be profitable for the farmer to do so if the amount received was more than $1 (the reduction in his return caused by switching crops). In fact, there would be room for a mutually satisfactory bargain in all cases in which change of crop would reduce the amount of damage by more than it reduces the value of the crop (excluding damage)—in all cases, that is, in which a change in the crop cultivated would lead to an increase in the value of production.

abandon the use of the neighbouring property. What payment would in fact be made would depend on the shrewdness of the farmer and the cattle-raiser as bargainers. But as the payment would not be so high as to cause the cattle-raiser to abandon this location and as it would not vary with the size of the herd, such an agreement would not affect the allocation of resources but would merely alter the distribution of income and wealth as between the cattle-raiser and the farmer.

I think it is clear that if the cattle-raiser is liable for damage caused and the pricing system works smoothly, the reduction in the value of production elsewhere will be taken into account in computing the additional cost involved in increasing the size of the herd. This cost will be weighed against the value of the additional meat production and, given perfect competition in the cattle industry, the allocation of resources in cattle-raising will be optimal. What needs to be emphasized is that the fall in the value of production elsewhere which would be taken into account in the costs of the cattle-raiser may well be less than the damage which the cattle would cause to the crops in the ordinary course of events. This is because it is possible, as a result of market transactions, to discontinue cultivation of the land. This is desirable in all cases in which the damage that the cattle would cause, and for which the cattle-raiser would be willing to pay, exceeds the amount which the farmer would pay for use of the land. In conditions of perfect competition, the amount which the farmer would pay for the use of the land is equal to the difference between the value of the total production when the factors are employed on this land and the value of the additional product yielded in their next best use (which would be what the farmer would have to pay for the factors). If damage exceeds the amount the farmer would pay for the use of the land, the value of the additional product of the factors employed elsewhere would exceed the value of the total product in this use after damage is taken into account. It follows that it would be desirable to abandon cultivation of the land and to release the factors employed for production elsewhere. A procedure which merely provided for payment for damage to the crop caused by the cattle but which did not allow for the possibility of cultivation being discontinued would result in too small an employment of factors of production in cattle-raising and too large an employment of factors in cultivation of the crop. But given the possibility of market transactions, a situation in which damage to crops exceeded the rent of the land would not endure. Whether the cattle-raiser pays the farmer to leave the land uncultivated or himself rents the land by paying the land-owner an amount slightly greater than the farmer would pay (if the farmer was himself renting the land), the final result would be the same and would maximise the value of production. Even when the farmer is induced to plant crops which it would not be profitable to cultivate for sale on the market, this will be a purely short-term phenomenon and may be expected to lead to an agreement under which the planting will cease. The cattle-raiser will remain in that location and the marginal cost of meat production will be the same as before, thus having no long-run effect on the allocation of resources.

## IV. The Pricing System with No Liability for Damage

I now turn to the case in which, although the pricing system is assumed to work smoothly (that is, costlessly), the damaging business is not liable for any of the damage which it causes. This business does not have to make a payment to those damaged by its actions. I propose to show that the allocation of resources will be the same in this case as it was when the damaging business was liable for damage caused. As I showed in the previous case that the allocation of resources was optimal, it will not be necessary to repeat this part of the argument.

I return to the case of the farmer and the cattle-raiser. The farmer would suffer increased damage to his crop as the size of the herd increased. Suppose that the size of the cattle-raiser's herd is 3 steers (and that this is the size of the herd that would be maintained if crop damage was not taken into account). Then the farmer would be willing to pay up to $3 if the cattle-raiser would reduce his herd to 2 steers, up to $5 if the herd were reduced to 1 steer and would pay up to $6 if cattle-raising was abandoned. The cattle-raiser would therefore receive $3 from the farmer if he kept 2 steers instead of 3. This $3 foregone is therefore part of the cost incurred in keeping the third steer. Whether the $3 is a payment which the cattle-raiser has to make if he adds the third steer to his herd (which it would be if the cattle-raiser was liable to the farmer for damage caused to the crop) or whether it is a sum of money which he would have received if he did not keep a third steer (which it would be if the cattle-raiser was not liable to the farmer for damage caused to the crop) does not affect the final result. In both cases $3 is part of the cost of adding a third steer, to be included along with the other costs. If the increase in the value of production in cattle-raising through increasing the size of the herd from 2 to 3 is greater than the additional costs that have to be incurred (including the $3 damage to crops), the size of the herd will be increased. Otherwise, it will not. The size of the herd will be the same whether the cattle-raiser is liable for damage caused to the crop or not.

It may be argued that the assumed starting point—a herd of 3 steers—was arbitrary. And this is true. But the farmer would not wish to pay to avoid crop damage which the cattle-raiser would not be able to cause. For example, the maximum annual payment which the farmer could be induced to pay could not exceed $9, the annual cost of fencing. And the farmer would only be willing to pay this sum if it did not reduce his earnings to a level that would cause him to abandon cultivation of this particular tract of land. Furthermore, the farmer would only be willing to pay this amount if he believed that, in the absence of any payment by him, the size of the herd maintained by the cattle-raiser would be 4 or more steers. Let us assume that this is the case. Then the farmer would be willing to pay up to $3 if the cattle-raiser would reduce his herd to 3 steers, up to $6 if the herd were reduced to 2 steers, up to $8 if one steer only were kept and up to $9 if cattle-raising were abandoned. It will be noticed that the change in the starting point has not altered the amount which would accrue to the cattle-raiser if he reduced the size of his herd by any given amount. It is still true that the cattle-raiser could receive an additional $3 from the farmer if he agreed to reduce his herd from 3 steers to 2 and that the $3 represents the value of the crop that would be destroyed by adding the third steer to the herd. Although a different belief on the part of the farmer (whether justified or not) about the size of the herd that the cattle-raiser would maintain in the absence of payments from him may affect the total payment he can be induced to pay, it is not true that this different belief would have any effect on the size of the herd that the cattle-raiser will actually keep. This will be the same as it would be if the cattle-raiser had to pay for damage caused by his cattle, since a receipt foregone of a given amount is the equivalent of a payment of the same amount.

It might be thought that it would pay the cattle-raiser to increase his herd above the size that he would wish to maintain once a bargain had been made, in order to induce the farmer to make a larger total payment. And this may be true. It is similar in nature to the action of the farmer (when the cattle-raiser was liable for damage) in cultivating land on which, as a result of an agreement with the cattle-raiser, planting would subsequently be abandoned (including land which would not be cultivated at all in the absence of cattle-raising). But such manoeuvres are preliminaries to an agreement and do not affect the long-run equilibrium position, which is the same whether or not the cattle-raiser is held responsible for the crop damage brought about by his cattle.

It is necessary to know whether the damaging business is liable or not for damage caused since without the establishment of this initial delimitation of rights there can be no market transactions to transfer and recombine them. But the ultimate result (which maximises the value of production) is independent of the legal position if the pricing system is assumed to work without cost.

## V. The Problem Illustrated Anew

The harmful effects of the activities of a business can assume a wide variety of forms. An early English case concerned a building which, by obstructing currents of air, hindered the operation of a windmill.[5] A recent case in Florida concerned a building which cast a shadow on the cabana, swimming pool and sunbathing areas of a neighbouring hotel.[6] The problem of straying cattle and the damaging of crops which was the subject of detailed examination in the two preceding sections, although it may have appeared to be rather a special case, is in fact but one example of a problem which arises in many different guises. To clarify the nature of my argument and to demonstrate its general applicability, I propose to illustrate it anew by reference to four actual cases.

Let us first reconsider the case of *Sturges v. Bridgman* [7] which I used as an illustration of the general problem in my article on "The Federal Communications Commission." In this case, a confectioner (in Wigmore Street) used two mortars and pestles in connection with his business (one had been in operation in the same position for more than 60 years and the other for more than 26 years). A doctor then came to occupy neighbouring premises (in Wimpole Street). The confectioner's machinery caused the doctor no harm until, eight years after he had first occupied the premises, he built a consulting room at the end of his garden right against the confectioner's kitchen. It was then found that the noise and vibration caused by the confectioner's machinery made it difficult for the doctor to use his new consulting room. "In particular . . . the noise prevented him from examining his patients by auscultation [8] for diseases of the chest. He also found it impossible to engage with effect in any occupation which required thought and attention." The doctor therefore brought a legal action to force the confectioner to stop using his machinery. The courts had little difficulty in granting the doctor the injunction he sought. "Individual cases of hardship may occur in the strict carrying out of the principle upon which we found our judgment, but the negation of the principle would lead even more to individual hardship, and would at the same time produce a prejudicial effect upon the development of land for residential purposes."

5. See Gale on *Easements* 237–39 (13th ed. M. Bowles 1959).
6. See *Fontainebleu Hotel Corp. v. Forty-Five Twenty-Five, Inc.*, 114 So. 2d 357 (1959).
7. 11 Ch. D. 852 (1879).
8. Auscultation is the act of listening by ear or stethoscope in order to judge by sound the condition of the body.

The court's decision established that the doctor had the right to prevent the confectioner from using his machinery. But, of course, it would have been possible to modify the arrangements envisaged in the legal ruling by means of a bargain between the parties. The doctor would have been willing to waive his right and allow the machinery to continue in operation if the confectioner would have paid him a sum of money which was greater than the loss of income which he would suffer from having to move to a more costly or less convenient location or from having to curtail his activities at this location or, as was suggested as a possibility, from having to build a separate wall which would deaden the noise and vibration. The confectioner would have been willing to do this if the amount he would have to pay the doctor was less than the fall in income he would suffer if he had to change his mode of operation at this location, abandon his operation or move his confectionery business to some other location. The solution of the problem depends essentially on whether the continued use of the machinery adds more to the confectioner's income than it subtracts from the doctor's.[9] But now consider the situation if the confectioner had won the case. The confectioner would then have had the right to continue operating his noise and vibration-generating machinery without having to pay anything to the doctor. The boot would have been on the other foot: the doctor would have had to pay the confectioner to induce him to stop using the machinery. If the doctor's income would have fallen more through continuance of the use of this machinery than it added to the income of the confectioner, there would clearly be room for a bargain whereby the doctor paid the confectioner to stop using the machinery. That is to say, the circumstances in which it would not pay the confectioner to continue to use the machinery and to compensate the doctor for the losses that this would bring (if the doctor had the right to prevent the confectioner's using his machinery) would be those in which it would be in the interest of the doctor to make a payment to the confectioner which would induce him to discontinue the use of the machinery (if the confectioner had the right to operate the machinery). The basic conditions are exactly the same in this case as they were in the example of the cattle which destroyed crops. With costless market transactions, the decision of the courts concerning liability for damage would be without effect on the allocation of resources. It was of course the view of the judges that they were affecting the working of the economic system—and in a desirable direction. Any other decision would have had "a prejudicial effect upon the development of land for residential purposes," an argument which was elaborated by examining the example of a forge operating on a barren moor, which was later developed

9. Note that what is taken into account is the change in income after allowing for alterations in methods of production, location, character of product, etc.

for residential purposes. The judges' view that they were settling how the land was to be used would be true only in the case in which the costs of carrying out the necessary market transactions exceeded the gain which might be achieved by any rearrangement of rights. And it would be desirable to preserve the areas (Wimpole Street or the moor) for residential or professional use (by giving non-industrial users the right to stop the noise, vibration, smoke, etc., by injunction) only if the value of the additional residential facilities obtained was greater than the value of cakes or iron lost. But of this the judges seem to have been unaware.

* * *

The reasoning employed by the courts in determining legal rights will often seem strange to an economist because many of the factors on which the decision turns are, to an economist, irrelevant. Because of this, situations which are, from an economic point of view, identical will be treated quite differently by the courts. The economic problem in all cases of harmful effects is how to maximise the value of production. In the case of *Bass v. Gregory* fresh air was drawn in through the well which facilitated the production of beer but foul air was expelled through the well which made life in the adjoining houses less pleasant. The economic problem was to decide which to choose: a lower cost of beer and worsened amenities in adjoining houses or a higher cost of beer and improved amenities. In deciding this question, the "doctrine of lost grant" is about as relevant as the colour of the judge's eyes. But it has to be remembered that the immediate question faced by the courts is *not* what shall be done by whom *but* who has the legal right to do what. It is always possible to modify by transactions on the market the initial legal delimitation of rights. And, of course, if such market transactions are costless, such a rearrangement of rights will always take place if it would lead to an increase in the value of production.

## VI. The Cost of Market Transactions Taken into Account

The argument has proceeded up to this point on the assumption (explicit in Sections III and IV and tacit in Section V) that there were no costs involved in carrying out market transactions. This is, of course, a very unrealistic assumption. In order to carry out a market transaction it is necessary to discover who it is that one wishes to deal with, to inform people that one wishes to deal and on what terms, to conduct negotiations leading up to a bargain, to draw up the contract, to undertake the inspection needed to make sure that the terms of the contract are being observed and so on. These operations are often extremely costly, sufficiently costly at any rate to prevent many transactions that would be carried out in a world in which the pricing system worked without cost.

In earlier sections, when dealing with the problem of the rearrangement of legal rights through the market, it was argued that such a rearrangement would be made through the market whenever this would lead to an increase in the value of production. But this assumed costless market transactions. Once the costs of carrying out market transactions are taken into account it is clear that such a rearrangement of rights will only be undertaken when the increase in the value of production consequent upon the rearrangement is greater than the costs which would be involved in bringing it about. When it is less, the granting of an injunction (or the knowledge that it would be granted) or the liability to pay damages may result in an activity being discontinued (or may prevent its being started) which would be undertaken if market transactions were costless. In these conditions the initial delimitation of legal rights does have an effect on the efficiency with which the economic system operates. One arrangement of rights may bring about a greater value of production than any other. But unless this is the arrangement of rights established by the legal system, the costs of reaching the same result by altering and combining rights through the market may be so great that this optimal arrangement of rights, and the greater value of production which it would bring, may never be achieved. The part played by economic considerations in the process of delimiting legal rights will be discussed in the next section. In this section, I will take the initial delimitation of rights and the costs of carrying out market transactions as given.

It is clear that an alternative form of economic organisation which could achieve the same result at less cost than would be incurred by using the market would enable the value of production to be raised. As I explained many years ago, the firm represents such an alternative to organising production through market transactions.[10] Within the firm individual bargains between the various cooperating factors of production are eliminated and for a market transaction is substituted an administrative decision. The rearrangement of production then takes place without the need for bargains between the owners of the factors of production. A landowner who has control of a large tract of land may devote his land to various uses taking into account the effect that the interrelations of the various activities will have on the net return of the land, thus rendering unnecessary bargains between those undertaking the various activities. Owners of a large building or of several adjoining properties in a given area may act in much the same way. In effect, using our earlier terminology, the firm would acquire the legal rights of all the par-

10. See Coase, "The Nature of the Firm," 4 *Economica,* New Series, 386 (1937). Reprinted in *Readings in Price Theory,* 331 (1952).

ties and the rearrangement of activities would not follow on a rearrangement of rights by contract, but as a result of an administrative decision as to how the rights should be used.

It does not, of course, follow that the administrative costs of organising a transaction through a firm are inevitably less than the costs of the market transactions which are superseded. But where contracts are peculiarly difficult to draw up and an attempt to describe what the parties have agreed to do or not to do (e.g. the amount and kind of a smell or noise that they may make or will not make) would necessitate a lengthy and highly involved document, and, where, as is probable, a long-term contract would be desirable,[11] it would be hardly surprising if the emergence of a firm or the extension of the activities of an existing firm was not the solution adopted on many occasions to deal with the problem of harmful effects. This solution would be adopted whenever the administrative costs of the firm were less than the costs of the market transactions that it supersedes and the gains which would result from the rearrangement of activities greater than the firm's costs of organising them. I do not need to examine in great detail the character of this solution since I have explained what is involved in my earlier article.

But the firm is not the only possible answer to this problem. The administrative costs of organising transactions within the firm may also be high, and particularly so when many diverse activities are brought within the control of a single organisation. In the standard case of a smoke nuisance, which may affect a vast number of people engaged in a wide variety of activities, the administrative costs might well be so high as to make any attempt to deal with the problem within the confines of a single firm impossible. An alternative solution is direct government regulation. Instead of instituting a legal system of rights which can be modified by transactions on the market, the government may impose regulations which state what people must or must not do and which have to be obeyed. Thus, the government (by statute or perhaps more likely through an administrative agency) may, to deal with the problem of smoke nuisance, decree that certain methods of production should or should not be used (e.g. that smoke preventing devices should be installed or that coal or oil should not be burned) or may confine certain types of business to certain districts (zoning regulations).

The government is, in a sense, a superfirm (but of a very special kind) since it is able to influence the use of factors of production by administrative decision. But the ordinary firm is subject to checks in its operations because of the competition of other firms, which might administer the same activities at lower cost and also because there is always the alternative of market transactions as against organisation

11. For reasons explained in my earlier article, see *Readings in Price Theory,* n. 14 at 337.

within the firm if the administrative costs become too great. The government is able, if it wishes, to avoid the market altogether, which a firm can never do. The firm has to make market agreements with the owners of the factors of production that it uses. Just as the government can conscript or seize property, so it can decree that factors of production should only be used in such-and-such a way. Such authoritarian methods save a lot of trouble (for those doing the organising). Furthermore, the government has at its disposal the police and the other law enforcement agencies to make sure that its regulations are carried out.

It is clear that the government has powers which might enable it to get some things done at a lower cost than could a private organisation (or at any rate one without special governmental powers). But the governmental administrative machine is not itself costless. It can, in fact, on occasion be extremely costly. Furthermore, there is no reason to suppose that the restrictive and zoning regulations, made by a fallible administration subject to political pressures and operating without any competitive check, will necessarily always be those which increase the efficiency with which the economic system operates. Furthermore, such general regulations which must apply to a wide variety of cases will be enforced in some cases in which they are clearly inappropriate. From these considerations it follows that direct governmental regulation will not necessarily give better results than leaving the problem to be solved by the market or the firm. But equally there is no reason why, on occasion, such governmental administrative regulation should not lead to an improvement in economic efficiency. This would seem particularly likely when, as is normally the case with the smoke nuisance, a large number of people are involved and in which therefore the costs of handling the problem through the market or the firm may be high.

There is, of course, a further alternative which is to do nothing about the problem at all. And given that the costs involved in solving the problem by regulations issued by the governmental administrative machine will often be heavy (particularly if the costs are interpreted to include all the consequences which follow from the government engaging in this kind of activity), it will no doubt be commonly the case that the gain which would come from regulating the actions which give rise to the harmful effects will be less than the costs involved in government regulation.

The discussion of the problem of harmful effects in this section (when the costs of market transactions are taken into account) is extremely inadequate. But at least it has made clear that the problem is one of choosing the appropriate social arrangement for dealing with the harmful effects. All solutions have costs and there is no reason to suppose that government regulation is called for simply because the problem is not well handled by the market or the firm. Satisfactory views on policy can only come from a patient study of how, in practice, the market, firms and governments handle the problem of harmful effects. Economists need to study the work of the broker in bringing parties together, the effectiveness of restrictive covenants, the problems of the large-scale real-estate development company, the operation of government zoning and other regulating activities. It is my belief that economists, and policy-makers generally, have tended to over-estimate the advantages which come from governmental regulation. But this belief, even if justified, does not do more than suggest that government regulation should be curtailed. It does not tell us where the boundary line should be drawn. This, it seems to me, has to come from a detailed investigation of the actual results of handling the problem in different ways. But it would be unfortunate if this investigation were undertaken with the aid of a faulty economic analysis. The aim of this article is to indicate what the economic approach to the problem should be.

## VII. The Legal Delimitation of Rights and the Economic Problem

The discussion in Section V not only served to illustrate the argument but also afforded a glimpse at the legal approach to the problem of harmful effects. The cases considered were all English but a similar selection of American cases could easily be made and the character of the reasoning would have been the same. Of course, if market transactions were costless, all that matters (questions of equity apart) is that the rights of the various parties should be well-defined and the results of legal actions easy to forecast. But as we have seen, the situation is quite different when market transactions are so costly as to make it difficult to change the arrangement of rights established by the law. In such cases, the courts directly influence economic activity. It would therefore seem desirable that the courts should understand the economic consequences of their decisions and should, insofar as this is possible without creating too much uncertainty about the legal position itself, take these consequences into account when making their decisions. Even when it is possible to change the legal delimitation of rights through market transactions, it is obviously desirable to reduce the need for such transactions and thus reduce the employment of resources in carrying them out.

A thorough examination of the presuppositions of the courts in trying such cases would be of great interest but I have not been able to attempt it. Nevertheless it is clear from a cursory study that the courts have often recognized the economic implications of their decisions and are aware (as many economists are not) of the reciprocal nature of the problem. Furthermore, from time to time, they take these economic implications into account, along with other factors, in arriving at their decisions. The American writers on this subject refer to the question in a more explicit fashion than do the British. Thus, to quote Prosser on Torts, a person may

> make use of his own property or . . . conduct his own affairs at the expense of some harm to his neighbors. He may operate a factory whose noise and smoke cause some discomfort to others, so long as he keeps within reasonable bounds. It is only when his conduct is unreasonable, *in the light of its utility and the harm which results* [italics added], that it becomes a nuisance. . . . As it was said in an ancient case in regard to candle-making in a town, "Le utility del chose excusera le noisomeness del stink."
>
> The world must have factories, smelters, oil refineries, noisy machinery and blasting, even at the expense of some inconvenience to those in the vicinity and the plaintiff may be required to accept some not unreasonable discomfort for the general good.[12]

The standard British writers do not state as explicitly as this that a comparison between the utility and harm produced is an element in deciding whether a harmful effect should be considered a nuisance. But similar views, if less strongly expressed, are to be found.[13] The doctrine that the harmful effect must be substantial before the court will act is, no doubt, in part a reflection of the fact that there will almost always be some gain to offset the harm. And in the reports of individual cases, it is clear that the judges have had in mind what would be lost as well as what would be gained in deciding whether to grant an injunction or award damages. Thus, in refusing to prevent the destruction of a prospect by a new building, the judge stated:

> I know no general rule of common law, which . . . says, that building so as to stop another's prospect is a nuisance. Was that the case, there could be no great towns; and I must grant injunctions to all the new buildings in this town. . . .[14]

12. See W. L. Prosser, *The Law of Torts* 398–99, 412 (2d ed. 1955). The quotation about the ancient case concerning candle-making is taken from Sir James Fitzjames Stephen, *A General View of the Criminal Law of England* 106 (1890). Sir James Stephen gives no reference. He perhaps had in mind *Rex. v. Ronkett,* included in Seavey, Keeton and Thurston, *Cases on Torts* 604 (1950). A similar view to that expressed by Prosser is to be found in F. V. Harper and F. James, *The Law of Torts* 67–74 (1956); *Restatement, Torts* §§ 826, 827 and 828.

13. See Winfield on *Torts* 541–48 (6th ed. T. E. Lewis 1954); Salmond on the *Law of Torts* 181–90 (12th ed. R. F. V. Heuston 1957); H. Street, *The Law of Torts* 221–29 (1959).

14. *Attorney General v. Doughty,* 2 Ves. Sen. 453, 28 Eng. Rep. 290 (Ch. 1752). Compare in this connection the statement of an American judge, quoted in Prosser, *op. cit. supra* n. 16 at 413 n. 54: "Without smoke, Pittsburgh would have remained a very pretty village," Musmanno, J., in *Versailles Borough v. McKeesport Coal & Coke Co.,* 1935, 83 Pitts. Leg. J. 379, 385.

* * *

. . . The problem which we face in dealing with actions which have harmful effects is not simply one of restraining those responsible for them. What has to be decided is whether the gain from preventing the harm is greater than the loss which would be suffered elsewhere as a result of stopping the action which produces the harm. In a world in which there are costs of rearranging the rights established by the legal system, the courts, in cases relating to nuisance, are, in effect, making a decision on the economic problem and determining how resources are to be employed. It was argued that the courts are conscious of this and that they often make, although not always in a very explicit fashion, a comparison between what would be gained and what lost by preventing actions which have harmful effects. But the delimitation of rights is also the result of statutory enactments. Here we also find evidence of an appreciation of the reciprocal nature of the problem. While statutory enactments add to the list of nuisances, action is also taken to legalize what would otherwise be nuisances under the common law. The kind of situation which economists are prone to consider as requiring corrective government action is, in fact, often the result of government action. Such action is not necessarily unwise. But there is a real danger that extensive government intervention in the economic system may lead to the protection of those responsible for harmful effects being carried too far.

## VIII. Pigou's Treatment in "The Economics of Welfare"

The fountainhead for the modern economic analysis of the problem discussed in this article is Pigou's *Economics of Welfare* and, in particular, that section of Part II which deals with divergences between social and private net products which come about because

> one person A, in the course of rendering some service, for which payment is made, to a second person B, incidentally also renders services or disservices to other persons (not producers of like services), of such a sort that payment cannot be exacted from the benefited parties or compensation enforced on behalf of the injured parties.[15]

Pigou tells us that his aim in Part II of *The Economics of Welfare* is

15. A. C. Pigou, *The Economics of Welfare* 183 (4th ed. 1932). My references will all be to the fourth edition but the argument and examples examined in this article remained substantially unchanged from the first edition in 1920 to the fourth in 1932. A large part (but not all) of this analysis had appeared previously in *Wealth and Welfare* (1912).

> to ascertain how far the free play of self-interest, acting under the existing legal system, tends to distribute the country's resources in the way most favorable to the production of a large national dividend, and how far it is feasible for State action to improve upon 'natural' tendencies.[16]

To judge from the first part of this statement, Pigou's purpose is to discover whether any improvements could be made in the existing arrangements which determine the use of resources. Since Pigou's conclusion is that improvements could be made, one might have expected him to continue by saying that he proposed to set out the changes required to bring them about. Instead, Pigou adds a phrase which contrasts "natural" tendencies with State action, which seems in some sense to equate the present arrangements with "natural" tendencies and to imply that what is required to bring about these improvements is State action (if feasible). That this is more or less Pigou's position is evident from Chapter I of Part II.[17] Pigou starts by referring to "optimistic followers of the classical economists" [18] who have argued that the value of production would be maximised if the government refrained from any interference in the economic system and the economic arrangements were those which came about "naturally." Pigou goes on to say that if self-interest does promote economic welfare, it is because human institutions have been devised to make it so. (This part of Pigou's argument, which he develops with the aid of a quotation from Cannan, seems to me to be essentially correct.) Pigou concludes:

> But even in the most advanced States there are failures and imperfections. . . . there are many obstacles that prevent a community's resources from being distributed . . . in the most efficient way. The study of these constitutes our present problem. . . . its purpose is essentially practical. It seeks to bring into clearer light some of the ways in which it now is, or eventually may become, feasible for governments to control the play of economic forces in such wise as to promote the economic welfare, and through that, the total welfare, of their citizens as a whole.[19]

Pigou's underlying thought would appear to be: Some have argued that no State action is needed. But the system has performed as well as it has because of State action. Nonetheless, there are still imperfections. What additional State action is required?

If this is a correct summary of Pigou's position, its inadequacy can be demonstrated by examining the first example he gives of a divergence between private and social products.

16. *Id.* at xii.
17. *Id.* at 127–30.
18. In *Wealth and Welfare,* Pigou attributes the "optimism" to Adam Smith himself and not to his followers. He there refers to the "highly optimistic theory of Adam Smith that the national dividend, in given circumstances of demand and supply, tends 'naturally' to a maximum" (p. 104).
19. Pigou, *op. cit. supra* n. 35 at 129–30.

> It might happen . . . that costs are thrown upon people not directly concerned, through, say, uncompensated damage done to surrounding woods by sparks from railway engines. All such effects must be included—some of them will be positive, others negative elements—in reckoning up the social net product of the marginal increment of any volume of resources turned into any use or place.[20]

The example used by Pigou refers to a real situation. In Britain, a railway does not normally have to compensate those who suffer damage by fire caused by sparks from an engine. Taken in conjunction with what he says in Chapter 9 of Part II, I take Pigou's policy recommendations to be, first, that there should be State action to correct this "natural" situation and, second, that the railways should be forced to compensate those whose woods are burnt. If this is a correct interpretation of Pigou's position, I would argue that the first recommendation is based on a misapprehension of the facts and that the second is not necessarily desirable.

Let us consider the legal position. Under the heading "Sparks from engines," we find the following in Halsbury's *Laws of England:*

> If railway undertakers use steam engines on their railway without express statutory authority to do so, they are liable, irrespective of any negligence on their part, for fires caused by sparks from engines. Railway undertakers are, however, generally given statutory authority to use steam engines on their railway; accordingly, if an engine is constructed with the precautions which science suggests against fire and is used without negligence, they are not responsible at common law for any damage which may be done by sparks. . . . In the construction of an engine the undertaker is bound to use all the discoveries which science has put within its reach in order to avoid doing harm, provided they are such as it is reasonable to require the company to adopt, having proper regard to the likelihood of the damage and to the cost and convenience of the remedy; but it is not negligence on the part of an undertaker if it refuses to use an apparatus the efficiency of which is open to bona fide doubt.

To this general rule, there is a statutory exception arising from the Railway (Fires) Act, 1905, as amended in 1923. This concerns agricultural land or agricultural crops.

> In such a case the fact that the engine was used under statutory powers does not affect the liability of the company in an action for the damage. . . . These provisions, however, only apply where the claim for damage . . . does not exceed £200 [£100 in the 1905 Act], and where written notice of the occurrence of the fire and the intention to claim has been sent to the company within seven days of the occurrence of the damage and particulars of the damage in writing showing the amount of the

20. *Id*. at 134.

claim in money not exceeding £200 have been sent to the company within twenty-one days.

Agricultural land does not include moorland or buildings and agricultural crops do not include those led away or stacked.[21] I have not made a close study of the parliamentary history of this statutory exception, but to judge from debates in the House of Commons in 1922 and 1923, this exception was probably designed to help the smallholder.[22]

Let us return to Pigou's example of uncompensated damage to surrounding woods caused by sparks from railway engines. This is presumably intended to show how it is possible "for State action to improve on 'natural' tendencies." If we treat Pigou's example as referring to the position before 1905, or as being an arbitrary example (in that he might just as well have written "surrounding buildings" instead of "surrounding woods"), then it is clear that the reason why compensation was not paid must have been that the railway had statutory authority to run steam engines (which relieved it of liability for fires caused by sparks). That this was the legal position was established in 1860, in a case, oddly enough, which concerned the burning of surrounding woods by a railway,[23] and the law on this point has not been changed (apart from the one exception) by a century of railway legislation, including nationalisation. If we treat Pigou's example of "uncompensated damage done to surrounding woods by sparks from railway engines" literally, and assume that it refers to the period after 1905, then it is clear that the reason why compensation was not paid must have been that the damage was more than £100 (in the first edition of *The Economics of Welfare*) or more than £200 (in later editions) or that the owner of the wood failed to notify the railway in writing within seven days of the fire or did not send particulars of the damage, in writing, within twenty-one days. In the real world, Pigou's example could only exist as a result of a deliberate choice of the legislature. It is not, of course, easy to imagine the construction of a railway in a state of nature. The nearest one can get to this is presumably a railway which uses steam engines "without express statutory authority." However, in this case the railway would be obliged to compensate those whose woods it burnt down. That is to say, compensation would be paid in the absence of Government action. The only circumstances in which compensation would not be paid would be those in which there had been Government action. It is strange that Pigou, who clearly thought it desirable that compensation should be

21. See 31 Halsbury, *Laws of England* 474–75 (3d ed. 1960), Article on Railways and Canals, from which this summary of the legal position, and all quotations, are taken.

22. See 152 H.C. Deb. 2622–63 (1922); 161 H.C. Deb. 2935–55 (1923).

23. *Vaughan v. Taff Railway Co.,* 3 H. and N. 743 (Ex. 1858) and 5 H. and N. 679 (Ex. 1860).

paid, should have chosen this particular example to demonstrate how it is possible "for State action to improve on 'natural' tendencies."

Pigou seems to have had a faulty view of the facts of the situation. But it also seems likely that he was mistaken in his economic analysis. It is not necessarily desirable that the railway should be required to compensate those who suffer damage by fires caused by railway engines. I need not show here that, if the railway could make a bargain with everyone having property adjoining the railway line and there were no costs involved in making such bargains, it would not matter whether the railway was liable for damage caused by fires or not. This question has been treated at length in earlier sections. The problem is whether it would be desirable to make the railway liable in conditions in which it is too expensive for such bargains to be made. Pigou clearly thought it was desirable to force the railway to pay compensation and it is easy to see the kind of argument that would have led him to this conclusion. Suppose a railway is considering whether to run an additional train or to increase the speed of an existing train or to install spark-preventing devices on its engines. If the railway were not liable for fire damage, then, when making these decisions, it would not take into account as a cost the increase in damage resulting from the additional train or the faster train or the failure to install spark-preventing devices. This is the source of the divergence between private and social net products. It results in the railway performing acts which will lower the value of total production—and which it would not do if it were liable for the damage. This can be shown by means of an arithmetical example.

Consider a railway, which is *not* liable for damage by fires caused by sparks from its engines, which runs two trains per day on a certain line. Suppose that running one train per day would enable the railway to perform services worth $150 per annum and running two trains a day would enable the railway to perform services worth $250 per annum. Suppose further that the cost of running one train is $50 per annum and two trains $100 per annum. Assuming perfect competition, the cost equals the fall in the value of production elsewhere due to the employment of additional factors of production by the railway. Clearly the railway would find it profitable to run two trains per day. But suppose that running one train per day would destroy by fire crops worth (on an average over the year) $60 and two trains a day would result in the destruction of crops worth $120. In these circumstances running one train per day would raise the value of total production but the running of a second train would reduce the value of total production. The second train would enable additional railway services worth $100 per annum to be performed. But the fall in the value of production elsewhere would be $110 per annum; $50 as a result of the employment of additional factors of production and $60 as a result of the destruction of crops. Since it would be better if the second train were not run and since it would not run if the railway were liable for damage caused to crops, the conclusion that the railway should be made liable for the damage seems irresistible. Undoubtedly it is this kind of reasoning which underlies the Pigovian position.

The conclusion that it would be better if the second train did not run is correct. The conclusion that it is desirable that the railway should be made liable for the damage it causes is wrong. Let us change our assumption concerning the rule of liability. Suppose that the railway is liable for damage from fires caused by sparks from the engine. A farmer on lands adjoining the railway is then in the position that, if his crop is destroyed by fires caused by the railway, he will receive the market price from the railway; but if his crop is not damaged, he will receive the market price by sale. It therefore becomes a matter of indifference to him whether his crop is damaged by fire or not. The position is very different when the railway is *not* liable. Any crop destruction through railway-caused fires would then reduce the receipts of the farmer. He would therefore take out of cultivation any land for which the damage is likely to be greater than the net return of the land (for reasons explained at length in Section III). A change from a regime in which the railway is *not* liable for damage to one in which it *is* liable is likely therefore to lead to an increase in the amount of cultivation on lands adjoining the railway. It will also, of course, lead to an increase in the amount of crop destruction due to railway-caused fires.

Let us return to our arithmetical example. Assume that, with the changed rule of liability, there is a doubling in the amount of crop destruction due to railway-caused fires. With one train per day, crops worth $120 would be destroyed each year and two trains per day would lead to the destruction of crops worth $240. We saw previously that it would not be profitable to run the second train if the railway had to pay $60 per annum as compensation for damage. With damage at $120 per annum the loss from running the second train would be $60 greater. But now let us consider the first train. The value of the transport services furnished by the first train is $150. The cost of running the train is $50. The amount that the railway would have to pay out as compensation for damage is $120. If follows that it would not be profitable to run any trains. With the figures in our example we reach the following result: if the railway is not liable for fire-damage, two trains per day would be run; if the railway is liable for fire-damage, it would cease operations altogether. Does this mean that it is better that there should be no railway? This question can be resolved by considering what would happen to the value of total production if it were decided to exempt the railway from liability for fire-damage, thus bringing it into operation (with two trains per day).

The operation of the railway would enable transport services worth $250 to be performed. It would also mean the employment of factors of production which would reduce the value of production elsewhere by $100. Furthermore it would mean the destruction of crops worth $120. The coming of the railway will also have led to the abandonment of cultivation of some land. Since we know that, had this land been cultivated, the value of the crops destroyed by fire would have been $120, and since it is unlikely that the total crop on this land would have been destroyed, it seems reasonable to suppose that the value of the crop yield on this land would have been higher than this. Assume it would have been $160. But the abandonment of cultivation would have released factors of production for employment elsewhere. All we know is that the amount by which the value of production elsewhere will increase will be less than $160. Suppose that it is $150. Then the gain from operating the railway would be $250 (the value of the transport services) minus $100 (the cost of the factors of production) minus $120 (the value of crops destroyed by fire) minus $160 (the fall in the value of crop production due to the abandonment of cultivation) plus $150 (the value of production elsewhere of the released factors of production). Overall, operating the railway will increase the value of total production by $20. With these figures it is clear that it is better that the railway should not be liable for the damage it causes, thus enabling it to operate profitably. Of course, by altering the figures, it could be shown that there are other cases in which it would be desirable that the railway should be liable for the damage it causes. It is enough for my purpose to show that, from an economic point of view, a situation in which there is "uncompensated damage done to surrounding woods by sparks from railway engines" is not necessarily undesirable. Whether it is desirable or not depends on the particular circumstances.

How is it that the Pigovian analysis seems to give the wrong answer? The reason is that Pigou does not seem to have noticed that his analysis is dealing with an entirely different question. The analysis as such is correct. But it is quite illegitimate for Pigou to draw the particular conclusion he does. The question at issue is not whether it is desirable to run an additional train or a faster train or to install smoke-preventing devices; the question at issue is whether it is desirable to have a system in which the railway has to compensate those who suffer damage from the fires which it causes or one in which the railway does not have to compensate them. When an economist is comparing alternative social arrangements, the proper procedure is to compare the total social product yielded by these different arrangements. The comparison of private and social products is neither here nor there. A simple example will demonstrate this. Imagine a town in which there are traffic lights. A motorist approaches an intersection and stops because the light is red. There are no cars approaching the intersection on the other street. If the motorist ignored the red signal, no accident would occur and the total product would increase because the motorist would arrive earlier at his destination. Why does he not do this? The reason is that if he ignored the light he would be fined. The private product from crossing the street is less than the social product. Should we conclude from this that the total product would be greater if there were no fines for failing to obey traffic signals? The Pigovian analysis shows us that it is possible to conceive of better worlds than the one in which we live. But the problem is to devise practical arrangements which will correct defects in one part of the system without causing more serious harm in other parts.

I have examined in considerable detail one example of a divergence between private and social products and I do not propose to make any further examination of Pigou's analytical system. But the main discussion of the problem considered in this article is to be found in that part of Chapter 9 in Part II which deals with Pigou's second class of divergence and it is of interest to see how Pigou develops his argument. Pigou's own description of this second class of divergence was quoted at the beginning of this section. Pigou distinguishes between the case in which a person renders services for which he receives no payment and the case in which a person renders disservices and compensation is not given to the injured parties. Our main attention has, of course, centred on this second case. It is therefore rather astonishing to find, as was pointed out to me by Professor Francesco Forte, that the problem of the smoking chimney—the "stock instance" [24] or "classroom example" [25] of the second case—is used by Pigou as an example of the first case (services rendered without payment) and is never mentioned, at any rate explicitly, in connection with the second case.[26] Pigou points out that factory owners who devote resources to preventing their chimneys from smoking render services for which they receive no payment. The implication, in the light of Pigou's discussion later in the chapter, is that a factory owner with a smokey chimney should be given a bounty to induce him to install smoke-preventing devices. Most modern economists would suggest that the owner of the factory with the smokey chimney should be taxed. It seems a pity that economists (apart from Professor Forte) do not seem to have noticed this feature of Pigou's treatment since a realisation that the problem could be tackled in either of these two ways would probably have led to an explicit recognition of its reciprocal nature.

In discussing the second case (disservices without compensation to those damaged), Pigou says that they are rendered "when the owner of a site in a residential quarter of a city builds a factory there and so de-

24. Sir Dennis Robertson, I *Lectures on Economic Principles* 162 (1957).
25. E. J. Mishan, "The Meaning of Efficiency in Economics," 189, *The Bankers' Magazine* 482 (June 1960).
26. Pigou, *op. cit. supra* n. 35 at 184.

stroys a great part of the amenities of neighbouring sites; or, in a less degree, when he uses his site in such a way as to spoil the lighting of the house opposite; or when he invests resources in erecting buildings in a crowded centre, which by contracting the air-space and the playing room of the neighbourhood, tend to injure the health and efficiency of the families living there." [27] Pigou is, of course, quite right to describe such actions as "uncharged disservices." But he is wrong when he describes these actions as "anti-social." [28] They may or may not be. It is necessary to weigh the harm against the good that will result. Nothing could be more "anti-social" than to oppose any action which causes any harm to anyone.

* * *

Indeed, Pigou's treatment of the problems considered in this article is extremely elusive and the discussion of his views raises almost insuperable difficulties of interpretation. Consequently it is impossible to be sure that one has understood what Pigou really meant. Nevertheless, it is difficult to resist the conclusion, extraordinary though this may be in an economist of Pigou's stature, that the main source of this obscurity is that Pigou had not thought his position through.

## IX. The Pigovian Tradition

It is strange that a doctrine as faulty as that developed by Pigou should have been so influential, although part of its success has probably been due to the lack of clarity in the exposition. Not being clear, it was never clearly wrong. Curiously enough, this obscurity in the source has not prevented the emergence of a fairly well-defined oral tradition. What economists think they learn from Pigou, and what they tell their students, which I term the Pigovian tradition, is reasonably clear. I propose to show the inadequacy of this Pigovian tradition by demonstrating that both the analysis and the policy conclusions which it supports are incorrect.

I do not propose to justify my view as to the prevailing opinion by copious references to the literature. I do this partly because the treatment in the literature is usually so fragmentary, often involving little more than a reference to Pigou plus some explanatory comment, that detailed examination would be inappropriate. But the main reason for this lack of reference is that the doctrine, although based on Pigou,

27. *Id.* at 185–86.

28. *Id.* at 186 n. 1. For similar unqualified statements see Pigou's lecture "Some Aspects of the Housing Problem" in B. S. Rowntree and A. C. Pigou, "Lectures on Housing," in 18 *Manchester Univ. Lectures* (1914).

must have been largely the product of an oral tradition. Certainly economists with whom I have discussed these problems have shown a unanimity of opinion which is quite remarkable considering the meagre treatment accorded this subject in the literature. No doubt there are some economists who do not share the usual view but they must represent a small minority of the profession.

The approach to the problems under discussion is through an examination of the value of physical production. The private product is the value of the additional product resulting from a particular activity of a business. The social product equals the private product minus the fall in the value of production elsewhere for which no compensation is paid by the business. Thus, if 10 units of a factor (and no other factors) are used by a business to make a certain product with a value of $105; and the owner of this factor is not compensated for their use, which he is unable to prevent; and these 10 units of the factor would yield products in their best alternative use worth $100; then, the social product is $105 minus $100 or $5. If the business now pays for one unit of the factor and its price equals the value of its marginal product, then the social product rises to $15. If two units are paid for, the social product rises to $25 and so on until it reaches $105 when all units of the factor are paid for. It is not difficult to see why economists have so readily accepted this rather odd procedure. The analysis focusses on the individual business decision and since the use of certain resources is not allowed for in costs, receipts are reduced by the same amount. But, of course, this means that the value of the social product has no social significance whatsoever. It seems to me preferable to use the opportunity cost concept and to approach these problems by comparing the value of the product yielded by factors in alternative uses or by alternative arrangements. The main advantage of a pricing system is that it leads to the employment of factors in places where the value of the product yielded is greatest and does so at less cost than alternative systems (I leave aside that a pricing system also eases the problem of the redistribution of income). But if through some God-given natural harmony factors flowed to the places where the value of the product yielded was greatest without any use of the pricing system and consequently there was no compensation, I would find it a source of surprise rather than a cause for dismay.

The definition of the social product is queer but this does not mean that the conclusions for policy drawn from the analysis are necessarily wrong. However, there are bound to be dangers in an approach which diverts attention from the basic issues and there can be little doubt that it has been responsible for some of the errors in current doctrine. The belief that it is desirable that the business which causes harmful effects should be forced to compensate those who suffer damage (which was exhaustively discussed in section VIII in connection with Pigou's railway sparks example) is undoubtedly the result of not comparing the total product obtainable with alternative social arrangements.

The same fault is to be found in proposals for solving the problem of harmful effects by the use of taxes or bounties. Pigou lays considerable stress on this solution although he is, as usual, lacking in detail and qualified in his support.[29] Modern economists tend to think exclusively in terms of taxes and in a very precise way. The tax should be equal to the damage done and should therefore vary with the amount of the harmful effect. As it is not proposed that the proceeds of the tax should be paid to those suffering the damage, this solution is not the same as that which would force a business to pay compensation to those damaged by its actions, although economists generally do not seem to have noticed this and tend to treat the two solutions as being identical.

Assume that a factory which emits smoke is set up in a district previously free from smoke pollution, causing damage valued at $100 per annum. Assume that the taxation solution is adopted and that the factory-owner is taxed $100 per annum as long as the factory emits the smoke. Assume further that a smoke-preventing device costing $90 per annum to run is available. In these circumstances, the smoke-preventing device would be installed. Damage of $100 would have been avoided at an expenditure of $90 and the factory-owner would be better off by $10 per annum. Yet the position achieved may not be optimal. Suppose that those who suffer the damage could avoid it by moving to other locations or by taking various precautions which would cost them, or be equivalent to a loss in income of, $40 per annum. Then there would be a gain in the value of production of $50 if the factory continued to emit its smoke and those now in the district moved elsewhere or made other adjustments to avoid the damage. If the factory owner is to be made to pay a tax equal to the damage caused, it would clearly be desirable to institute a double tax system and to make residents of the district pay an amount equal to the additional cost incurred by the factory owner (or the consumers of his products) in order to avoid the damage. In these conditions, people would not stay in the district or would take other measures to prevent the damage from occurring, when the costs of doing so were less than the costs that would be incurred by the producer to reduce the damage (the producer's object, of course, being not so much to reduce the damage as to reduce the tax payments). A tax system which was confined to a tax on the producer for damage caused would tend to lead to unduly high costs being incurred for the prevention of damage. Of course this could be avoided if it were possible to base the tax, not on the damage caused, but on the fall in the value of production (in its widest sense) resulting from the emission of smoke. But to do so would require a detailed knowledge of individual prefer-

29. *Id.* 192–4, 381 and *Public Finance* 94–100 (3d ed. 1947).

ences and I am unable to imagine how the data needed for such a taxation system could be assembled. Indeed, the proposal to solve the smoke-pollution and similar problems by the use of taxes bristles with difficulties: the problem of calculation, the difference between average and marginal damage, the interrelations between the damage suffered on different properties, etc. But it is unnecessary to examine these problems here. It is enough for my purpose to show that, even if the tax is exactly adjusted to equal the damage that would be done to neighboring properties as a result of the emission of each additional puff of smoke, the tax would not necessarily bring about optimal conditions. An increase in the number of people living or of businesses operating in the vicinity of the smoke-emitting factory will increase the amount of harm produced by a given emission of smoke. The tax that would be imposed would therefore increase with an increase in the number of those in the vicinity. This will tend to lead to a decrease in the value of production of the factors employed by the factory, either because a reduction in production due to the tax will result in factors being used elsewhere in ways which are less valuable, or because factors will be diverted to produce means for reducing the amount of smoke emitted. But people deciding to establish themselves in the vicinity of the factory will not take into account this fall in the value of production which results from their presence. This failure to take into account costs imposed on others is comparable to the action of a factory-owner in not taking into account the harm resulting from his emission of smoke. Without the tax, there may be too much smoke and too few people in the vicinity of the factory; but with the tax there may be too little smoke and too many people in the vicinity of the factory. There is no reason to suppose that one of these results is necessarily preferable.

I need not devote much space to discussing the similar error involved in the suggestion that smoke producing factories should, by means of zoning regulations, be removed from the districts in which the smoke causes harmful effects. When the change in the location of the factory results in a reduction in production, this obviously needs to be taken into account and weighed against the harm which would result from the factory remaining in that location. The aim of such regulation should not be to eliminate smoke pollution but rather to secure the optimum amount of smoke pollution, this being the amount which will maximise the value of production.

## X. A Change of Approach

It is my belief that the failure of economists to reach correct conclusions about the treatment of harmful effects cannot be ascribed simply to a few slips in analysis. It stems from basic defects in the current approach to problems of welfare economics. What is needed is a change of approach.

Analysis in terms of divergencies between private and social products concentrates attention on particular deficiencies in the system and tends to nourish the belief that any measure which will remove the deficiency is necessarily desirable. It diverts attention from those other changes in the system which are inevitably associated with the corrective measure, changes which may well produce more harm than the original deficiency. In the preceding sections of this article, we have seen many examples of this. But it is not necessary to approach the problem in this way. Economists who study problems of the firm habitually use an opportunity cost approach and compare the receipts obtained from a given combination of factors with alternative business arrangements. It would seem desirable to use a similar approach when dealing with questions of economic policy and to compare the total product yielded by alternative social arrangements. In this article, the analysis has been confined, as is usual in this part of economics, to comparisons of the value of production, as measured by the market. But it is, of course, desirable that the choice between different social arrangements for the solution of economic problems should be carried out in broader terms than this and that the total effect of these arrangements in all spheres of life should be taken into account. As Frank H. Knight has so often emphasized, problems of welfare economics must ultimately dissolve into a study of aesthetics and morals.

A second feature of the usual treatment of the problems discussed in this article is that the analysis proceeds in terms of a comparison between a state of laissez faire and some kind of ideal world. This approach inevitably leads to a looseness of thought since the nature of the alternatives being compared is never clear. In a state of laissez faire, is there a monetary, a legal or a political system and if so, what are they? In an ideal world, would there be a monetary, a legal or a political system and if so, what would they be? The answers to all these questions are shrouded in mystery and every man is free to draw whatever conclusions he likes. Actually very little analysis is required to show that an ideal world is better than a state of laissez faire, unless the definitions of a state of laissez faire and an ideal world happen to be the same. But the whole discussion is largely irrelevant for questions of economic policy since whatever we may have in mind as our ideal world, it is clear that we have not yet discovered how to get to it from where we are. A better approach would seem to be to start our analysis with a situation approximating that which actually exists, to examine the effects of a proposed policy change and to attempt to decide whether the new situation would be, in total, better or worse than the original one. In this way, conclusions for policy would have some relevance to the actual situation.

A final reason for the failure to develop a theory adequate to handle the problem of harmful effects stems from a faulty concept of a factor of production. This is usually thought of as a physical entity which the businessman acquires and uses (an acre of land, a ton of fertiliser) instead of as a right to perform certain (physical) actions. We may speak of a person owning land and using it as a factor of production but what the land-owner in fact possesses is the right to carry out a circumscribed list of actions. The rights of a land-owner are not unlimited. It is not even always possible for him to remove the land to another place, for instance, by quarrying it. And although it may be possible for him to exclude some people from using "his" land, this may not be true of others. For example, some people may have the right to cross the land. Furthermore, it may or may not be possible to erect certain types of buildings or to grow certain crops or to use particular drainage systems on the land. This does not come about simply because of government regulation. It would be equally true under the common law. In fact it would be true under any system of law. A system in which the rights of individuals were unlimited would be one in which there were no rights to acquire.

If factors of production are thought of as rights, it becomes easier to understand that the right to do something which has a harmful effect (such as the creation of smoke, noise, smells, etc.) is also a factor of production. Just as we may use a piece of land in such a way as to prevent someone else from crossing it, or parking his car, or building his house upon it, so we may use it in such a way as to deny him a view or quiet or unpolluted air. The cost of exercising a right (of using a factor of production) is always the loss which is suffered elsewhere in consequence of the exercise of that right—the inability to cross land, to park a car, to build a house, to enjoy a view, to have peace and quiet or to breathe clean air.

It would clearly be desirable if the only actions performed were those in which what was gained was worth more than what was lost. But in choosing between social arrangements within the context of which individual decisions are made, we have to bear in mind that a change in the existing system which will lead to an improvement in some decisions may well lead to a worsening of others. Furthermore we have to take into account the costs involved in operating the various social arrangements (whether it be the working of a market or of a government department), as well as the costs involved in moving to a new system. In devising and choosing between social arrangements we should have regard for the total effect. This, above all, is the change in approach which I am advocating.

# 7

RALPH TURVEY

# *On Divergences between Social Cost and Private Cost*

*Ralph Turvey is Joint Deputy Chairman of the National Board for Prices and Incomes in Great Britain.*

The notion that the resource-allocation effects of divergences between marginal social and private costs can be dealt with by imposing a tax or granting a subsidy equal to the difference now seems too simple a notion. Three recent articles have shown us this. First came Professor Coase's "The Problem of Social Cost", then Davis and Whinston's "Externalities, Welfare and the Theory of Games" appeared, and, finally, Buchanan and Stubblebine have published their paper "Externality".[1] These articles have an aggregate length of eighty pages and are by no means easy to read. The following attempt to synthesise and summarise the main ideas may therefore be useful. It is couched in terms of external diseconomies, i.e. an excess of social over private costs, and the reader is left to invert the analysis himself should he be interested in external economies.

The scope of the following argument can usefully be indicated by

*"On Divergences between Social Cost and Private Cost," by Ralph Turvey, from* Economica, *August 1963.*

1. *Journal of Law and Economics,* Vol. III, October, 1960, *Journal of Political Economy,* June, 1962, and *Economica,* November, 1962, respectively. [Professor Coase's article is reprinted in this volume, Selection 6.]

starting with a brief statement of its main conclusions. The first is that if the party imposing external diseconomies and the party suffering them are able and willing to negotiate to their mutual advantage, state intervention is unnecessary to secure optimum resource allocation. The second is that the imposition of a tax upon the party imposing external diseconomies can be a very complicated matter, even in principle, so that the *a priori* prescription of such a tax is unwise.

To develop these and other points, let us begin by calling *A* the person, firm or group (of persons or firms) which imposes a diseconomy, and *B* the person, firm or group which suffers it. How much *B* suffers will in many cases depend not only upon the *scale* of *A*'s diseconomy-creating activity, but also upon the precise *nature* of *A*'s activity and upon *B*'s *reaction* to it. If *A* emits smoke, for example, *B*'s loss will depend not only upon the quantity emitted but also upon the height of *A*'s chimney and upon the cost to *B* of installing air-conditioning, indoor clothes-dryers or other means of reducing the effect of the smoke. Thus to ascertain the optimum resource allocation will frequently require an investigation of the nature and costs both of alternative activities open to *A* and of the devices by which *B* can reduce the impact of each activity. The optimum involves that kind and scale of *A*'s activity and that adjustment to it by *B* which maximises the algebraic sum of *A*'s gain and *B*'s loss as against the situation where *A* pursues no diseconomy-creating activity. Note that the optimum will frequently involve *B* suffering a loss, both in total and at the margin.[2]

If *A* and *B* are firms, gain and loss can be measured in money terms as profit differences. (In considering a social optimum, allowance has of course to be made for market imperfections.) Now assuming that they both seek to maximise profits, that they know about the available alternatives and adjustments and that they are able and willing to negotiate, they will achieve the optimum without any government interference. They will internalize the externality by merger,[3] or they will make an agreement whereby *B* pays *A* to modify the nature or scale of its activity.[4] Alternatively,[5] if the law gives *B* rights against *A, A* will pay *B* to accept the optimal amount of loss imposed by *A*.

If *A* and *B* are people, their gain and loss must be measured as the amount of money they respectively would pay to indulge in and prevent *A*'s activity. It could also be measured as the amount of money they respectively would require to refrain from and to endure *A*'s activity, which will be different unless the marginal utility of income is constant. We shall assume that it is constant for both *A* and *B,* which is reason-

2. Buchanan-Stubblebine, pp. 380–1.
3. Davis-Whinston, pp. 244, 252, 256; Coase, pp. 16–17.
4. Coase, p. 6; Buchanan-Stubblebine agree, p. 383.
5. See previous references.

able when the payments do not bulk large in relation to their incomes.[6] Under this assumption, it makes no difference whether *B* pays *A* or, if the law gives *B* rights against *A, A* compensates *B*.

Whether *A* and *B* are persons or firms, to levy a tax on *A* which is *not* received as damages or compensation by *B* may prevent optimal resource allocation from being achieved—still assuming that they can and do negotiate.[7] The reason is that the resource allocation which maximises *A*'s *gain less B*'s *loss* may differ from that which maximises *A*'s *gain less A*'s *tax less B*'s *loss*.

The points made so far can usefully be presented diagrammatically (Figure 1). We assume that *A* has only two alternative activities, I and

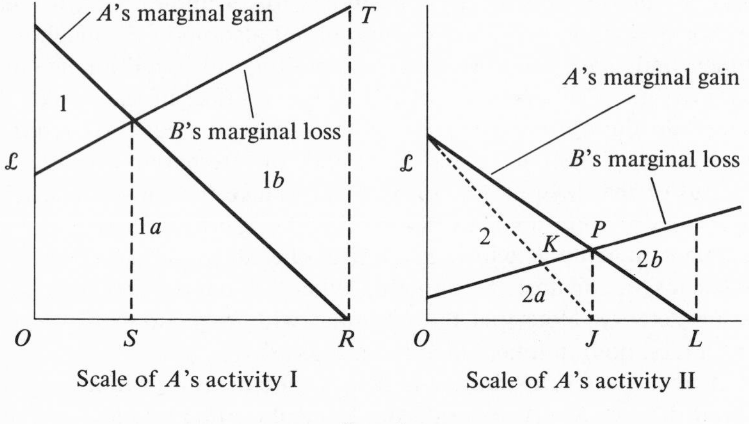


FIG. 1

II, and that their scales and *B*'s losses are all continuously variable. Let us temporarily disregard the dotted curve in the right-hand part of the diagram. The area under *A*'s curves then gives the total gain to *A*. The area under *B*'s curves gives the total loss to *B* after he has made the best adjustment possible to *A*'s activity. This is thus the direct loss as reduced by adjustment, plus the cost of making that adjustment.

If *A* and *B* could not negotiate and if *A* were unhampered by restrictions of any sort, *A* would choose activity I at a scale of *OR*. A scale of *OS* would obviously give a larger social product, but the optimum is clearly activity II at scale *OJ,* since area 2 is greater than area 1. Now *B* will be prepared to pay up to $(1a+1b-2a)$ to secure this result, while *A* will be prepared to accept down to $(1+1a-2-2a)$ to assure

6. Dr. Mishan has examined the welfare criterion for the case where the only variable is the scale of *A*'s activity, but where neither *A* nor *B* has a constant marginal utility of income. Cf. his paper "Welfare Criteria for External Effects," *American Economic Review*, September, 1961.

7. Buchanan-Stubblebine, pp. 381–3.

it. The difference is $(1b-1+2)$, the maximum gain to be shared between them, and this is clearly positive.

If $A$ is liable to compensate $B$ for actual damages caused by either activity I or II, he will choose activity II at scale $OJ$ (i.e. the optimum allocation), pay $2a$ to $B$ and retain a net gain of 2. The result is the same as when there is no such liability, though the distribution of the gain is very different: $B$ will pay $A$ up to $(1a+1b-2a)$ to secure this result. Hence whether or not we should advocate the imposition of a liability on $A$ for damages caused is a matter of fairness, not of resource allocation. Our judgment will presumably depend on such factors as who got there first, whether one of them is a non-conforming user (e.g. an establishment for the breeding of maggots on putrescible vegetable matter in a residential district), who is richer, and so on. Efficient resource allocation requires the imposition of a liability upon $A$ only if we can show that inertia, obstinacy, etc. inhibit $A$ and $B$ from reaching a voluntary agreement.[8]

We can now make the point implicit in Buchanan-Stubblebine's argument, namely that there is a necessity for any impost levied on $A$ to be paid to $B$ when $A$ and $B$ are able to negotiate. Suppose that $A$ is charged an amount equal to the loss he imposes on $B$; subtracting this from his marginal gain curve in the right-hand part of the diagram gives us the dotted line as his marginal net gain. If $A$ moves to point $J$ it will then pay $B$ to induce him to move back to position $K$ (which is sub-optimal) as it is this position which maximises the *joint* net gain to $A$ and $B$ together.

There is a final point to be made about the case where $A$ and $B$ can negotiate. This is that if the external diseconomies are reciprocal, so that each imposes a loss upon the other, the problem is still more complicated.[9]

We now turn to the case where $A$ and $B$ cannot negotiate, which in most cases will result from $A$ and / or $B$ being too large a group for the members to get together. Here there are certain benefits to be had from resource re-allocation which are not privately appropriable. Just as with collective goods,[10] therefore, there is thus a case for collective action to achieve optimum allocation. But all this means is that *if* the state can ascertain and enforce a move to the optimum position at a cost less than the gain to be had, and *if* it can do this in a way which does not have unfavourable effects upon income distribution, then it should take action.

These two "ifs" are very important. The second is obvious and requires no elaboration. The first, however, deserves a few words. In

8. Cf. the comparable argument on pp. 94–8 of my *The Economics of Real Property*, 1957, about the external economy to landlords of tenants' improvements.
9. Davis-Whinston devote several pages of game theory to this problem.
10. Buchanan-Stubblebine, p. 383.

order to ascertain the optimum type and scale of *A*'s activity, the authorities must estimate all of the curves in the diagrams. They must, in other words, list and evaluate all the alternatives open to *A* and examine their effects upon *B* and the adjustments *B* could make to reduce the loss suffered. When this is done, if it can be done, it is necessary to consider how to reach the optimum. Now, where the nature as well as the scale of *A*'s activity is variable, it may be necessary to control both, and this may require two controls, not one. Suppose, for instance, that in the diagram, both activities are the emission of smoke: I from a low chimney and II from a tall chimney. To induce *A* to shift from emitting *OR* smoke from the low chimney to emitting *OJ* smoke from the tall chimney, it will not suffice to levy a tax of *PJ* per unit of smoke.[11] If this alone were done, *A* would continue to use a low chimney, emitting slightly less than *OR* smoke. It will also be necessary to regulate chimney heights. A tax would do the trick alone only if it were proportioned to losses imposed rather than to smoke emitted, and that would be very difficult.

These complications show that in many cases the cost of achieving optimum resource allocation may outweigh the gain. If this is the case, a second-best solution may be appropriate. Thus a prohibition of all smoke emission would be better than *OR* smoke from a low chimney (since 1 is less than 1*b*) and a requirement that all chimneys be tall would be better still (giving a net gain of 2 less 2*b*). Whether these requirements should be imposed on existing chimney-owners as well as on new ones then introduces further complications relating to the short run and the long run.

There is no need to carry the example any further. It is now abundantly clear that any general prescription of a tax to deal with external diseconomies is useless. Each case must be considered on its own and there is no *a priori* reason to suppose that the imposition of a tax is better than alternative measures or indeed, that any measures at all are desirable unless we assume that information and administration are both costless.[12]

To sum up, then: when negotiation is possible, the case for government intervention is one of justice not of economic efficiency; when it is not, the theorist should be silent and call in the applied economist.

11. Note how different *PJ* is from *RT*, the initial observable marginal external diseconomy.

12. Coase, pp. 18, 44.

# 8

ALLEN V. KNEESE AND
BLAIR T. BOWER

# *Causing Offsite Costs to Be Reflected in Waste Disposal Decisions*

*Allen V. Kneese is Director, and Blair T. Bower is Associate Director of the Quality of the Environment Program at Resources for the Future.*

Under a free market economy municipalities and private firms can escape certain costs associated with waste disposal by passing the problem along to other parties, and may find it to their economic advantage to do so. The ultimate costs throughout a whole watershed are not reduced—in fact, they are increased if offsite costs exceed the costs of reducing waste discharges—but they are borne by someone other than the dischargers. As we have already seen, the incentive to neglect offsite costs would disappear if a single competitive firm managed all phases of water supply and use throughout an entire basin, and if there were no public goods involved, because the external diseconomies would then become internal. The basin-wide firm idea, though useful for illuminating the character and role of external diseconomies, is not a satisfactory solution. Public intervention, however, can cause offsite costs to be reflected in the waste disposal decisions of individual firms and of local government units, and is needed for optimal resources allocation.

Some possible means of public intervention—effluent charges, incentive payments, and enforcement of quality standards—are considered in this chapter from the economic viewpoint.

In areas of considerable urban and industrial development the offsite costs associated with unregulated waste discharge are likely to be great enough to justify public regulation, despite the costs and distortions which may be entailed in the regulation itself.[1] Here the focus is on the economic aspects of dealing with the external costs of waste discharge, and the assumption is made that regulatory public agencies exist and that they have sufficient geographical scope to internalize the major external costs associated with waste discharge in their areas. It is also assumed that the objective of these agencies, generally referred to as "river basin authorities," is to obtain the maximum net benefit from the water resources.

How can the basin authority induce the individual municipality or industry to consider offsite costs in its decisions involving waste generation, treatment, and discharge? This question is discussed, first, in terms of the method to be used. Emphasis is placed on a device whereby the basin authority would levy charges on effluents discharged. The alternative possibility of framing a system of payments for waste discharge reduction to achieve optimum water quality is also explored.

Attention is then directed to a variety of considerations involved in actually determining the costs associated with a given waste discharge.

## Economic Incentives for Reducing Waste Discharge

### *Charges*

Economists have long held that technological spillovers can be counteracted by levying a tax on the unit responsible for the diseconomy and by paying a subsidy to the damaged party. Some have even demonstrated that under certain conditions the appropriate tax is just large

1. As judged in terms of the criteria of economic welfare theory. See James M. Buchanan, "Politics, Policy and the Pigovian Margin," *Economica*, Vol. 29 (February 1962).

enough to pay the appropriate subsidy.[2] This idea makes a good bit of sense when one views the uneconomic effects of spillovers as resulting basically from a maldistribution of costs. However, in the case of waste loads, and from the point of view of resources allocation, it is not necessary both to levy a tax and pay a subsidy if the waste discharger and the damaged party do not themselves bargain about the externality. In principle, either a charge on effluents or a payment to reduce discharge will serve to induce the combination of measures that will minimize the costs associated with waste disposal in a region. But if bargaining takes place, a unilateral fee or subsidy will not produce an efficient result. To see this, assume that a public authority is created and directed to impose the cost of incremental damages upon the waste discharger. In other words, the authority endeavors to compute the relevant portion of a function like $YX'$ in Figure 1 and to confront the waste discharger

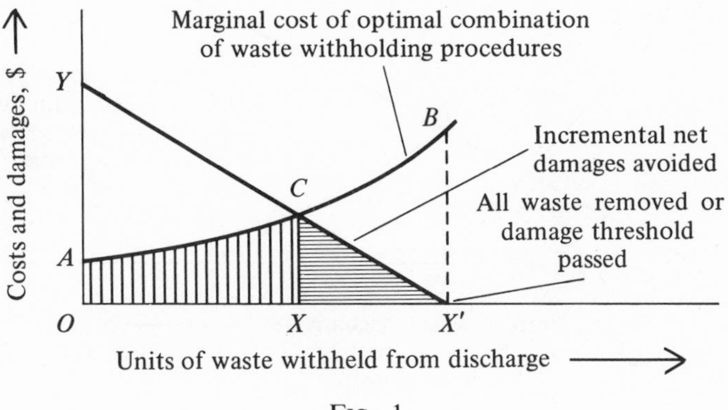


FIG. 1

with it. The horizontal axis measures units of waste withheld from discharge by the waste discharger and the $YX'$ function indicates the incremental effluent charge which the waste disposer must pay for every unit of residual waste discharged.

If damages are properly determined and assessed, does optimization require that the loss to the "damaged" party be compensated? If the authority were in fact to levy a charge equal to the damages associated with each increment of waste discharge, it would collect just enough in tolls to cover the residual damages (area $XCX'$).[3]

From the viewpoint of efficiency (as contrasted with equity) compen-

2. James E. Meade, "External Economies and Diseconomies in a Competitive Situation," *Economic Journal,* March 1952.

3. If, on the other hand, a single charge were imposed at a level which would cause incremental damage and reduction costs to be equated, the authority would collect more than the residual damages unless the total damage cost function were linear (i.e., resulting in a flat $YX'$). The excess is basically an economic rent. More will be said about this later.

sation must be paid if the parties can and do negotiate. Otherwise, optimal resource allocation will be prevented both in the short run and the long run, because the level of waste reduction that minimizes the sum of party 1's cost for waste reduction plus party 2's residual damages does not necessarily minimize party 1's cost plus party 1's charge plus party 2's damage. Consequently, a schedule of charges equal to marginal damages imposed and not used to pay compensation will not eliminate what have been termed "Pareto-relevant" externalities at the optimum level of waste reduction, i.e., from the viewpoint of the parties further gains from trade are possible.[4] Figure 2 illustrates the reasoning involved.

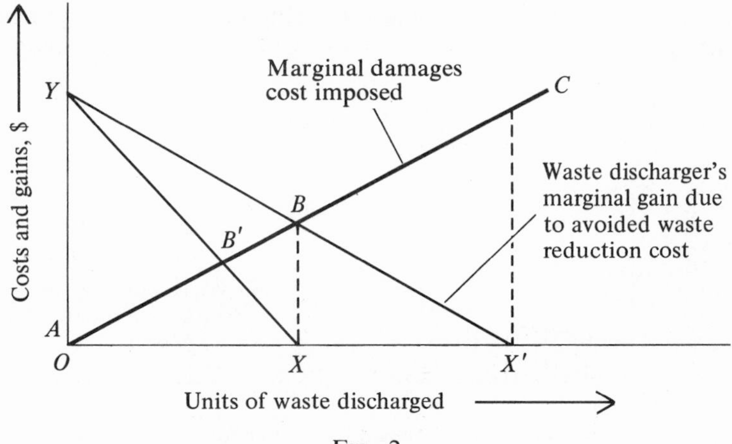


FIG. 2

Increments to waste discharge are measured along the $x$-axis rather than, as in Figure 1, reductions in waste discharge. Function $AC$ indicates the incremental damage costs imposed and $YX'$ indicates the marginal gain to the waste discharger due to not having to reduce his waste discharge (say by treatment). $YX$ is the excess of the marginal gain from discharging waste over the marginal damage cost imposed at each level of waste discharge. If the marginal cost of offsite damages is levied upon the waste discharger, his net gain becomes function $YX$. At the optimum point $X$, his marginal gains are cancelled by the marginal damages imposed so that his net gain is zero. This is another way of saying that $X$ is the optimum point. Acting by himself, if marginal offsite damage costs are imposed upon him, the waste discharger would tend to move to this point. As noted in the previous chapter, an exter-

4. A particularly clear exposition of this point as it relates to externalities generally is found in Ralph Turvey, "On Divergencies Between Social Cost and Private Cost," *Economica,* Vol. 30 (August 1963). [Reprinted in this volume, Selection 7.

nal effect continues to exist at the optimum point. This simply indicates that the benefit to the discharger from not reducing the waste load further is greater than the damage imposed on other parties.

Unless, however, the damaged party is compensated so that he experiences no damage costs at the rate of discharge indicated by point *X,* he will be willing to pay the waste discharger to reduce his waste discharge further to the point where the discharger's marginal net gain equals the marginal damage imposed. This is indicated by point *B′* in Figure 2. On the other hand, if he is compensated and his compensation is contingent on engaging in the pollution damaged activity, this will tend to lead to inefficient longer-term adjustments.

If the parties can not or do not negotiate, the social optimum can be attained by taxing the waste discharger and not compensating the damaged party. We therefore conclude that achievement of an optimum by means of effluent charges would not in general involve compensation of parties adversely affected by waste discharges. In the remainder of the chapter we assume that the parties involved cannot negotiate with each other and that no compensation is paid.

The discussion up to this point has been conducted in terms of effluent charges representing incremental offsite or external costs imposed. It is not difficult to demonstrate that, *in principle,* the same result could be achieved by making payments for the reduction of waste discharge—even under conditions where longer-run adjustments such as plant location and process change are involved. The distribution of income would of course be different under the two arrangements.

### *Payments*

We believe that a system of payments, or "bribes" as they have recently been termed in the literature, could in principle achieve the same result as an optimal charges scheme, despite some recent statements to the contrary.[5]

Assume that a profit-maximizing firm has an incremental production cost curve as indicated by *MC* in Figure 3, that the price at which the firm can sell the commodity it produces is given, as indicated by the curve *D,* that the *only* way the firm can diminish the amount of a residual substance which it discharges into a stream is to reduce production,

5. See D. F. Bramhall and E. S. Mills, "A Note on the Asymmetry Between Fees and Payments," *Water Resources Research* (Third Quarter 1966), pp. 615–16; M. J. Kamien, N. L. Schwartz, and F. T. Dolbear, Jr., "Asymmetry Between Bribes and Charges," *Water Resources Research* (First Quarter 1966), pp. 147–57; and A. Myrick Freeman III, "Bribes and Charges: Some Comments," *Water Resources Research* (First Quarter 1967), pp. 287–98. See also F. T. Dolbear, Jr., "On the Theory of Optimum Externality," *American Economic Review* (March 1967).

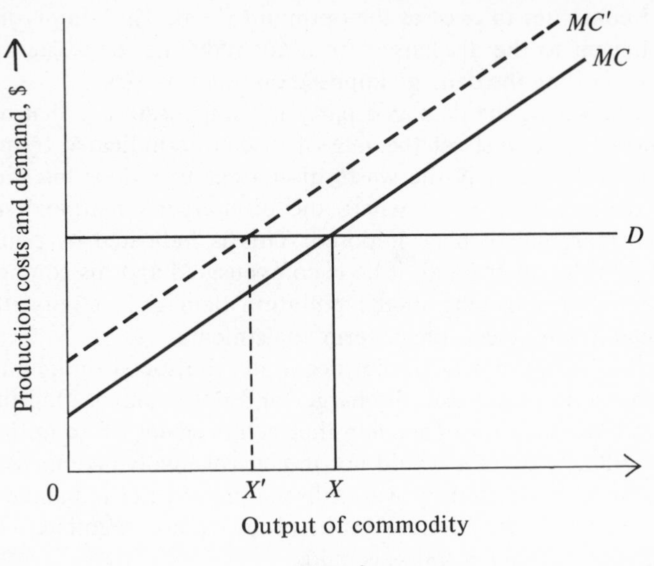


FIG. 3

and that residual waste per unit of output is a constant. If a regulatory authority imposes a unit charge on the effluent of the firm, the incremental production cost function will shift upward by the amount of the charge per unit of output (i.e., charge per unit of effluent times effluent per unit of output) from *MC to MC'*.

On the other hand, if the regulatory authority offers to pay the same amount per unit for reducing waste discharge, the incremental cost function will still be *MC'*. A firm rationally trying to maximize its profits will view the payment as an opportunity cost of production because waste discharge is, by assumption, a straightforward function of production. This proposition can perhaps be clarified by the numerical example below:

| Output of product | Total cost of production | Incremental cost of production | Effluent charge on waste discharge per unit of production | Payment for reduction of waste discharge per unit of production | Minimum product price necessary to induce indicated no. of units of production with charge or payment |
|---|---|---|---|---|---|
| 0 | 10 | . . . | . . . | . . . | $. . . |
| 1 | 11 | 1 | 1 | 1 | 2 |
| 2 | 13 | 2 | 1 | 1 | 3 |
| 3 | 16 | 3 | 1 | 1 | 4 |
| 4 | 20 | 4 | 1 | 1 | 5 |
| 5 | 25 | 5 | 1 | 1 | 6 |

If, for example, the incremental cost of producing the fifth unit (neglecting the cost of waste discharge) is $5.00, and a public authority stands ready to pay $1.00 if the manufacturer does not discharge the waste associated with that unit, he will produce that unit only if the price is at least $6.00. If the price were less than $6.00, his *net* income would be higher if he simply accepted the $1.00 payment. The $1.00 payment thus becomes an *opportunity cost* even though it is not a direct outlay. The point is that in the sense of *opportunity* cost, which is the relevant concept of cost for decision purposes, the two procedures have entirely the same effect on incremental costs.[6]

The question may well be raised whether in terms of longer-term adjustments, i.e., a firm's decision to enter or leave an industry or to expand or contract production capacity, the effect may not be different. The answer is in principle "no," but great informational and administrative difficulties emerge if the payments route is adopted. These are discussed later.

In Figure 4, *AC* indicates the long-run average cost curve of a plant. This curve indicates the average costs of producing various levels of

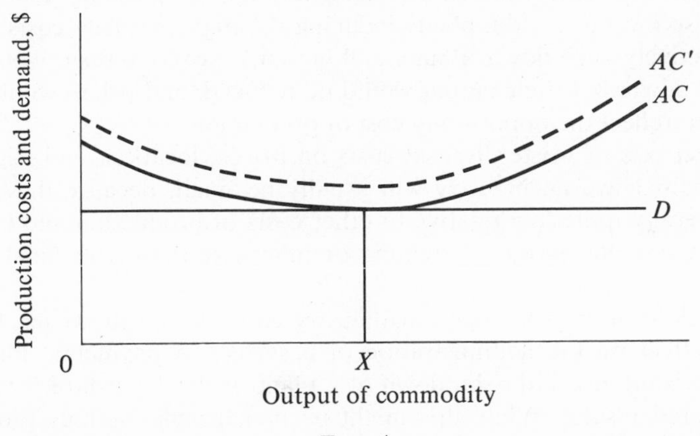


FIG. 4

output (including an average return on investment) under conditions where a plant can be designed to produce a given output at least overall cost. If the price is as indicated by *D*, the firm would construct a plant for which costs were lowest at output level *X*. This is the relationship which would tend to prevail in equilibrium in a competitive industry. A charge on effluent per unit would raise the average cost function by the same amount for each unit of output. Consequently, the new average cost curve would be a simple vertical displacement of the old, as shown by *AC′*. A payment would have the same effect because it is an opportunity cost. The price of output must be high enough to cover the

6. For a mathematical development of this point, see Kamien *et al., op. cit.*

amount of the available payment as well as costs of production (including a normal return on investment), if it is to be worthwhile for the firm to produce an additional unit. In effect, the plant's cost curve (reflecting the value of all foregone opportunities if production is carried on) is raised by the amount of the payment if the payment is available to it whether it is producing or not.

Assume that all plants in the industry are initially earning no "excess" return (i.e., they are earning only the minimum necessary to keep their resources employed in the industry). Assume further that they have their cost curves raised by a charge or payment by the amount indicated in Figure 4. The new long-term equilibrium adjustment would then find the industry price higher by the amount of the subsidy or the payment, fewer firms in the industry, and fewer resources devoted to producing the output of the industry. All this would reflect the fact that costs imposed elsewhere in the system, and previously neglected by management, have been internalized to individual firms.

A system of charges or payments reflecting downstream costs would, however, probably not raise the opportunity costs of all firms by the same amount. Under these circumstances—and assuming costs were otherwise the same—the plants inducing the highest offsite costs would be selectively shut down. Plants making an "excess" return, would not be shut down, but their output would be reduced, and prices would tend to rise to reflect the opportunity cost of production.

The effects of waste disposal costs on prices, location, and decisions to enter or leave an industry will usually be small, because these costs are generally quite low relative to other costs of production and because the firm has the option of treating or otherwise modifying final waste output.

Decisions on location and / or industry entry or exit, however, have a major effect on the administration of a system of payments. Payment must be continued after shutdown of a plant, if the procedure is to have the desired results. While this might be manageable, serious problems would occur if a shift in demand for the product should increase the potential profitability of the plant, or if other dynamic adjustments should take place.[7] Moreover, plants might introduce processes that produce a

7. Major administrative and informational problems arise in both the short and long run if cost and revenue functions of the waste discharging firm shift. In general, the reason is that there is no "natural origin" for the bribe payment. In the charges case the basin agency simply collects for every unit of waste discharge which causes external cost. The authority need not and does not concern itself with how much would have been discharged by an industrial firm in the absence of the charge. In the case of a bribe, however, a total amount of payment to the firm is involved which is contingent upon the amount the firm would have discharged in the absence of the bribe—the dotted area in Figure 1. In order to determine this quantity when cost or revenue functions shift, thus altering the privately optimum level of production of the firm and associated waste discharge,

greater quantity of waste in order to obtain payment for reducing waste.

Even more perplexing would be the handling of proposals for new industrial locations. The administrative authority would have to stand ready to make payments to industrial plants which never do locate in the area but which would do so if a payment reflecting the costs their effluents would impose were not made.[8] Payments on this basis would of course be an open invitation to extortion. If charges were levied, however, the authority would only have to provide the prospective firm with an estimate of the unit charge to be placed on its effluent.[9]

To recapitulate, the basic point about the payments-charges controversy is that only the payments scheme requires information about a "status quo" point or rate of discharge. Obtaining such information would be difficult; if industries enter or leave, it might be impossible. Moreover, the payments technique would make it easy for the unscrupulous operator to benefit by exaggerating his potential waste load.

Thus far, it has been assumed, for the sake of simplicity, that the only way in which waste loads can be reduced is by decreasing production. Waste loads can, of course, be reduced by other means including industrial process changes, materials recovery, by-product production, and treatment. That a payment per unit reduction of waste load will produce the same balance between production cutback and treatment (or other methods) as a charge of equivalent amount can be conveniently shown by means of an isoquant diagram (Figure 5). The isoquants $CC'$ and $KK'$ indicate along their lengths a constant amount of waste reduction (say, pounds of BOD) which can be achieved by alternative combinations of treatment or production decrease.

The lines $AA'$ and $BB'$ indicate the number of units of treatment capacity and reduction in output that can be purchased at equal cost. In principal they could relate to any combinations of alternatives. In many industries, internal process adjustments will be economical substitutes for external treatment. If more than two alternatives are involved, dia-

---

the authority must have full knowledge of the firm's altered cost and revenue functions. For a rigorous development of this point, see the excellent article by Kamien *et al.*, *op. cit.* See also the following discussion, "Asymmetry Between Bribes and Charges: A Comment," by Gordon Tullock and "Asymmetry Between Bribes and Charges: Reply," by Kamien *et al.*, *Water Resources Research* (Fourth Quarter 1966), pp. 854–57.

For a municipality, this problem is not so severe because the amount of waste discharge without treatment would be closely related to the sewered population, although it must be remembered that sizable amounts of industrial wastes are discharged to many municipal waste handling systems.

8. This is to avoid making the payment contingent on actually engaging in the activity—an arrangement which, we have already seen, would lead to inefficiency.

9. This procedure is followed by many municipalities that levy charges on industrial wastes discharged to their waste-handling systems.

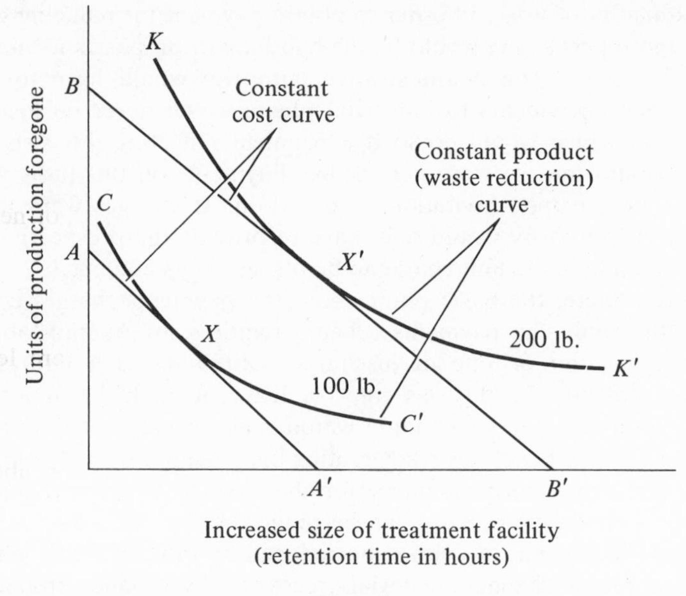


FIG. 5

grammatic presentation becomes cumbersome if not impossible, and mathematics must be used.

Point $A'$ in Figure 5 is established by spending all of a given amount of funds on increasing waste treatment. Point $A$ shows the number of units of production which can be foregone at the expense of an amount of net revenue equivalent to the outlay necessary to reach $A'$ (disregarding waste disposal costs). The line $AA'$ represents all the alternative combinations of production foregone and outlay for increased treatment which can be obtained at the same cost. The line $BB'$ represents a similar set of alternatives for a higher cost level. For simplicity these functions are pictured as straight lines although this is not essential to derive the optimization criteria outlined below. An infinite number of curves like $AA'$ and $BB'$ corresponding to all possible levels of cost could be drawn.

Expenditure on treatment and loss of revenue on foregone production of incremental units may be in the interest of the firm either to avoid a charge on effluent or to obtain a payment for reducing discharge. A central point is that for any level of cost (outlay or foregone revenue) the firm will seek that combination of measures which will achieve the greatest reduction in waste discharge. That combination is attained for the cost level $BB'$ at point $X'$ where the constant cost curve is tangent to the constant product curve $KK'$.

An infinite number of curves like $KK'$ and $CC'$ could also be drawn corresponding to each level of waste reduction. The convex-to-the-origin slope of these curves indicates a diminishing marginal rate of substitution between the two alternatives. As treatment capacity is expanded, it substitutes less well for output reduction. Another way of saying this is that there are diminishing physical returns to increased retention time in the treatment plant.[10]

Points like $X$ and $X'$ indicate greatest effect at a given cost as no higher isoquant or constant product curve (in this case product refers to waste discharge reduction) is attainable given the combinations of input that can be purchased. At such points the combination of the two alternatives pictured is economically optimum—the rate at which one can be "traded off" for the other at the margin in order to produce a given output is equal to the ratio of their respective prices, and we can conveniently refer to the cost of output reduction and of increased treatment input as their respective "prices." The rate of tradeoff or marginal technical rate of substitution is equal to the slope of the isoquant. It is also readily demonstrable [11] that the slope of this curve equals the marginal physical product of treatment (i.e., the physical output result produced by a small increment of treatment) over the marginal physical output reduction. Since the slope of the price line (like $AA'$) equals the ratio of the prices of the two alternatives we can write the following optimizing rule:

$$\frac{P_T}{MPP_T} = \frac{P_O}{MPP_O}.$$

A marginal cost function analogous to the one pictured in Figure 3 but relating the marginal cost of an optimum combination of waste reduction measures to the degree of waste reduction achieved can readily be constructed from the information contained in an isoquant diagram by establishing numerous points such as $X$ and reading off corresponding costs and outputs. The optimal waste reduction level can then be established by equating the marginal cost of the optimum combination of output reduction and treatment (or other methods) with the charge or payment representing the downstream incremental cost. When this condition exists the marginal costs of all relevant alternatives (including treatment, process change, output reduction, and pollution damages) are equated and the level of waste production and treatment is optimized. The major point here is that since the costs of alternatives are evaluated equivalently under the charges and payments procedures, both the combination and level of use of alternatives will be the same.

10. For reasons of symmetry, the assumption that waste loads decline linearly with reduction in output is retained.

11. See William Baumol, *Economic Theory and Operations Analysis* (Prentice Hall, 1965).

In actuality, lines like $AA'$ and $BB'$ would frequently be curved. They would probably be convex to the origin, since with rising marginal costs the cost per unit of reducing output would tend to rise. (There is an important distinction here between a rising price and rising incremental costs.) Straight lines (given $P$'s) were assumed in order to simplify the derivation of the optimizing rules, and the demonstration that effluent charges and payments for waste reduction lead to the same optimizing rules. Conclusions with respect to charges and payments remain the same even if equal cost lines like $AA'$ curve, provided certain constraints on their curvature are met.[12]

The basic and quite reasonable rule remains that in order to minimize costs the marginal costs of all relevant ways of achieving a given result must be equalized. Moreover, the result will be the same whether the cost is an actual outlay or the foregone opportunity to receive a payment.

This section has illustrated that, under idealized conditions, payments and charges serve equally well to achieve optimal amounts of waste discharge reduction for both the short-run or the long-run situation and for multiple alternatives to reduce waste as well as for output reduction. But it has also illustrated the extensive informational and administrative requirements of the payments scheme. Accordingly, the discussion in the remainder of the book is almost entirely in terms of "putting the costs on the waste discharger." If costs are properly defined, this procedure tends to produce an optimal allocation of resources and optimal residual or final waste loads. However, water quality management is such a complex matter that any procedure is bound to be accompanied by problems of adjustment, equity, or inefficiency, or perhaps all three. The effluent charge procedure is no exception. It presents problems of adjustment and equity—especially when locations have been made on the assumption that a certain amount of waste can be discharged freely.

## Problems in Determining Charges

### *The Nature of Damage Functions*

Basic to an optimum system of effluent charges, or for that matter any procedure aimed at achieving an economically optimal level of water quality, is delineation of the "damage cost function," which is the func-

12. For an example of the application of isoquant analysis to a design problem where the equal cost lines curve, see Maynard M. Hufschmidt, "Application of Basic Concepts: Graphic Techniques," in Maass *et al.*, *Design of Water Resource Systems* (Harvard University Press, 1962).

TABLE 1. Simple Illustration of Damage Distribution

| Plant no. (*serially located along stream*) | Chloride load discharged (*1,000 lb. per day*) | Chloride load at plant intake (*1,000 lb. per day*) | Flow Condition I: Streamflow (*mill. gpd*) | Flow Condition I: Chloride concentration (*1,000 lb. per mill. gpd*) | Flow Condition I: Damage per day (*$1,000*) | Flow Condition I: Total damage per day (*$1,000*) | Flow Condition II: Streamflow (*mill. gpd*) | Flow Condition II: Chloride concentration (*1,000 lb. per mill. gpd*) | Flow Condition II: Damage per day (*$1,000*) | Flow Condition II: Total damage per day (*$1,000*) |
|---|---|---|---|---|---|---|---|---|---|---|
| 1 | 1.0 | | | | | | | | | |
| 2 | 0.5 | 1.0 | 1.0 | 1.0 | 1.00 | 1.00 | 0.5 | 2.0 | 2.00 | 2.00 |
| 3 | 1.5 | 1.5 | 1.0 | 1.5 | 3.00 | 4.00 | 0.5 | 3.0 | 6.00 | 8.00 |
| 4 | 1.0 | 3.0 | 2.0 | 1.5 | 3.00 | 7.00 | 1.0 | 3.0 | 6.00 | 14.00 |
| 5 | 0.5 | 4.0 | 2.0 | 2.0 | 1.00 | 8.00 | 1.0 | 4.0 | 2.00 | 16.00 |

| Damages caused at | Flow Condition I: Damage caused by: Plant 1 | Plant 2 | Plant 3 | Plant 4 | Sum of damages caused to | Flow Condition II: Damage caused by: Plant 1 | Plant 2 | Plant 3 | Plant 4 | Sum of damages caused to |
|---|---|---|---|---|---|---|---|---|---|---|
| | ($1,000) | | | | | | | | | |
| Plant 1 | 0.00 | 0.00 | 0.00 | 0.00 | 0.00 | 0.00 | 0.00 | 0.00 | 0.00 | 0.00 |
| Plant 2 | 1.00 | 0.00 | 0.00 | 0.00 | 1.00 | 2.00 | 0.00 | 0.00 | 0.00 | 2.00 |
| Plant 3 | 2.00 | 1.00 | 0.00 | 0.00 | 3.00 | 4.00 | 2.00 | 0.00 | 0.00 | 6.00 |
| Plant 4 | 1.00 | 0.50 | 1.50 | 0.00 | 3.00 | 2.00 | 1.00 | 3.00 | 0.00 | 6.00 |
| Plant 5 | 0.25 | 0.125 | 0.375 | 0.25 | 1.00 | 0.50 | 0.25 | 0.75 | 0.50 | 2.00 |
| Sum of damages caused by: | 4.25 | 1.625 | 1.875 | 0.25 | 8.00 | 8.50 | 3.25 | 3.75 | 0.50 | 16.00 |

tional relationship between the amount of a waste discharged and damages.

In this discussion we do not distinguish between short-run and long-run damage functions. Also we ignore the stochastic character of the function resulting from environmental variability—primarily temperature and streamflow.[13] This permits us to first discuss basic principles without introducing duration and frequency as affecting the magnitude of damages.

The simplest situation occurs when this function is linear, i.e., each additional unit of waste discharge results in an equal increment of damage. In general, the present exposition will assume linear damage functions. Such a situation is illustrated in Table 1 where it is assumed that five plants are arrayed along a stream, that streamflow increases along the course of the stream (say, because a tributary enters), and that the waste it contains is nondegradable (say, chloride). The nondegradability assumption does not change the analysis in any way, but it simplifies the example.

Since damage per day is assumed to be in direct proportion to concentration, a level of charges equal to incremental damage costs can be worked out for each level of flow and for each plant. For example, at flow level one, the charge for plant 1 is $4.25 per pound of waste discharged which is the sum of the damages caused by plant 1 to plants 2, 3, 4, and 5. (At flow level two, the charge for plant 1 is $8.50 per pound, because the same waste discharge results in a doubling of the concentration.) The charge for plant 1 is the same regardless of the level of discharge of the other plants. If plant 1 reduced its discharge by half in response to the charge levied on it, its assessment would drop by half to $2.125—the amount by which downstream damages are reduced. This "separability" characteristic of linear damage functions is very important since it greatly reduces the amount of information which a regulatory authority must have to implement an efficient effluent charge system.

The manner in which a plant maximizing its profits (or minimizing its losses) will respond to an effluent charge levied on it at a given level of streamflow and the effect of this response on costs associated with its waste disposal are shown in Figure 6. Under the circumstances pictured, the plant will reduce its waste discharge from point *D* to point *E*, thus minimizing the costs associated with its waste disposal.

The marginal cost of withholding wastes in an optimal manner includes, in the case of chlorides, such alternatives as temporary storage, process adjustments, and reducing production. The marginal cost function rises after some point because it becomes progressively more ex-

13. Editors' note: Both of these refinements are introduced later in the book by Kneese and Bower.

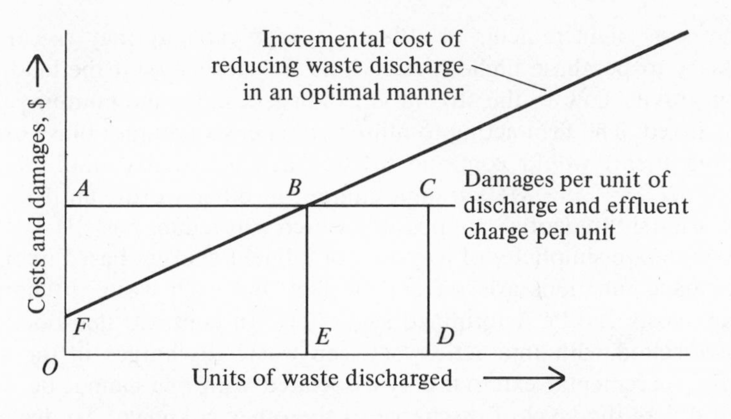


*OD* = Units of waste discharge if no charge levied on effluent.
*OA* = Damages per unit of waste discharge and effluent charge per unit.
*OACD* = Total damages associated with unrestricted waste discharge, i.e., no effluent charge levied.
*OE* = Reduction of waste discharge with effluent charge *OA*.
*OFBE* = Total cost of reducing effluent discharge to *ED*.
*OFBDC* = Total cost associated with waste disposal with *ED* waste discharge, i.e., residual damage costs plus cost of reducing discharge.
*OABE* = Total damages avoided.
*ABF* = Net reduction in waste disposal associated costs by reducing waste discharge by *OE*, i.e., *OABE* minus *OFBE*. This also equals the total cost of *not* reducing the effluent discharge.

FIG. 6

pensive to withhold production (say, as inventory is run down), or to adjust production processes.[14]

14. If the costs of an optimal combination of waste reduction procedures have the shapes conventionally attributed to cost functions (text illustrations show only the relevant range of such functions), marginal costs and the effluent charge will be equal at more than one point. This is illustrated in the figure below:

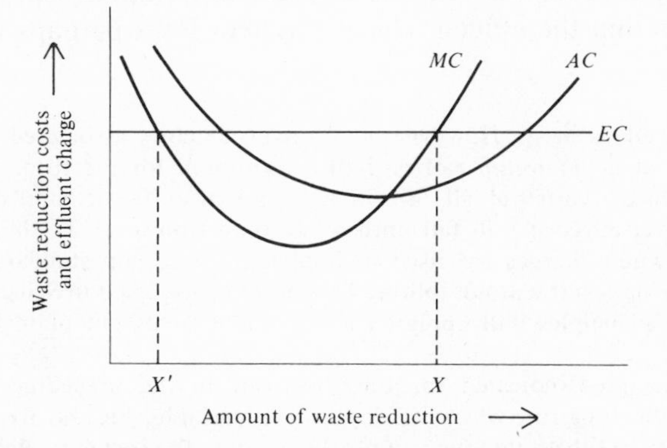


Marginal cost for function *EC* equals the effluent charge at levels of waste reduc-

The costs of increments of effluent storage capacity may rise if it is necessary to purchase higher-priced land for storage or if the land providing gravity flow to the stream is exhausted and some pumping costs are incurred. The firm acting to minimize its costs (charges plus costs of avoiding them) would continue to use each alternative until $MC_A = MC_B = MC_C = \ldots MC_N =$ unit charge; in other words, until no further marginal "tradeoffs" are possible which will reduce costs.[15]

The relative simplicity of a system of effluent charges based upon linear damage functions arises from the fact that each waste discharger's damage costs can be determined separately. In contrast, non-linearities are associated with interaction between waste discharges in the sense that the incremental external cost associated with one cannot be determined unless the level of discharge of the other is known. To deal with this problem we have to distinguish between "separable" and "nonseparable" damage functions. A pollution damage function is separable if there are several waste dischargers and the pollution damage imposed at any downstream point is simply the sum of the damages that would be caused by the individual waste dischargers, each acting in isolation. If, as is often the case, the individual wastes interact in a more complicated way and the total damage caused by the several dischargers is not the simple sum of their individual effects, then the damage function is nonseparable. This complicates matters, because, if the damage function is nonseparable, the control authority cannot tell how much damage any waste discharger is causing and could appropriately be charged for, unless the authority knows the amounts of pollutants being emitted by other dischargers at the same time. Instead of needing to know merely the individual damage functions and being able to achieve an optimal solution by imposing damage costs on the waste dischargers and letting them respond, the agency must now know the cost functions for waste reduction at each interdependent point of waste discharge. It must determine the optimum level of waste reduction for each point *before* charges are assessed on effluents in order to obtain an optimal solution. This means that the effluent charges system loses perhaps its major ad-

---

tion $X'$ as well as at $X$. However, at $X'$ over-all costs associated with its waste disposal are at a *maximum* rather than a *minimum* for the firm. Fortunately, if the firm reduces wastes at all, it will have no incentive to stop at $X'$, for if it continues, over-all costs will fall until waste reduction level $X$. Under certain circumstances when charges are used to implement a stream standards system, this cost minimizing result will not follow. This point is discussed in Selection 9.

The same principles will apply to a municipal treatment plant having similar cost curves.

15. The principles indicated for the short-run in the preceding footnote hold equally for the long-run where all inputs are variable, i.e., no fixed plants. It is only necessary to substitute long-run for short-run in the cost curve labels.

vantage over other systems of control such as payments or effluent standards.

In some cases it might be preferable to assume linearity even when it is known that the discharge-damage relation is actually somewhat more complex because taking account of greater complexity may rapidly increase costs and yield strongly diminishing returns. In other words, among the incremental costs that should not be neglected are the costs of refining the system of charges.[16]

## *Some Additional Theoretical Problems in Determining Optimal Charges*

Even when damage functions are linear, "separability" problems may arise if waste disposers impose damages on a series of downstream units and the waste discharge of one firm affects the marginal production cost of another. A separable externality is defined as one that occurs when the total cost of a productive process is affected by the level of output in another process (and hence the level of upstream waste discharge), but marginal cost is not.[17] The separability case requires that the mar-

16. The use of linear separable functions may not be a good approximation of the actual conditions should any of the following situations hold to a significant degree: (1) Damage costs definitely do not increase in direct proportion to the increase in waste concentration; (2) The relationship between different types of wastes is either synergistic or cancelling rather than additive; (3) A downstream treatment process simultaneously removes two or more types of wastes coming from different sources. In the third case, an inevitable element of arbitrariness is introduced into allocating the costs to individual waste-discharging firms which emit different wastes or the same wastes in different proportions. When the costs of removing residual materials from intake water are strictly "joint," there is literally no non-arbitrary way of determining who is responsible for what. In some or many cases, however, the costs are probably not truly joint. In these cases, for any given level of concentration the cost of dealing with an increment of the waste can be estimated.

In the first two cases, the complexity occurs because the cost for which plant 2's discharge can be held responsible becomes in part a function of how much plant 1 discharges.

There is currently much uncertainty concerning the character of damage functions but there is evidence that some of them are non-linear. Research is needed to define how pervasive this situation is and whether the additional data and analytical requirements of taking non-linearities into account could be justified in devising actual management programs.

17. Otto A. Davis and Andrew H. Whinston, "Externalities, Welfare, and the Theory of Games," *Journal of Political Economy* (June 1962), pp. 241 ff. For an excellent brief article relating Davis and Whinston to the previously cited articles by Coase and Buchanan and Stubblebine, see Ralph Turvey, "On Divergences between Social Cost and Private Cost," *Economica,* Vol. 30 (August 1963), 309. [Reprinted in this volume, Selection 7.]

ginal production cost of each firm linked to another by a technological externality be given entirely in terms of its own output. This would follow, for example, if factory A discharged a waste which caused factory B to put in a water supply treatment plant which thereafter would be operated in an invariant fashion. The new total cost function becomes a vertical displacement of the old and marginal costs are not changed, as is shown in the following example.

| Plant B output | Total cost before factory A | Marginal cost | Total cost after factory A | Marginal cost |
|---|---|---|---|---|
| 1 | 1.0 | | 2.0 | |
| | | 1.5 | | 1.5 |
| 2 | 2.5 | | 3.5 | |
| | | 2.0 | | 2.0 |
| 3 | 4.5 | | 5.5 | |

Since marginal costs of one firm are not changed by the output decisions of the other firm, the externality does not cause output to be different from what it would have been in the absence of the externality presuming it continues to be in the affected firm's interest to produce at all. The effect of the externality is intra-marginal and thus merely affects the firm's profit position and the decision whether to continue production or to stop. In this instance, the appropriate effluent fee is simply a lump sum equal to the increase in total cost.[18] This is a fee levied for the privilege of discharging at all. The only question involved is whether to discharge or not.

A nonseparable externality is defined to occur when the marginal cost in a productive process is affected by the level of output in another process. For example, assume that a petroleum refinery A, which expels

18. Except where the externalities are mutually or reciprocally imposed on each other by plants. Such a case could arise for example if two thermal power plants both used a lake for cooling water and by raising the temperature of the lake reduced each other's cooling efficiency. In this case the cost functions of the two would be related in the following manner:

$$C_1 = C_1(q_1, q_2) = f_1(q_1) + g_1(q_2)$$
$$C_2 = C_2(q_1, q_2) = f_2(q_2) + g_2(q_1)$$

where subscripts refer to the respective plants, $C$ indicates costs, and $q$ the output level. These externalities are non-separable because each cost function can be written as the sum of two functions, both of which involve only one variable as its argument. To obtain an optimal solution in this instance the basin agency would need to know the cost functions of both plants and their interrelationships. In this as in other cases of non-separability the use of charges loses its information requirements advantage over direct controls. For an elaboration of this point see Otto A. Davis and Andrew B. Whinston, "On Externalities, Information and the Government-Assisted Invisible Hand," *Economica*, Vol. 33 (August 1966).

a hot effluent, locates upstream from a petroleum refinery B, which is operated by a different firm and which uses the stream for cooling water. Cooling efficiency for plant B will drop, and for the purposes of our example we assume it will have to pump more water for cooling per unit of output. The marginal costs of plant B *for a given level of output* will be affected by the externality, shown below.

| Plant A discharge | Total pumping costs before plant A | Marginal cost | Total pumping costs after plant A | Marginal cost | Total external cost of plant A discharge | Marginal external cost |
|---|---|---|---|---|---|---|
| 1 | 1.0 | | 2.0 | | 1 | |
| | | 1.5 | | 2.5 | | 1 |
| 2 | 2.5 | | 4.5 | | 2 | |
| | | 2.0 | | 3.0 | | 1 |
| 3 | 4.5 | | 7.5 | | 3 | |

The reader will note that, so far as we have gone, the damage function is linear in the illustrated case. But the response of plant B would be to adjust output (assuming other less costly alternatives such as precooling of influent are unavailable).

A particular marginal cost function for plant B corresponds to each level of discharge of plant A. (A set of such cost functions is shown in Figure 7.) Plant A can determine its own profit-maximizing level of output, and at this rate of production a specific level of output (say, $x_1$

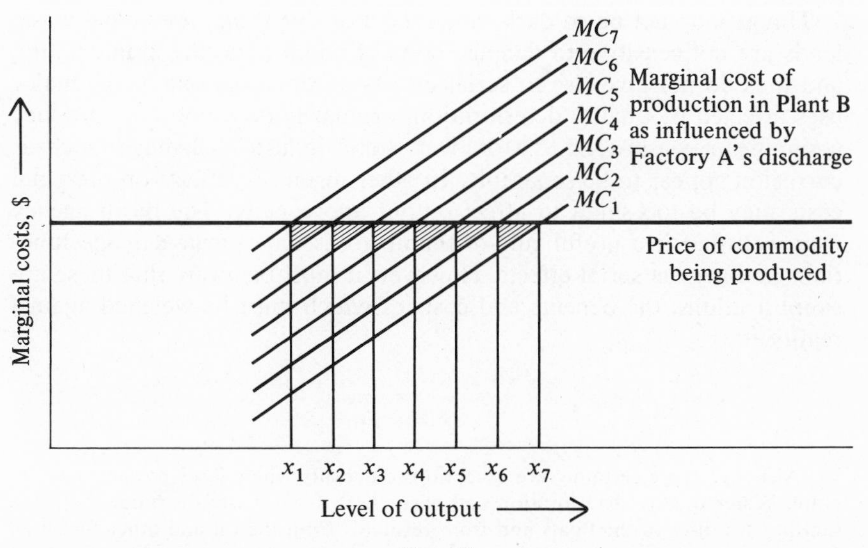


FIG. 7

corresponding to $MC_1$) will also maximize profits for plant B (although this will in general not be the level of output which will maximize joint profits for both plants taken together). However, it is possible to compute the amount of a levy on the output of plant A which will maximize joint profits. The appropriate level of charge will be the one which causes the marginal net loss imposed on firm B to be considered an opportunity cost in firm A. The net losses corresponding to different levels of waste discharge are shown in the shaded areas of Figure 7. The net damage or net loss function may also turn out to be linear as indeed in Figure 7 it is. But even so we are not out of the woods if different waste dischargers and water users impose external costs on multiple production units.

Assume an initial equilibrium along a stream where a number of water users and waste dischargers impose external diseconomies on one another seriatim. Assume the regulatory authority starts by imposing the incremental costs of the last one of these ($Z$) upon the previous one ($Y$) and so on back up the line. When the charge is levied upon $Y$ the outputs of both $Y$ and $Z$ change. As the cost imposed by $X$ on $Y$ is levied on $X$ the output of $Y$ changes again; accordingly so does that of $Z$. We have an interdependent system, and a simultaneous solution is necessary for the entire system. Not only is the problem analytically complex, but it would require detailed information on the costs of reducing waste discharges as well as damages imposed, and once again we lose the two major advantages of the effluent charges system as a means of water quality management. Decentralized decision making is no longer possible, and the information needed by the basin agency is no longer limited to damage functions.

Things may not be so dark, however. For one thing, municipal waste loads are not sensitive to damage costs imposed upon the municipality, and thus do not give rise to serial effects. Recreation, one of the major uses affected by quality deterioration, ordinarily does not itself produce a significant waste load.[19] Moreover, some industrial damages such as corrosion appear to be separable. In other instances, effects on marginal costs may be too small to affect output appreciably. The basin agency may well find it a useful rule-of-thumb to assume linear damage functions and neglect serial effects. However, it must be aware that these are simplifications, the benefits and costs of which must be weighed against reality.

19. Although there certainly are exceptions especially where large power boats operate. Water quality deterioration can occur both from waste discharge from the sanitary facilities of the boats and from residuals from the oil and other fuel used for power.

# III

# *POLICIES FOR ENVIRONMENTAL PROTECTION*

The preceding section set the economic framework within which environmental policies are to be analyzed. We turn here to a less formal examination, generally speaking, of the elements of sound policy and some specific recommendations. All the contributors in this section would, no doubt, accept the economic basis for environmental disruption laid out in the formal analysis, but their prescriptions vary. These prescriptions include changes in the legal structure as well as in strictly economic policies and some question whether governmental intervention is likely to be worth the cost.

Kneese and Bower and Dales argue strongly in favor of levying some sort of effluent or damage charge on polluters as the least costly means of achieving abatement. The two approaches differ, however, in a fundamental respect. Kneese and Bower recommend the conventional policy of fixing a charge to which potential polluters are expected to respond by adjusting their individual emission levels. Dales' alternative is to fix the total amount of pollution that is to be permitted and allow the price charged for polluting to adjust by means of an auction of pollution rights. The two approaches will lead to the same results only if the regulatory agency is successful in selecting either the price or the quantity, as the case may be, which corresponds to the equilibrium or crossing point of the community's demand and supply curves for environmental improvement. If the agency in question is in possession of sufficient market information to select the ideal price or quantity, it can, for that matter, obtain the same results with direct regulation of pollution.

In the practically universal absence of perfect market information, however, any method is likely to require a process of trial and error in order to approximate the optimal level of pollution, and this process may itself be a costly one. A choice between the two proposals might well depend, then, on which provides the least room for error in the initial decision and which permits the least costly adaptation in the event of error and of changing circumstances.

Dales defends his proposal on grounds that, whereas it is, he feels, impossible to estimate how much various amounts of abatement are worth to a community, it is possible to decide, as a political matter, how much abatement the community wants, presumably irrespective of price. Others would argue, contrarily, that whereas a community cannot properly decide on the total amount of abatement it wants without some idea of how much it will cost, it is possible to calculate the value the community places on incremental amounts of abatement. The superiority of one policy over another will depend in large part on the elasticity of the community's demand for pollution abatement. If demand is extremely inelastic, a decision to set some limit on pollution will not go far astray regardless of the price that pollution rights turn out to command in the market. If, on the other hand, demand is highly elastic, it will be easier to estimate the price that ought to be charged for damaging the environment than to predict the amount of pollution that will turn out to be ideal in the light of polluters' responses.

E. J. Mishan and Guido Calabresi are both concerned with adapting the legal structure to remove impediments that stand in the way of a more nearly optimal use of the environment. In particular, they confront the question of where legal liability for environmental damage should be placed. Each would grant Coase's point that optimal allocation would result if negotiation among affected parties were free, but they emphasize that it is not free. Mishan points out that, in view of the obstacles to negotiation or recourse through the courts, the status quo will be vastly different under a system that places the burden of proof on those whose amenities are damaged than under one that places it on the polluter. He argues for the latter, making—unlike most economists—no pretense of neutrality between those who wish to enjoy the amenities and those whose actions are likely to damage them. Calabresi, a lawyer, draws attention to some of the subtleties involved in establishing liability rules designed to minimize the social cost of moving toward an optimal resource allocation, emphasizing that adaptation will always be costly.

Milton Friedman is the leading spokesman of a school of thought that insistently reminds us that "Every act of government intervention limits the area of individual freedom directly and threatens the preservation of freedom indirectly. . . ." This aspect of governmental intervention applies as fully to activities in the environmental sphere as to any others. Thus, Friedman feels, a substantial advantage over other possibilities should be shown before the government interferes with economic activities, including those with "neighborhood" or environmental effects. He does not maintain that the drawbacks of bigger government are always decisive but insists that they should always be taken into account.

The selection by Robert Dorfman and Henry D. Jacoby is rather different from the others and serves two purposes. First, it describes, as realistically as possible in brief compass, the complexity of the issues to be faced in making a specific decision designed to protect or improve the environment. This complexity arises because any such decision will affect many different participants in many different ways. All the interests concerned are likely to be both legitimate and dearly held. A principal thesis of the article is that no inclusive "social welfare function" or "public interest" incorporates all the considerations that have to be taken into account. Thus it emphasizes the distributional aspect of environmental policy. Second, the paper sketches a quantitative model of the political process by which conflicting claims on the use of the environment are adjudicated. The authors claim predictive power for their model, but this claim has never been tested adequately.

Harold Thomas, a thoughtful environmental engineer, raises the practical problem of setting standards of environmental quality. He concludes that this is essentially an economic problem, not a technical or engineering one, and that a poor country or community should adopt lower standards than a rich one. This insight is particularly pertinent to the conflicting attitudes of rich and poor countries toward environmental policy: India is not nearly so willing to forego the use of DDT as the United States, and with reason.

# 9

ALLEN V. KNEESE AND
BLAIR T. BOWER

# *Standards, Charges, and Equity*

*Allen V. Kneese is Director and Blair T. Bower Associate Director of the Quality of The Environment Program at Resources for the Future in Washington, D.C.*

## Efficient Quality Control and Stream Standards

Since the level of water quality to be achieved in each of the nation's watercourses cannot be directly established on economic grounds—because all of the relevant benefits from water quality improvement cannot be computed—and since the Congress, the executive, and various state officials have expressed an urgent desire to improve water quality, it appears that some form of watercourse standards will be the mechanism used to establish water quality levels, at least for the time being. These standards will be based on some, usually vague, consideration of damage costs vs. costs of quality improvement. In the following discussion we use the term "stream standards" even though the standards might relate to any type of watercourse: stream, lake, estuary, or underground.

One form such a standard could take for a river is to set an upper limit on the concentration of a given waste, say, a material toxic to fish in the stream. Assuming that damage functions are approximately linear and additive, a system of charges can be established which provides incentives for achieving this standard at lowest cost. The following general statements are of significance here:

1. The symmetry of the *stream standards* analysis with that of effluent charges can perhaps be most readily seen if the standard is viewed as representing a highly inelastic damage function. Then, the general principles governing the adjustment of charges to damage functions become relevant.[1]

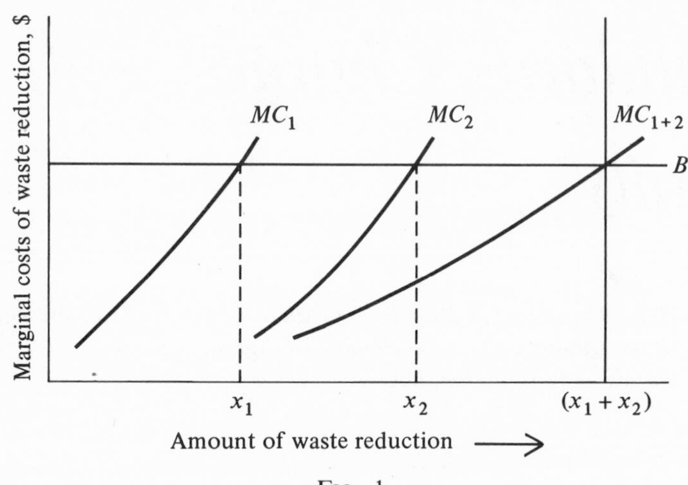


FIG. 1

To see this, consider Figure 1 where $MC_1$ indicates the marginal cost of waste reduction at outfall 1 and $MC_2$ is the marginal cost of waste reduction at outfall 2. The waste involved is assumed to be nondegradable. Assume that waste reduction level $(x_1+x_2)$ corresponds to the amount of removal needed to achieve the standard in the watercourse. The vertical damage function at $(x_1+x_2)$, representing the watercourse standard, implies that up to that point the benefits accruing from waste reduction are infinitely large, while thereafter they are zero. This is one of the implications of stream standards that is almost always false in reality.

A charge, set at level $B$ (which the authority finds by calculating the incremental costs of waste reduction at all relevant points of discharge or by experimentation), will be viewed as a perfectly elastic damage function by each of the two waste dischargers individually; accordingly,

1. For a discussion of these principles see Selection 8.

the separate adjustments of the waste dischargers will lead to equalization of marginal waste reduction costs at both points. This means that the standard is achieved at minimum cost. A charge higher or lower than *B* would exceed or fail to meet the standard, but would tend to achieve whatever level is reached at least cost.

2. If a standard is set for a critical point on a stream and discharge of a given waste must be cut back to achieve the standard, a charge per unit of nondegradable waste discharge can be worked out which will achieve the standard. The same is true of degradable wastes, but in this case the charge will depend upon the degree to which the wastes are degraded in the stream between the outfalls and the critical point. From the viewpoint of this study, credit must be given for degradation in the stream, not because it is equitable, for no criteria of equity have yet been offered, but because it is necessary for the minimization of costs. To see this, imagine the extreme case where a waste is fully degraded before it reaches the critical point in the stream. If a charge were imposed on the waste discharged at the outfalls, waste reduction outlays would be induced unnecessarily.

3. A level of effluent charges that would achieve the stated water quality standard could be arrived at by experimentation but the initial charge would have to be substantially correct to induce the least-cost system. Any substantial error in the initial charge [2] would be built into subsequent investment decisions involving durable plant and equipment, as well as into any further adjustments. Under these circumstances the effluent charge needed to achieve the standard might be quite different from what would have emerged had the initial charge level been the correct one, i.e., the one that achieves the quality standard at least cost.[3]

A strategy for taking advantage of the decentralized decision making inherent in effluent charges without inducing serious non-optimal adjustments must be devised. A suitable approach would appear to be to do a cost study in sufficient detail to insure that the charge initially set is not far from the minimum required to meet the standard. Changes in waste load reduction within a limited range can often be achieved by adjusting operation of treatment plants or production processes without substantial losses in efficiency. Changes in the necessary charge, as demand for assimilative capacity shifts, will probably be gradual, reflecting economic growth and changing technology.

4. The effluent charges procedure would have the advantage over other possible techniques of permitting each waste discharger to adjust in the most efficient way for his particular circumstances. Individual

2. Or in the estimates of the expected value and duration of the charges if charges are to be varied to reflect time variation in the assimilative capacity of the watercourses.

3. The MC curves in Figure 1 were drawn on the assumption that the effluent charge was correctly specified in the initial instance.

dischargers could withhold wastes in temporary storage, adjust production processes, change raw materials, treat wastes, cut back on production, change the character of their output, pay the charge, or use a combination of these procedures.

The charge required to achieve the standard indicates the incremental cost of achieving the standard. This means that without specific knowledge of the costs of waste reduction to upstream dischargers, the responsible authority can estimate the marginal social cost of the standard. This is important for judging its appropriateness. Presume, for example, that the standard relates to the killing of fish by, say, a toxic substance, and that it is not possible to establish the value of the "damage costs" of varying concentrations of the waste. The charges necessary to achieve the standard give an indication of what a small change in the standard and the accompanying physical effects *must at least be worth*. For example, if the charge is $1.00 per unit of waste, a unit reduction in the standard must reduce damage costs by at least $1.00 if the standard is worth maintaining. This type of information facilitates a decision on whether to raise or lower standards.

## Complementarities in Treatment or Other Measures for Reduction of Waste Discharges

A thoroughgoing water quality control system in a river basin with reasonably complex development would probably result in a whole array of charges relating to different types of wastes. But measures to reduce one waste may simultaneously reduce others at a given point of discharge. Examples can be cited both with regard to municipal and industrial waste reduction processes. Standard physical, or primary, treatment (principally sedimentation) removes suspended solids but also some portion of the degradable organics. Standard biological treatment removes the bulk of the degradable organics, and has some effect on persistent organics, plant nutrients, bacteria, and other substances as well. Cooling towers may reduce BOD and phenols, as well as the waste heat discharged to watercourses.

Since joint costs cannot be allocated in a way that assists rational decisions, the discharger endeavoring to minimize his waste disposal costs would have to calculate the amount by which the joint effect of his action would reduce his charges, and weigh it against his specific cost of introducing and operating waste reduction measures. In a system of charges designed both to reflect downstream costs and meet the established "standards," the level of charges needed to accomplish the "standards" could still be established in the experimental manner outlined above. For example, if it is found that a charge can be reduced without having water quality drop below the level specified in the standard, this means that: (1) the charge is too high in terms of those dischargers responding directly to it; and / or (2) one waste is jointly treated with another by at least some dischargers and the charge levied on the first is high enough to induce higher than necessary joint removal of the second. The charge levied on the first would then be reduced until the standard is just met. If the influence of joint treatment is strong, the charge might be dropped to zero.

In either case, the charge which just meets the standard reflects the cost of an incremental change in the standard. An "ideal" system of charges would reflect the costs imposed downstream, and all charges levied to meet the standard would be at the lowest level consistent with the standard.

## Effluent Standards

Consideration of *effluent* standards (as distinct from stream standards) —perhaps the most discussed method of achieving some coordination between the water quality desired in watercourses and individual waste discharges—has been purposely postponed until now so that the problem of external economies could be examined first in terms of charges and payments. These measures are the ones that have been traditionally advocated by economists for dealing with externalities; they are also consistent with the operation of the price mechanism in private markets.

Before pursuing this discussion further it should be noted that many professionals in water quality management who have advocated the standards approach have appreciated that stream and effluent standards must be viewed as potential complements in a rational program of management. Nevertheless, confused and confusing debate has been going on over the years about stream *vs.* effluent standards, as though the two were always alternatives. This is a relevant issue if a single waste discharge affecting quality in a water body is involved. Then the waste discharger can monitor the stream and adjust his discharge so as to maintain a specified standard in it. This would provide improved flexibility over a simple effluent standard which would presumably be based on some critical condition in the stream. There are a number of cases where industrial operations do precisely this in rural and small town areas. But, if there are multiple waste dischargers, achieving the standard by independent decisions based on a *stream* standard will be impossible. The problem here is another illustration of "inseparability," because what one waste discharger can do is contingent on what another does. A central agency must therefore provide information and incentives (via charges or effluent standards) which will produce co-ordinated behavior. Under these conditions, stream and effluent standards must be viewed as potential complements and not alternatives, i.e., effluent standards are meaningful only in the context of water quality goals or standards in the watercourse.

In this section, effluent standards are compared with a system of charges and payments as a tool for water quality management. A first and basic point is that *if a regional authority had full information concerning all the costs associated with existing and potential waste generation and disposal, it could establish a set of effluent standards that would have the same effect on resource allocation as an ideal system of charges*. A charge on each individual outfall tends to produce a certain quality of effluent. Standards could be set to obtain the same effluent quality at each outfall and prevent potential discharges which would come into existence only if the full cost associated with them, including all external costs, did not have to be paid. It is important to note, however, that the standard would have to be stated in terms of amount of wastes discharged (pounds, tons, etc.) rather than in terms of concentrations, which are sometimes used. In order to impose such effluent standards, it would be necessary to forecast the response of each actual and potential discharger at each level of flow to a system of charges which reflected the damage costs and / or point-of-use standards in the area. This would mean a whole system of individually tailored effluent standards. Minimization of the sum of waste reduction and damage costs *could* be achieved by this system as well as by an ideal system of charges or subsidies. Neither can claim an advantage on the basis of superior performance under *ideal* conditions of information and authority on the part of a responsible regional agency. The differences relate to ease of administration and to income distribution, or equity.

A system of charges has the advantage of requiring less information than other approaches when the objective is the minimization of the costs associated with water quality management. This is true where a stream standard is set, or where charges reflect damage costs and the costs are approximately linear functions of concentrations for individual wastes and approximately additive for different wastes.

In this case, charges based upon damage costs will tend toward a minimization of damage and waste reduction costs even though reduction costs are not explicitly known. The charges technique also requires less information than effluent standards when it is used to obtain the least-cost system for achieving a stream standard.

By a series of approximations, a charge can be set that will achieve a specified stream standard and that will tend to produce the equimarginal costs for alternatives that are necessary for cost minimization. Sample studies could produce a forecast of the appropriate level of charges suitable for planning purposes.

Because land values, production processes, and other significant variables differ from case to case, precise information concerning "marginal tradeoffs" between alternative reduction procedures at different outfalls would be necessary to establish cost-minimizing combinations by the use of effluent standards. But, to set a charge, the water quality management agency does not need to know the cost of waste reduction for each individual waste discharger; the agency only has to be able to estimate the average of the dischargers' marginal costs. Suppose that three dischargers have marginal cost curves for waste reduction as shown in Figure 2.

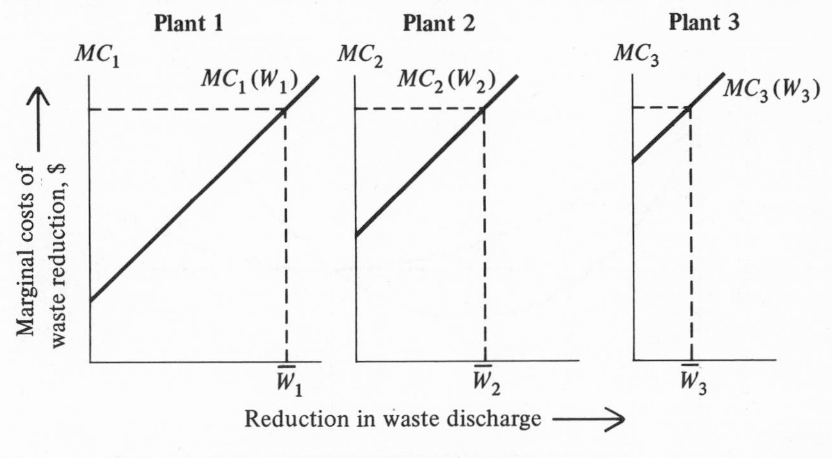


FIG. 2

These curves have been drawn so that $MC_2$ is an average of $MC_1$ and $MC_3$, i.e., for any level of waste reduction $W$, by each individual firm,

$$MC_2(W) = \tfrac{1}{2}\{MC_1(W) + MC_3(W)\}.$$

Being linear, the curves also have the property that $W_2$ associated with any level of $MC$ is equal to the average of $W_1$ and $W_3$ associated with that $MC$. Suppose now that a stream standard is established which requires that the total combined waste reduction by the three firms equals $3\overline{W}_2$. The least cost allocation to achieve this level of waste reduction is $W_1 = \overline{W}_1$, $W_2 = \overline{W}_2$, and $W_3 = \overline{W}_3$. For the water quality agency to achieve this least-cost solution through a system of effluent standards it would have to know the cost curve for each discharger. However, with a system of effluent charges all it needs to know is the cost curve for an average or typical firm such as the second plant in Figure 2.[4]

There is one circumstance where effluent standards could in principle achieve minimum cost in meeting a stream standard while charges could

4. This formulation was suggested by Louis M. Falkson.

not. This is where a stream standard affects only one or a few dischargers and requires only low level treatment for its attainment. To see this consider the long-run average and marginal waste reduction costs for a waste treatment plant indicated in Figure 3. Say that the pictured plant is the only one affected by a stream standard, and that the plant must have reduction capacity $x_1$ to meet the standard. No single effluent charge would induce waste reduction to that level. The lowest effluent charge which would induce any waste reduction is *B*. A lower charge would induce no waste reduction because investment costs could not be recovered. But a charge at level *B* induces treatment to point $x_2$, which

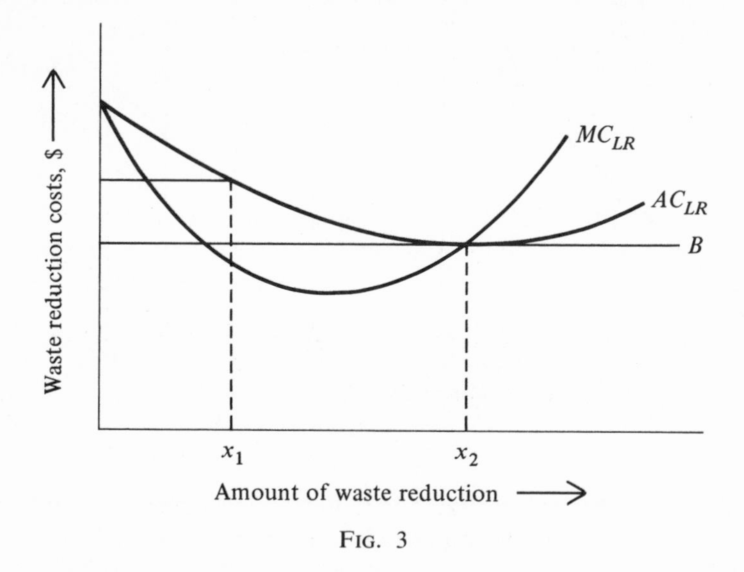


FIG. 3

is higher than needed and therefore excessively costly. This may be a problem in the case of a large stream with great assimilative capacity, comparatively few points of waste discharge, and a relatively large number of points where specified standards must be met. This could turn out to be a *significant* problem in using effluent charges to obtain efficient results where stream standards apply to all points along the stream as they frequently do. Empirical study of this matter is needed.[5]

## Some Further Comments on Charges

Where actual conditions are more complicated than those outlined above, detailed information is needed on specific alternatives and their

5. See J. C. Liebman's case study of the Willamette in *The Optimal Allocation of Stream Dissolved Oxygen Resources* (Cornell University Water Resources Center, 1965).

related marginal costs at each outfall for forecasting, and later establishing, either appropriate effluent charges or appropriate effluent standards. In fact, if there is substantial direct interaction among waste discharges, implementing either system becomes very complex, and reasonable rules of thumb will have to be developed for the regulation of individual waste discharges.

The charges alternative still seems to have certain advantages although these are harder to specify than in the previous cases. For example, a charge offers some incentive to take action even to the lowest level of waste discharge, while a standard (even though stated in terms of quantity rather than concentration) provides no incentive to curb waste discharge beyond the required level even though it might be possible to do so quite inexpensively. To put it another way, for any given level of information, the charges alternative would cost less in terms of resources because charges automatically tend toward an equalization of marginal costs.

If the complexities in an actual situation result in the specification of a stream standard rather than an actual effort to assess and distribute damage costs, the advantages of the charges procedure are those already described. And for any given degree of approximation of optimum results, the information or administrative requirements for charges are no greater than for effluent standards. If the burden of proof concerning the quality of effluents can be put on the discharger for one method, it can be for the other as well.

In a dynamic context, charges have the advantage of exerting a continuous pressure on the waste discharger to improve his waste-handling technology. Under an effluent standards system, the waste discharger has no incentive to do more than meet the standard. In the case of an effluent charge it is always in the interest of the waste discharger to develop and employ cost-reducing technology to diminish his discharge by keeping his (falling) incremental costs of waste reduction in balance with the charge.

A final point—and one that may be extremely important in implementing large-scale regional measures—is that the effluent charge yields revenue, whereas effluent standards and subsidies do not. Even if the revenues are not used for water quality management—and whether they should be is a matter for economic analysis—the revenue yielding aspect is desirable because the effluent charge does not have the resource misallocation effects of most other taxes.

## Some Equity Questions

This study is built largely upon the concept of economic efficiency, but obtaining the maximum net value from the use of a resource is not the only objective relevant to decision making. Equity may be a significant consideration in water quality management policy. Any change in the rules on which decisions are based will result in some rather arbitrary gains and losses. For example, if an industrial plant is built on the banks of a stream with the expectation that no effective limitation on waste discharge would ever be required, and then such limitation is imposed, the result will be a severe decline of capital values—perhaps even shutdown of the plant. Any effective control program faces this problem.

Another context in which questions of equity arise is the distribution of costs of a program even where no actual cases of hardship occur. We offer no criterion of equity in this study because we can find no analytical basis for doing so. We can only record what we hope will be reasonable impressions. The idea that the waste discharger should bear the external costs of his discharge will strike most persons as equitable. This is analogous to requiring a firm or municipality to pay the full cost of the materials, labor, and energy it uses, which does not seem to offend most people's sense of equity.[6]

If effluent standards or charges are used to implement a stream water quality standard, the equity issues are considerably more complicated. At first glance it may seem equitable to require all waste dischargers in the relevant area to meet the same effluent standard or to cut back proportionally when a stream standard is being violated. However, our theory tells us that the result will be inefficient because marginal costs of reducing waste discharge will not be equated except by accident. Moreover, empirical research has shown that the inefficiency of such approaches may be substantial.[7] This result, of course, suggests that there must be some distribution of costs of an efficient program which will leave everybody better off than the equal reduction alternative. It is possibly the most important characteristic of the economist's efficiency concept that gains in efficiency produce net benefits which can be distributed in such a way that everyone is made better off. When rearrangements can no longer be made which make this possible, an efficient point has been reached. Is it really equitable to make all waste dischargers adhere to the same effluent standard when there can be large differences in the costs of achieving it?

Perhaps the scheme that would strike one as most inequitable is the

6. Compare the statement by a Dow Chemical Company official: "We are convinced that pollution control is part of the cost of doing business and we have always treated it that way. We firmly believe that industry can be clean and profitable at the same time, and we have amply proven this to our satisfaction." C. Gerstacker, "Management's Role in Pollution Control," *Industrial Water Engineering,* Vol. 3, No. 4 (1966), 39.

7. See the case study of the Delaware estuary in A. V. Kneese and B. T. Bower, *Managing Water Quality* (Baltimore: The Johns Hopkins Press, 1968), chap. 8. Also J. C. Liebman, *op. cit.*, pp. 29 ff.

efficiently programmed effluent standards approach because it emphasizes waste reduction by dischargers with low marginal waste reduction costs. In effect, it could penalize those with low costs and present the high-cost waste discharger with a free property right to cntinue his waste discharge unmodified. It would put a premium on costly waste disposal methods.[8]

The preceding discussion of equity has focused on the distribution of costs of water quality management. Equally relevant are equity considerations with respect to who gains from water quality improvement. If, in fact, much of the allocation of resources to water quality improvement has to be "justified" in terms of recreation and aesthetic benefits, whether or not the resulting distribution of benefits is in accord with social policy regarding income distribution must also be considered.

## Conclusion

The effluent charge can be used to achieve stream water quality standards efficiently, i.e., at lowest economic or resource cost. It tends to induce the least costly combination of measures for waste reduction at each outfall, and the least-cost distribution of waste reduction among outfalls, thereby minimizing the real resources cost of attaining a stream water quality standard. In the process, it will normally yield a net revenue as well.

What happens is that each waste discharger is charged in proportion to the use he makes of a resource—the waste assimilative and transport capacity of the watercourse. The waste discharger can compare his marginal costs and marginal charges and decide whether it pays him to reduce his waste, and to what degree. The revenue that accrues to the agency from the charges can be viewed as a rental return on a *natural resource*. How much this will be depends on the scarcity of the resource, i.e., how small the natural flow is, how high the standard is, how highly developed the basin is. It is as though parties wishing to use other parts of the public domain—public land, for instance—were competing for its use and thereby establishing its rental value, and the rental return was accruing to the public. This strikes us as equitable.

We have focused in our theoretical development on charges on actual waste discharge. In principle, of course, charges should be levied on all activities which reduce water quality sufficiently to impose external costs. This may not be administratively feasible in some instances, but it should be possible to assess a hydropower plant which draws water from the oxygen-depleted strata of reservoirs in the summer months,

8. It might also induce some longer-term adjustments that are unfavorable from an efficiency point of view.

discharges it into a stream, and causes a reduction in the dissolved oxygen content of the stream. Similarly, other activities which divert or deplete water in such a way as to reduce assimilative capacity should be subject to charges.

Our analysis so far leads us to the conclusion that systems of charges may properly be viewed as a central tool of water quality management. We feel that regional water quality management agencies should be provided with this tool. We feel also, however, that other devices such as litigation between waste dischargers and affected parties and between individual parties and the regional agency and the selective enforcement of effluent standards can be useful complements.

# 10

J. H. DALES *

# *Land, Water, and Ownership*

*J. H. Dales is Professor of Economics at the University of Toronto.*

## I

Increasing public concern about the pollution of natural water systems in North America has confronted governments with a new problem in resource administration, and challenged economists to devise an artificial pricing system for water that will itself promote wise use of the resource, thereby greatly simplifying the lives of water administrators. The pricing problem turns out, not unexpectedly, to be a deliciously complex tangle of joint uses, externalities, and peak-load problems. The administrative problem of approximating optimum shadow prices by actual user charges promises to be a nightmare.

The economic and administrative complexity of water problems is commonly explained as being inherent in the nature of a fluid resource. Because of the self-mixing quality of a fluid, one use of water at a given point may affect other uses at the same point; and because water flows

*"Land, Water, and Ownership," by J. H. Dales, from the* Canadian Journal of Economics, *November 1968. Reprinted by permission of the* Canadian Journal of Economics *and J. H. Dales.*

* I am grateful to colleagues for comments on earlier versions of this paper. My main debt, however, is to Mr. J. C. McManus, who has spent much time discussing the paper with me and has done his best to prevent me from making errors. The paper has been written during a sabbatical year financed in part by the Canada Council.

through space, use at one point may also affect uses at other points. Opportunity cost pricing is accordingly very complex because of the number of alternative opportunities that may be affected by any one use at any one point, not to mention the complications introduced by time of use, varying stream flow, different rates of self-regeneration of different stretches of water, and the chemical interactions of different types of waste after they have been discharged into a natural water system.

Before we submit to this incubus of complexity, however, we might seek comfort in the reflection that the great virtue of a pricing system is that it solves, avoids, mediates, or somehow manages to dispel, all sorts of complexities, particularly those that arise from various interdependencies between uses and users of goods. Yet the existence of a natural pricing system depends crucially on the institution of ownership. What is not owned cannot be priced since prices are payments for property rights or rights to the use of an asset.[1] In the course of allocating property rights to assets among different owners, the price system in fact transforms most potential "technological externalities" into "pecuniary externalities," a synonym for prices. Thus we hear very little about externalities of land use, precisely because property rights to land use are well established and allocated by the price system. It is quite otherwise where water is concerned.

We can now re-formulate the water problem and blame its complexity not on nature and the laws of fluids, but on man and his failure to devise property rights to the use of natural water systems. Economists tend to assume implicitly that it is impossible to own water and therefore seek to devise artificial price systems that are identical to what prices "would be" if ownership were possible. The alternative strategy is to devise an ownership system and then let a price system develop. The purpose of this article is to suggest that there are very considerable advantages to attacking our water problems by means of a system of explicit ownership rather than by a system of shadow prices.

A geographical reflection is also in order. Despite the large numbers of people who live on the St. Lawrence, the Fraser, and the St. John, most Canadians live on lakes, or on rivers that flow into lakes, rather than into oceans, whereas most Americans, despite the large population around the southern rim of the Great Lakes, live on river systems that

1. One never owns physical assets, but only the rights to use physical assets. Professor Ronald Coase writes that a factor of production "is usually thought of as a physical entity which the businessman acquires and uses (an acre of land, a ton of fertilizer) instead of as a right to perform certain (physical) actions. We may speak of a person owning land and using it as a factor of production but what the land-owner in fact possesses is the right to carry out a circumscribed list of actions. The rights of a land-owner are not unlimited." See his "The Problem of Social Cost," *Journal of Law and Economics,* Oct. 1960, 1–44. [Reprinted in this volume, Selection 6.]

discharge into salt water. Lakes are much less "mobile" and much less "self-mixing" than rivers. Most of the water in lakes *stays* there for prolonged periods, and recent research has shown that during much of the year the shallow, inshore waters of large lakes are effectively isolated from the very large volumes of water in their deep centres.[2] (Similar propositions apparently apply to oceans; otherwise the serious pollution problems of coastal cities such as San Francisco, New York, and Vancouver would not exist.)

People who live on river systems, as most Americans do, tend to pass on their pollution to the next downstream community, thereby creating vexing externality problems. American literature on pollution has been strongly influenced by the uni-directional flow of rivers, which makes it relatively easy to solve the identification problem of who pollutes whom. River pollution therefore lends itself to economic analysis in terms of externalities, and shadow-pricing schemes to offset them. People who live on lake systems (most Canadians) tend to pollute themselves; because inshore lake water, far from being uni-directional, tends to slosh up and down the shoreline, lake pollution tends to be a sort of Hobbesean war of all against all. It therefore requires economic analysis in terms of social decision-making and social welfare functions rather than in terms of the effects of autonomous upstream communities on autonomous downstream communities. The economics of Canadian water pollution is therefore quite different from the economics of American water pollution.

## II

Rent theory has been the traditional vehicle for studying the economics of natural resources, and a review of some of the effects of the ownership-rental system as applied to land highlights the opposite effects induced by the absence of an ownership-rental system as applied to water. Following normal practice, we shall speak of the supply of land (or water) available to any society as fixed by nature. Though fixed in supply when measured in natural units (acres or gallons), the *quality* of land and water can be changed by human action. (Were we to measure

2. See G. K. Rodgers and D. V. Anderson, "The Thermal Structure of Lake Ontario," *Proceedings Sixth Conference Great Lakes Research,* 1963, University of Michigan publication 10, 59–69; P. F. Hamblin and G. K. Rodgers, *The Currents in the Toronto Region of Lake Ontario,* publication (PR 29) of the Great Lakes Institute of the University of Toronto, 1967; and G. K. Rodgers, "Thermal Regime and Circulation in the Great Lakes," in Claude E. Dolman, ed., *Water Resources of Canada,* Royal Society of Canada, studia varia 11, (Toronto, 1967), 87–95.

the quantity of land in efficiency units of a given quality, its supply would be variable; we conduct the present argument, however, in terms of natural units, fixed supplies, and variable qualities.) Imagine a society where all land is being used and even the poorest of it commands a positive rent (this assumption allows us to avoid those not very illuminating discussions about no-rent land and the relationships between the extensive and intensive margins) and suppose that an initial state of equilibrium exists, and in particular that at existing land values and rents there is neither investment nor disinvestment in the quality of the soil. Population growth when superimposed on this initial state will lead to increases in land values and rents; these increases in turn will lead to economies in the use of soil by means of the substitution of manufactured fertilizers and other intensive farming practices against inputs of natural soil fertility. The process may be described as one of investing in soil fertility, and the equilibrium stock of soil fertility will accordingly rise (or its rate of decline will fall). In a general form, the conclusion is that *the level of rent determines the quality of the soil that it is economic to maintain.* It is also clear, as Ricardo showed, that when man-made inputs are substituted for natural inputs in the food-producing industry the real cost of food increases and the standard of living in terms of food falls. Rising rents, therefore, tend to slow down population growth and lessen the population pressure that produces them.

The working out of these processes in the historical development of the United States has been brilliantly described by Bunce.[3] In the early history of the country there was a high ratio of soil fertility to the human demands on it, and rent was accordingly low; "high farming" practices on the European model were therefore rejected, and economic use necessitated soil-depleting practices. In the course of time the man-soil fertility ratio rose, as a result both of population growth and of soil depletion; rents rose; more intensive farming practices became economic, and the rate of soil depletion was thereby reduced. A slowing down in the rate of population growth after 1900 further reduced pressure on the land; it is possible that in general soil depletion has now been brought to a halt in the United States and Canada, and that soil erosion and soil-depleting practices in some areas are balanced by soil-building investments in others.

The contrast between the history of land and of water use on this continent is eloquent. Property rights were established in land, with rent being the payment for the right to use the soil fertility; there were no water rents because property rights to water use were not established. Rising land rents led to more intensive land use and after 350 years there is no problem of population pressure on the land in North America. Water rents were zero; over-use of the water led to continuous re-

3. A. C. Bunce, *The Economics of Soil Conservation* (Ames, Iowa, 1945).

duction in water quality; and there is now a growing problem of population pressure on North American water resources. If we accept a simple dynamic extension of rent theory and assume a direct relationship between the level of rent and the development of improved technologies, the land-water comparison is again suggestive. Rising land rents have been associated with phenomenal improvements in land-use technology; zero rents for water have been associated with virtually zero improvement in water-use technology so far as quality-depleting uses are concerned.

The short-run function of land rent, of course, is to allocate parcels of land among different users and different uses. What is interesting is that potential externalities of land use, for example, the operation of a pig farm in the centre of a choice residential area, seldom materialize. Land being immobile, the ownership-rental system seems to work in such a way as to produce "natural zoning" in land use. Differential rents provide the mechanism for such zoning, and the result is that potential technological externalities are continuously transformed into pecuniary externalities, or prices. It should also be noted that a formal economic description of this process depends on a recognition of space, and particularly of the socially "insulating" quality of space. So long as space exists—and we must remember that in most economic analysis it does not—"zoning" solutions to externalities, or what Mishan has recently called "separate facilities" solutions,[4] are possibilities. Given space, there is no need for pig farmers and business executives to live as neighbours, and therefore no need to devise a system of bribes to compensate one or the other party for damages suffered.

The absence of an ownership-rental system for water has meant that water use has in fact been determined by such things as historical priority, gall, and force and fraud; it cannot be otherwise when property rights do not exist and when the price for the use of a valuable asset is zero. When no pricing process exists, there is no mechanism to transform technological externalities into pecuniary externalities. Accordingly we *do* observe striking examples of externalities in water use; stinking streams flow through choice residential areas, and anglers experience a mixture of rage and resignation as their favourite streams are polluted by industrial wastes. And then there are the externalities of all against all—householders help to destroy swimming beaches by their use of detergents (which promote algal growth), motorists pollute the air they breathe, and we all promote municipal and industrial pollution by insisting on cheap products and low taxes.

These considerations suggest the enormous social benefits that have resulted from applying an ownership-rental system to land, and, by contrast, the enormous social friction and economic waste that result from

4. E. J. Mishan, *The Costs of Economic Growth* (New York, 1967), chap. 8.

not applying an ownership-rental system to water. It has, of course, been relatively "easy" to apply property rights to land because land is both divisible and immobile. The awkward problem remains: is it *possible* to apply an ownership-rental system to the use of our water resources?

## III

To speak of owning an asset is to use a convenient abbreviation for a complex interaction between a legal concept and an economic concept. An asset may be thought of as "a bundle of potential utility-yielding services that can be used in alternative ways." In the same vein, ownership consists of "a bundle of legally defined user rights to an asset." As Coase has pointed out, it is rights, never objects, that are owned, and the rights themselves are always limited by law; "outright" ownership can never, by definition, extend to the use of an asset for illegal purposes.[5]

From the whole spectrum of possible ownership arrangements, we shall pick four major types for brief comment. What we shall call *common-property* ownership is, from an economic point of view, virtually non-ownership. A common-property asset is one that can be used by everyone, for almost any purpose, at zero cost. Examples are the medieval commons, the high seas, wild game, freeways, and (until recently in this country) air and water. Common-property ownership is justified economically *only* when the costs of enforcing a more restricted form of property-rights would be greater than the benefits of doing so. H. S. Gordon has shown that, neglecting enforcement costs, common-property ownership of an asset is economically inefficient in that the asset will be over-used by comparison with assets that are subject to more restrictive property rights.[6]

Empirically it is clear that if the asset is depletable it will be continuously depleted on the grounds that "everybody's property is nobody's property": medieval commons were overstocked; modern freeways (but not toll-ways) quickly become congested; wild animals (but never domestic animals) become scarce or extinct; and the deteriorating quality of our air and water resources has become a matter of widespread concern. The concept of a free good has always been a contradiction in terms; it is time we appreciated its sardonic overtones, for anything that is treated as a free good is indeed likely to become a valueless thing.

5. See Coase, "The Problem of Social Cost." Reprinted in this volume, pp. 100–129.
6. H. S. Gordon, "The Economic Theory of a Common-Property Resource: The Fishery," *Journal of Political Economy,* April 1954, 124–42. Reprinted in this volume, pp. 88–99.

In general common-property assets are nominally owned by some public body, usually a government, and the owner may restrict use of the property in a variety of ways. Some roads may be used by motorists but not by cyclists or pedestrians; some wild animals may be photographed, but not shot; on some lakes canoes and sailboats may be used, but not motor-boats. It seems reasonable to refer to such property as *restricted common-property;* though the type of use is restricted, it is still common-property in the sense that everyone can use it for designated purposes at zero cost. If uses that deplete the asset in a physical sense are banned, the quality of the asset can be maintained, though "congestion" problems may reduce its value to other users.

When the use of an asset is restricted by law to particular persons, or a particular person, we have what can conveniently be called *status-tenure* or *fixed-tenure* ownership. Such ownership guarantees exclusivity of use to the parties authorized to use the property, but these user rights are not transferable. Though secure right of access to an asset by a limited group of people is valuable, the absence of the right of transferability prevents an explicit price system from developing. Nevertheless implicit prices are likely to appear. If the right to send one's children to a particularly good school is limited to those who live in a particular area, the value of the rights is likely to become reflected in the value of real estate in the area concerned. The "regulatory" branches of modern governments create an enormous variety of valuable property rights that are imperfectly transferable, and that tend to be capitalized and monetized in ways that are usually unsuspected by their creators. The value of tariff protection, a quota to grow tobacco, a licence to transport milk or to operate a taxicab, are reflected in the values of tariff-protected businesses, tobacco farms, milk routes, and taxi fleets.[7] Though the indirect monetization of such rights is seldom illegal, contemporary populations choose to be as hypocritical about the process as medieval populations were about the evasions of prohibitions on the payment of interest; social inhibitions about a rational approach to property and prices have outlived social inhibitions about rational approaches to astronomy and sex.

From status-tenure to full *ownership,* in the usual contemporary sense of the term, is but a short step. Once the property right is separated from the person, it becomes transferable, and transfers of assets (rights) then take place at explicit prices. *Transferable* property rights stand in a one-to-one relationship to prices; everything that is owned is priced, and everything that is priced is owned—which is to say nothing about either the form of ownership (transferable property rights to assets may be owned by individuals, corporations, or governments) or

7. On this general question, see Charles A. Reich, "The New Property," *Yale Law Journal,* 73, no. 5 (April 1964), 733–87.

about the precise functional relationship between ownership and prices. Ideological hang-ups on concepts of property rights and ownership are understandable because such concepts touch the very roots of society. We have not yet learned to discuss such matters unemotionally. Though we are inclined to take a condescending view of medieval man's distrust of full property rights to land, we tend to become quite agitated when valuable government-granted rights (licences to import, for example) are traded in the market place, or when suggestions to extend property rights to air and water are put forward for discussion. Property and prices still raise ancient fears that "the rich will eat out the poor."

## IV

Since the right to use water is valuable, and since ownership consists of user rights, it should in principle be possible to devise an ownership-rental system for water. As is well known, however, certain characteristics of a natural water system create special problems in ownership.

The characteristics of an ownership system reflect in part the "divisibility" of the asset to which it is applied. Let us define an *asset-unit* as the smallest physical amount of the asset to which it is practicable to apply property rights, i.e., for which it is practicable to enforce exclusivity of use. In land, the asset-unit is very small, perhaps a few square yards; when the asset-unit is small compared to the quantity of the asset available, the asset can be held by a large number of individual owners. In such cases a "private property" form of ownership is likely to work well; decisions about the use of the asset will be decentralized among many owners, and a reasonably competitive market in asset-units will emerge.

In water, the asset-unit is very large. If water were completely "self-mixing," no one would pay anything to own Lake Ontario unless he could also own the whole Great Lakes drainage basin above the St. Lawrence River. As we have seen, however, water, especially in large lakes, mixes only slowly and imperfectly; because of this, and because of the self-purifying characteristic of water, the quality of water in the eastern end of Lake Ontario may be effectively independent of the uses made of the water at the western end of the lake. Even so, it is clear that the asset-unit is very large. It might be possible to divide the Great Lakes water system into, say, a dozen "regions" each of which would be self-contained for practical purposes, but it would certainly be impossible to divide them into a thousand such regions. In a democratic society it would be unacceptable to allow as few as a dozen, or even a score, of owners to control such an immense property as the Great Lakes drainage system. The only sensible alternative is the one actually adopted, namely, monopoly ownership by government. The reverse side of this coin is that the government must decide how its property is to be used and must enforce its decisions—assuming that it wishes to avoid the horrors of the common-property approach to resource management.

The decision about how water shall be used must be an arbitrary one from the standpoint of economics. Let me argue this point on the basis of a simple (but seemingly realistic) classification of water uses.

If we ignore such uses as navigation and the generation of hydroelectric power, which have insignificant effects on water quality, it seems reasonable to classify other uses into two categories: waste disposal and "all other" uses, which we shall call amenity use. These two uses are competitive. Though it is not true that fishermen, swimmers, industries that use water for processing purposes, and municipal authorities responsible for residential water supplies all have the same quality demands, it is true that some of these users would be benefited, and none would be harmed, by an improvement in water quality. Waste disposers, on the other hand, would be harmed by such an improvement since it could only occur if less waste were discharged into the water. We thus reduce the many uses of water to two: amenity use and waste disposal. The social problem is then to decide on the division of water services between these two conflicting uses. In principle, the division should be made in such a way that the value of a marginal increment in the one good is equal to the value of a marginal decrement in the other. But since the value of a marginal change in amenity use cannot be measured, the optimum amount of waste disposal cannot be identified. In practice, the decision is made on a political rather than an economic calculus. Once there is a political demand for "pollution control," anti-pollution measures tend to be instituted incrementally until complaints about their cost outweigh complaints about pollution! That sort of solution, applied also to such things as education, road systems, and various social welfare schemes, seems to me to be eminently sensible, *faute de mieux*.

In water quality problems, however, it is important to keep in mind that, within limits, water can be "regionalized" for practical purposes, and that "zoning" solutions to quality problems are therefore possible in some cases. In practice it would probably be wise to provide for different ratios of amenity use to waste disposal use in different water "regions"; the socially insulating quality of space should be utilized wherever possible. People are mobile, and if they can consume the amenity services of water in the upper reaches of a river and the waste disposal services of the same river in its lower reaches, there is no need to force them to decide on the optimum division between amenity uses and waste disposal uses of the water in both the upper and lower parts of the river. But again, alas, economics has little to say about a feasible or desirable delimitation of water "regions"; a sensible "mapping" of water must be left to the good judgment of physical scientists and politicians.

The contention that there exists no economically optimum division between amenity and pollution uses of water will be resisted by exponents of damage-cost pricing.[8] In the classic example of an upstream community polluting a downstream community, an allegation of damage to the downstream user seems to rest on three assumptions: that the downstream community owns its water and in particular owns the right not to have its water polluted by others; that the downstream community gains no advantage from the upstream pollution, i.e., that its residents buy no goods from their upstream neighbours at prices that are lower than they would be if the upstream community were forced to reduce its pollution; and that the upstream residents suffer no disadvantages from the downstream pollution because they never visit the downstream area for fishing, swimming, or other recreational purposes. The property rights assumption has not generally been true in the past in North America, and even to-day it is far from clear that a downstream community has any more right to use the river water for swimming and drinking than the upstream community has to use it for waste disposal purposes. The other two assumptions about the inter-community immobility of goods and people are, in general, untenable. The "polluter-pollutee" view of the problem that underlies the recommendation of damage-cost pricing derives from the apparently easy identification of the two parties on a river. Once the mobility of goods and people up and down a river is taken into account, however, identification becomes much more difficult and the problem appears much like the "war of all against all" that is characteristic of lake and ocean pollution.

Even if everyone is at once a polluter and a pollutee, however, the optimum amount of pollution could be achieved if the value of a marginal dose of pollution could be measured. But it cannot be measured, because its value is the value of the amenity use forgone, which cannot be measured. Attempts have been made to measure the recreational value of particular land and water areas, but all such measurements are made on the partial equilibrium assumption that the recreational use of neighbouring areas is held constant. In general, however, a reduction in the amenity capacity of one river or lake will result in increased pressure on the amenity capacities of other rivers and lakes in the same general area. So far as I know, no one has been able to measure the amenity value of an acre-foot of water under general equilibrium assumptions. All we can be reasonably sure about is that the recreational value of water rises as population grows and the standard of living increases.

8. A good exposition of damage-cost pricing is to be found in Allen V. Kneese, *The Economics of Regional Water Quality Management* (Baltimore, 1964). [See Kneese and Bower in Section II of this volume (*Editors*).]

In brief, it seems to me that it is unrealistic to view water management as a problem in externalities, and that the question of how water should be used is purely a matter of collective decision-making. Economics cannot be of any significant help in making this decision. Even the principle that property rights should be set so as to maximize social product is of no use in the case of water because the values of amenity uses of water—recreation, and the simple aesthetic satisfaction that most of us gain from looking at, or even merely contemplating the existence of, clean water—cannot be measured, though such values are certainly part of any society's gross national welfare. Social welfare functions, community indifference curves, and benefit-cost analysis are ways of visualizing the social decision-making problem, but not of solving it.

What is special about the ownership of water, therefore, is that the owner must decide, without the benefit of economics, how his asset is to be divided among different uses. (When asset-units are small relative to the amount of the asset available, as in land, decentralized ownership is possible and the amount of the asset devoted to different uses is, for practical purposes, determined by market forces.) But this special quality of water (and air) ownership does not make it impossible to apply a rental system to water management.

## V

If economics has nothing useful to say about the ownership decision of how water should be used, it has a great deal to say about how the decision, once made, should be implemented. What the government-owner of a natural water system must decide is how many equivalent tons of wastes may be discharged into the waters of each water region. The decision has at least three arbitrary components. Since in given circumstances a ton of one waste is likely to be more injurious than a ton of another, some equivalence must be established between different waste products, and since circumstances differ widely I assume that some average equivalence is chosen for each region in order to simplify the problem and reduce administrative difficulties. The other two sources of arbitrariness from the economic point of view, the mapping of regions and the choice of the amount of pollution in each region, have already been discussed. Let us now suppose that the owner has decided that during the next five years no more than $x$ equivalent tons of waste per year are to be dumped into the waters of region $A$, and that $x$ represents a 10 per cent reduction from the amount of waste that is currently being discharged into the region's waters. How can the government-owner enforce this decision?

The government can enforce its decision in one of six main ways. It can *regulate:* (1) a waste quota can be assigned to each waste discharger and set so that the sum of the quotas does not exceed *x;* or (2) an across-the-board regulation that each discharger must reduce his waste discharge by 10 per cent may be promulgated. It can *subsidize:* (3) dischargers can be subsidized to reduce their wastes, either individually or (4) on an across-the-board basis of so much per ton of waste discharge reduced. It can *charge:* (5) an effluent charge can be levied on dischargers, either individually, or (6) on an across-the-board basis of so much per ton of waste discharged.

I suggest that it is intuitively obvious that the *individual,* or point-by-point, procedures would involve staggering administrative costs. Yet it should be noted that politicians and civil servants seem to favour point-by-point *regulation,* and that economists who recommend damage-cost pricing favour point-by-point *charging* schemes. It seems intuitively obvious that in practice no point-by-point procedure could distribute the cost of reducing pollution among polluters in an economically optimal way, i.e., in a way that would minimize the total cost of reducing pollution by 10 per cent. To suppose that optimality in this sense is possible is to suppose that the administrative authority is able to solve a set of thousands of simultaneous equations, when the information required to write the equations in numerical form is not only not available, but also often unobtainable. It is also obvious that an across-the-board regulation to the effect that all dischargers must reduce their wastes by 10 per cent would result in a non-optimal distribution of the cost burden.[9]

Let us then examine the across-the-board schemes of subsidization and charging. Both possess the advantages of low administrative costs relative to the point-by-point schemes, and both would result in an optimum distribution of costs among dischargers; all dischargers would reduce their wastes up to the point where the marginal cost of doing so equalled the subsidy provided, or the charge levied. Both schemes have two disadvantages: a certain amount of experimentation would be necessary to establish the level of subsidy, or charge, that would produce a 10 per cent reduction in waste discharge; and the levels would have to be varied annually to take account of industrial and demographic growth (or decline) in the region in order to keep to the target of *x*

9. Paul A. Bradley, "Producers' Decisions and Water Quality Control" (Background Paper D 29–3 in *Pollution and Our Environment,* papers presented at a conference held in Montreal, Oct. 31 to Nov. 4, 1966, by the Canadian Council of Resource Ministers), discusses various possible reactions of individual firms to the regulation of effluent standards and to a system of effluent charges. Standards, charges, and subsidies are discussed extensively in Kneese, *Economics of Regional Water Quality Management,* chaps. 4 and 8. [See Bower and Kneese in this volume, Selection 8.]

equivalent tons of waste discharge. The subsidy scheme, however, has two disadvantages that the charging scheme does not have. First, if a subsidy of so much per ton of waste reduced is set, extra profits will accrue to those firms that can reduce their wastes at a cost per ton that is less than the subsidy provided, and no change in relative prices of goods is necessary. In the charging scheme excess profits will not be generated, and there will necessarily be a change in relative prices of goods, which in turn will result in a socially desirable adjustment of consumption patterns to reflect the differential costs of waste disposal as between different goods. Second, the subsidization scheme provides no incentive to choose production methods that reduce the amount of waste generated (and may indeed have the opposite effect!), whereas the charging scheme provides incentives both to reduce waste and improve the technology of treating waste before it is discharged. The across-the-board charging scheme is therefore clearly the best of the six possible ways of implementing the government's decision.[10]

Its victory is made decisive by the fact that it lends itself easily to a market mechanism, whereas the subsidy scheme does not. The government's decision is, let us say, that for the next five years no more than $x$ equivalent tons of waste per year are to be discharged into the waters of region $A$. Let it therefore issue $x$ pollution rights and put them up for sale, simultaneously passing a law that everyone who discharges one equivalent ton of waste into the natural water system during a year must hold one pollution right throughout the year. Since $x$ is less than the number of equivalent tons of waste being discharged at present, the rights will command a positive price—a price sufficient to result in a 10 per cent reduction in waste discharge. The market in rights would be continuous. Firms that found that their actual production was likely to be less than their initial estimate of production would have rights to sell, and those in the contrary situation would be in the market as buyers. Anyone should be able to buy rights; clean-water groups would be able to buy rights and not exercise them. A forward market in rights might be established. The rights should be for one year only, the price of one right for one year representing the annual rental value of the water for waste disposal purposes.[11] (There is no reason, though, why speculators should not gamble in one year on the price of rights in later years.) The virtues of the market mechanism are that no person, or agency, has to *set* the price—it is set by the competition among buyers

10. Editors' note: Since foregoing a subsidy represents an opportunity cost of production just as does a charge, it is difficult to see why its effects on either relative prices or production methods should differ from those of a charge. See Kneese and Bower in this volume, Selection 8, on the same point.

11. Professor Neufeld has suggested that it would be desirable to issue rights of different durations; more complicated schemes than the one outlined in the text could easily be arranged.

and sellers of rights—and that the price in the market automatically "allows for" the regional growth (or decline) factor. If the region experiences demographic or industrial growth the price of rights will automatically rise and induce existing dischargers to reduce their wastes in order to make room for the newcomers. The government should make it clear that it reserves the right to alter the allowable level of pollution (the number of rights it issues) at stated time intervals (say, every five or ten years). All that is required to make the market work is the inflexible resolve of the government not to change the rights issue during the interval, no matter what the political pressures to do so may be, and to enforce rigidly the requirement that a ton-year of waste discharge *must* be paid for by the holding of one pollution right for one year. Pollution rights are fully transferable property rights, and any welching on the enforcement of the right would be a breach of trust.

The automaticity of the market mechanism reduces administrative costs by relieving administrators of the necessity of setting the charge for rights and changing it periodically to reflect economic growth or decline. The administrative costs of enforcement would remain, of course, but they would be no greater than the costs of enforcing any of the other implementation schemes that we have considered. Technological change in the form of automatic monitoring devices to measure the volume of effluents from discharge points promises to reduce the costs of policing for all anti-pollution schemes.

Compliance with any point-by-point regulatory or subsidization scheme of pollution control establishes a sort of *status-tenure* property right. The right inheres to the discharger that earns it, and is only transferable (at the capitalized value of its implicit price) when the property to which it applies is sold. The market mechanism of the across-the-board charging scheme separates the property right to water use from the other assets of the discharger, and thereby makes the property right *fully transferable*. Full transferability and explicit prices are, as has been noted, considered preferable to status tenure and implicit prices by contemporary populations in Western democratic societies.

## VI

Having puffed the merits of the across-the-board *cum* market mechanism scheme of pollution control, I must now take note of its deficiencies. There are four arbitrary elements: the mapping of water regions; the setting of waste equivalents; the choice of the allowable amount of waste discharge; and the choice of time interval during which the number of pollution rights is fixed. By comparison with some ideal, Pareto-optimal scheme laid up in Heaven, each of these decisions is bound to introduce elements of non-optimality into the arrangements I have proposed. In each case, however, I suggest that the saving in administrative costs is likely to outweigh the loss in terms of resource misallocation, measured from some theoretical optimum that ignores administrative and other transactions costs—notably the cost of acquiring enough information to administer an optimal pricing system.[12]

The question of the possible effects of pollution charges on the location of industries (and population) requires special comment. It is often suggested in the literature that waste discharged into a large, lightly populated river system does less damage than if it is discharged into a small, thickly populated river system, and that accordingly pollution charges for use of the former ought to be lower than for use of the latter. This reasoning assumes that the only costs of waste discharge are the objective, measurable, costs to residents in the area, or more generally—if people are allowed to live in one area and vacation in another—that the damage done to amenities by an extra ton of waste is everywhere the same. A system of charging that equalized marginal measurable costs as between water systems would then minimize the objective costs of disposing of a given tonnage of wastes over all the water systems in an area. But this argument does *not* hold if the goal is to minimize *total* costs of disposing of a given tonnage of wastes.

In general, as one river system (or one part of a lake) becomes more polluted, the amenity value of neighbouring unpolluted waters rises. Moreover, when pollution reaches a level that is inconsistent with all recreational uses, added waste discharge has no recreational cost, while added pollution (that destroys swimming even if not, say, boating) in a popular vacation area probably has a very high recreational cost. The demand for amenity uses of water is certainly not a continuous function of water quality. Not enough is known, or perhaps knowable, about the demand for amenity uses of water to devise a fully optimal use of water in an "*n*-region system." In general, though, when congestion problems arise—when people begin to realize the existence of a spatially generalized pollution problem—it is clear that as pollution levels in one area rise the amenity value of relatively clear water in neighbouring areas rises; thus the opportunity cost of using such waters for waste disposal purposes also rises. This consideration by itself, therefore, suggests that pollution charges should be *higher* in areas where pollution is currently at low levels than in areas where it is at high levels—the reverse of the pricing system usually recommended. The system of low pollution charges for a low pollution level tends to spread pollution evenly over

12. A referee for this paper wrote that my scheme requires "that the questions of how much pollution, where pollution is to be allowed, how it is to be measured . . . etc., are all answered beforehand. But *these* are the really big questions." I agree. I don't think that economic analysis can answer these questions; it can, however, point to the best means of implementing the given answers.

the countryside. I prefer the opposite system of high pollution charges for a low pollution level; it tends to create the separate facilities recommended by Mishan.

In any event, in the present state of knowledge about amenity values of water, it is obvious that the spatial pattern of pollution, or the price differentials between regions for pollution rights, will reflect an arbitrary decision by government. In the scheme outlined in this paper initial differentials in the prices of pollution rights would probably not be large if waste disposal were to be reduced by 10 per cent in each region. As time goes on, however, price differentials will tend to change as other forces lead to the centralization or decentralization of industries and populations. These tendencies can be offset, or encouraged, by the government's decision about the absolute and relative numbers of rights made available for sale in different regions. Thus the government-owner of a water (or air) system must decide not only the over-all quality of his asset, but also the quality of the asset in each region.

It should be noted, finally, that the market in pollution rights is not a "true" or "natural" market. In natural markets price creates two-way communication between sources of supply and demand and affects amounts supplied as well as amounts demanded. (Where supply is fixed in natural units, as in the land market, price affects the equilibrium quality of the asset, and mediates between the users of land on the one hand and the users of the products of land on the other.) My market provides only for one-way communication. It transmits the government-owner's decisions about the use of water to the users of the asset, but there is no feedback from the users to the owner. A rise in the price of a pollution right signals that the waste disposal use of water is becoming more valuable; but this does *not* mean that the supply of allowable waste disposal capacity should be increased, for the value of the competing amenity use of water is also likely to be increasing under the impact of the same growth forces that make the waste disposal use more valuable. The price signals that the government gets from the market are "false," in the sense that they are largely echoes of its own arbitrary decision about the supply of rights. The market proposed in this paper is therefore nothing more than an administrative tool. But administrative tools that have some *prima facie* claim to efficiency should not be ignored in an increasingly administered society.

# 11

E. J. MISHAN

# Property Rights and Amenity Rights

*E. J. Mishan is Reader in Economics at the London School of Economics.*

## I

The competitive market has long been recognized by economists as an inexpensive mechanism for allocating goods and services with tolerable efficiency. Once it is observed that the production of 'bads', or noxious spillover effects, have begun, increasingly, to accompany the production of goods, one might be excused for talking about a serious failure of the market mechanism. In fact the failure is not to be attributed to the market itself, but to the legal framework within which it operates. In particular, we must remind ourselves that what constitutes a cost to commercial enterprise depends upon the existing law. If the law recognized slavery the costs of labour would be no greater than the costs involved in capturing a man and maintaining him thereafter at subsistence level.

How, then, can the law be altered so as to remove the existing inequities?

*"Property Rights and Amenity Rights," by E. J. Mishan. From* Technology and Growth: The Price We Pay, *by E. J. Mishan.*  *Chap. 5, pp. 36–42.*

In so far as the activities of private industry are in question, the alteration required of the existing law is clear. For private industry, when it troubles at all to justify its existence to society, is prone to do so just on the grounds that the value of what it produces exceeds the costs it incurs—gains exceed losses, in short. But what *are* costs under the law and what *ought* to be counted as costs is just what is in issue. A great impetus would doubtless take place in the expansion of certain industries if they were allowed freely to appropriate, or freely to trespass upon, the land or property of others. Even where they were effectively bought off by the property-owning victims, the owners of these specially licensed industries would certainly become richer. And one could be sure if, after the elapse of some years, the Government sought to revoke this license there would be a tremendous outcry followed by a determined campaign of opposition alleging that such arbitrary infringement of liberties would inevitably stifle progress, jeopardize employment and, of course, 'lose the country valuable export markets'.

Such an example, though admittedly far-fetched, is distinctly relevant. For private property in this country has been regarded as inviolate for centuries. Even if the Government, during a national emergency or in fulfilment of some radical piece of legislation, takes over the ownership of private property it is obliged to compensate the owners. It may well be alleged that in any instance the Government paid too little or too much. But it would not occur to a British Government merely to confiscate private property.

In extending this principle of compensation, largely on the grounds of equity, the law should explicitly recognize also the facts of allocation. Privacy, quiet, and clean air are scarce goods, far scarcer than they were before the war and sure to become scarcer in the foreseeable future. There is no warrant, therefore, for allowing them to be treated as though they were free goods; as though they were so abundant that a bit more or a bit less made not the slightest difference to anyone. Indeed, if the world were so fashioned that clean air and quiet took on a physically identifiable form, and one that allowed it to be transferred as between people, we should be able to observe whether a man's quantum of the stuff had been appropriated, or damaged, and institute legal proceedings accordingly. The fact that the universe has not been so accommodating in this respect does not in any way detract from the principle of justice involved, or from the principle of economy regarding the allocation of scarce resources. One has but to imagine a country in which men were invested by law with property rights in privacy, quiet, and clean air—simple things, but for many indispensable to the enjoyment of the good life—to recognize that the extent of the compensatory payments that would perforce accompany the operation of industries, of motorized traffic and airlines, would constrain many of them to close down—or else to operate at levels far below those which would prevail in the absence of such legislation, until such time as industry and transport discovered inexpensive ways of controlling their own noxious by-products.

Thus, if the law were altered so that private airport authorities were compelled to fully compensate victims of aircraft noise it is more than possible—even though the decision costs would be very much lower than those which would have to be incurred by the victims under the present law—that most airports would be quite unable to cover such costs with their profits. They would be recognized as uneconomic and would have to close down.[1]

## II

The consequence of recognizing such rights in one form or another, let us call them *amenity rights,* would be far-reaching. Such innovations as the invisible electronic bugging devices currently popular in the United States among people eager to 'peep in' on other people's conversations could be legally prohibited in recognition of such rights.[2] The case

1. Recent calculations of the differences in the market value of houses, alike in all other relevant respects, at different distances from an airport, understate the loss suffered from aircraft noise for at least two reasons:

(1) They represent an estimate of the maximum loss that the house-owner in the noisier area is able and willing to bear in order to move out of the area, not the larger estimate of the minimum sum they would accept to put up with the inconvenience. Moreover, as alternative quiet zones become harder to find, this minimum acceptable sum grows relative to the maximum sum they are willing to pay to move. Even if there were several currently quiet areas into which a family might move without great expense or inconvenience the absence of an announced government plan of maintaining these areas free of aircraft noise over the future leaves open a risk that reduces the attraction of such areas.

(2) If the Government's existing policies continue and, therefore, noise-free inhabitable areas soon disappear, the increased level of noise throughout the country as a whole is accompanied by a narrowing of differentials between areas. To regard such consequent differentials as an index of disamenity is absurd. For it will reveal zero disamenity for any area as the whole country becomes subject to the same amount of aerial disturbance, no matter how great.

2. According to *Life International* (13th June, 1966): 'As manufacturers leapfrog each other turning out ingenious new refinements, the components they sell have been getting smaller and more efficient. . . . So rapidly is the field developing that today's devices may be soon outmoded by systems using microcircuits so tiny that a transmitter made of them would be thinner and smaller than a postage stamp, and could be slipped undetected virtually anywhere. . . . How to safeguard individual rights in a world suddenly turned into a peep-hole and listening-post has become the toughest legal problem facing the US today.'

Whether the law could be made effective is, of course, a problem. To the extent it could not, one would have to recognize a loss of welfare arising directly from technological progress.

against their use would rest simply on the fact that the users of such devices would be unable to compensate the victims, including all the potential victims, to continue living in a state of unease or anxiety. So humble an invention as the petrol-powered lawn-mower, and other petrol-driven garden implements would come also into conflict with such rights. The din produced by any one man is invariably heard by dozens of families who, of course, may be enthusiastic gardeners also. If they are all satisfied with the current situation or could come to agreement with one another, well and good. But once amenity rights were enacted, at least no man could be forced against his will to absorb these noxious byproducts of the activity of others. Of course, compensation that would satisfy the victim (always assuming he tells the truth) may exceed what the offender could pay. In the circumstances, the enthusiast would have to make do with a hand lawn-mower until the manufacturer discovered means of effectively silencing the din. The manufacturer would, of course, have every incentive to do so, for under such legislation the degree of noise-elimination would be regarded as a factor in the measurement of technical efficiency. The commercial prospects of the product would then vary with the degree of noise-elimination achieved.

Admittedly there are difficulties whenever actual compensation payments have to be made, say, to thousands of families disturbed by aircraft noise. Yet once the principle of amenity rights is recognized in law, a rough estimate of the magnitude of compensation payments necessary to maintain the welfare of the number of families affected would be entered as a matter of course into the social cost calculus.[3] And unless these compensatory payments could also be somehow covered by the proceeds of the air service there would be no *prima facie* case for maintaining the air service. If, on the other hand, compensatory payments could be paid (and their payment costs the company less than any technical device that would effectively eliminate the noise) some method of compensation must be devised.

It is true that the courts, from time to time, have enunciated the doctrine that in the ordinary pursuit of industry a reasonable amount of inconvenience must be borne with. The one defect of this otherwise judicious doctrine lies in the clear implication that the costs of such inconvenience be borne by the victim. In a world where the inconve-

3. Like collective goods, most collective 'bads' (such as aircraft noise) would, under amenity legislation, apparently place each person affected in a monopoly position: such a person could ask what price he wished in order to surrender the amenity in question, and could do so without fear of some other person's consent being 'substituted' for his. But this situation is no different from that of, say, landowners whose consent is required in order to lay a railway line. Just as a body of law has evolved to determine just compensation for the surrender of rights of way, so will the courts discover ways of giving effect to citizens' rights under amenity legislation.

niences are becoming increasingly intolerable the legal recognition of amenity rights has the virtue of imposing an economic interpretation on the word reasonable, and therefore also on the word unreasonable, by transferring the cost of the inconvenience on to the shoulders of those who cause it. If, by actually compensating the victims (or by paying to eliminate the disamenity by the cheapest technical method available) an existing service cannot be continued, the inconvenience that was generated is deemed to be unreasonable. And since those who cause the inconvenience are now compelled to shoulder the increased costs associated with it there should be no trouble in convincing them that the inconvenience is unreasonable or persuading them to withdraw the activity in question.

Governments may continue to claim that, say, a certain airline service should be maintained even though it cannot cover its social costs under such new legislation for reasons connected with the defence of the realm or the national interest. But it would now have to think twice about such popular formulae, since it would be required to vindicate its claims about the high value to the nation of this particular service by paying a direct subsidy to the operators of the service, from the taxpayers' money, in order to cover the costs of fully compensating the victims.

## III

What is of prior importance, however, is that the ethical and economic principles served by amenity rights be accepted by law in the first instance. Once accepted, it will not overtax the wit of man to devise over time the machinery necessary to implement the law. But there should be no mistake about it; such a law will have the most drastic effects on private enterprise which, for too long, has neglected the damage inflicted on society at large in producing its wares. For many decades now, private firms have, without giving it a thought, polluted the air we breathe, poisoned lakes and rivers with their effluence, and produced gadgets that have destroyed the quiet of millions of families, gadgets that range from motorized lawn-mowers and motor-cycles to transistors and private planes. What is being proposed therefore may be regarded as an *alteration of the legal framework within which private firms operate in order to direct their enterprise towards ends that accord more closely with the interests of modern society*.

More specifically, it would provide industry with the pecuniary incentive necessary to undertake prolonged research into methods of removing the potential and existing amenity-destroying features of so many of today's products and services.

The social advantage of enacting legislation embodying amenity rights is further reinforced by a consideration of the regressive nature of the chief spillover effects. The rich have legal protection of their property and have less need, at present, of protection from the disamenity created by others. The richer a man is the wider is his choice of neighbourhood. If the area he happens to choose appears to be sinking in the scale of amenity he can move, if at some inconvenience, to a quieter area. He can select a suitable town house, secluded perhaps, or made soundproof throughout, and spend his leisure and pleasure in the country or abroad, and at times of his own choosing. In contrast, the poorer a family the less opportunity it has for moving from its present locality. To all intents it is stuck in the area and must put up with whatever disamenity is inflicted upon it.

And generalizing from the experience of the last ten years or so, one may depend upon it that it will be the neighbourhoods of the working and lower middle classes that will suffer most from the increased construction of fly-overs and fly-unders and road-widening schemes that inevitably tend to concentrate the traffic and thicken the pollution. Thus the recognition of amenity rights would have favourable distributive effects on the welfare of society. It would promote not only a rise in the standards of environment generally from which all would benefit, it would raise them most for the lower income groups that have suffered more than any other group from unchecked 'development' since the war.

## IV

Finally, in any practical appraisal of the range of consequences following such an innovation, one must be aware that the existence of decision costs in making new economic arrangements does build inertia into the *status quo*. As mentioned above, the existence of potential economic improvements—that is, new economic arrangements whose value to some exceeds the losses incurred by others, thereby enabling everyone affected to be made better off—does not imply that they will or should be realized. It is necessary also that the decision costs entailed in effecting the change be smaller than the value of the potential economic improvement. Such decision costs, to say nothing of the need for initiative, act as a cost barrier to many potential economic improvements irrespective of the state of the law concerning the responsibility for amenity damages (although, as indicated earlier, the decision-cost barrier is likely to be very much greater for the existing law than for one recognizing amenity rights).

None the less, if we restrict ourselves to the number of spillover activities with potential economic improvements that would, because of their associated decision costs, be excluded under *either* law, the *status quo* resulting from the existing law will be markedly different from that resulting from a law recognizing amenity rights. And since there is this marked difference, it goes without saying that the law should be chosen with an eye to the status to be perpetuated. Neither state is 'optimal', in the economist's sense, while such inertia persists. But since the 'suboptimal' states are quite different, we should certainly want to choose the less intolerable suboptimal state.

In sum, under the existing law, a proliferation of adverse spillover effects continues to take refuge behind the barrier of decision costs. Under the proposed law, it is amenity which is sheltered behind this barrier. Using words loosely: the magnitude of decisions costs implies that under the *status quo* there will be 'too much' spillover; under the proposed status, on the other hand, there will be 'too little'.

Over time, of course, changes in population, in tastes, and in technology may reduce some of the decision costs and may raise the value of the potential economic improvements under either law (though the converse seems just as likely). And wherever this occurs, wherever the value of the potential economic gain of some mutual arrangement exceeds the decision costs, the new arrangement will be brought into effect and we shall get the 'right amount' of spillover and the 'right amount' of amenity—an optimal outcome in fact. But bearing in mind that spillovers are likely to grow rapidly over the future, and that many of them, such as the destruction of natural beauty and the poisoning of the earth's atmosphere, cause irrevocable damage, the interest of society, certainly of posterity, is better served by 'too little' spillover rather than 'too much'. And, since in practice we have to choose between them, our continued acceptance of the *status quo* (rather than amenity legislation) implies an acceptance of the 'too much' spillover option.

# 12

GUIDO CALABRESI

# *Transaction Costs, Resource Allocation, and Liability Rules*

*Guido Calabresi is a Professor in the Law School at Yale University.*

In his article on "The Problem of Social Cost" Professor Coase argued that (assuming no transaction costs) the same allocation of resources will come about regardless of which of two joint cost causers is initially charged with the cost, in other words regardless of liability rules.[1] Various writers—including me—accepted that conclusion for the short run, but had doubts about its validity in the long run situation. The argument was that even if transactions brought about the same short run allocation, liability rules would affect the relative wealth of the two joint cost causing activities, and in the long run this would af-

*"Transaction Costs, Resource Allocation, and Liability Rules—A Comment," by Guido Calabresi, from* The Journal of Law and Economics, *April 1968.*

1. Coase, The Problem of Social Cost. 3 J. Law & Econ. 1 (1960). Reprinted in this volume, pp. 100–129.

fect the relative number of firms and hence the relative output of the activities.[2]

Further thought has convinced me that if one assumes no transaction costs—including no costs of excluding from the benefits the free loaders, that is, those who would gain from a bargain but who are unwilling to pay to bring it about—and if one assumes, as one must, rationality and no legal impediments to bargaining, Coase's analysis must hold for the long run as well as the short run. The reason is simply that (on the given assumptions) the same type of transactions which cured the short run misallocation would also occur to cure the long run ones. For example, if we assume that the cost of factory smoke which destroys neighboring farmers' wheat can be avoided more cheaply by a smoke control device than by growing a smoke resistant wheat, then, even if the loss is left on the farmers they will, under the assumptions made, pay the factory to install the smoke control device. This would, in the short run, result in more factories relative to farmers and lower relative farm output than if the liability rule had been reversed. But if, as a result of this liability rule, farm output is too low relative to factory output those who lose from this "misallocation" would have every reason to bribe farmers to produce more and factories to produce less. This process would continue until no bargain could improve the allocation of resources.

The interesting thing about this analysis, however, is that there is no reason whatsoever to limit it to joint cost causers. Thus, if one assumes rationality, no transaction costs, and no legal impediments to bargaining, *all* misallocations of resources would be fully cured in the market by bargains.[3] Far from being surprising, this statement is tautological, at least if one accepts any of the various classic definitions of misallocation. These ultimately come down to a statement akin to the following: A misallocation exists when there is available a possible reallocation in which all those who would lose from the reallocation could be fully compensated by those who would gain, and, at the end of this compensation process, there would still be some who would be better off than before.

This and other similar definitions of resource misallocation merely mean that there is a misallocation when a situation can be improved by bargains. If people are rational, bargains are costless, and there are no legal impediments to bargains, transactions will *ex hypothesis* occur to the point where bargains can no longer improve the situation; to the

2. See, for example, Calabresi, The Decision for Accidents: An Approach to Nonfault Allocation of Costs, 78 Harv. L. Rev. 713, 730 n.28, 731 n.30 (1965) and Calabresi, Fault, Accidents and the Wonderful World of Blum and Kalven, 75 Yale L.J. 216, 231–232 and accompanying footnotes (1965).

3. See note 5 *infra*.

point, in short, of optimal resource allocation.[4] We can, therefore, state as an axiom the proposition that all externalities can be internalized and all misallocations, even those created by legal structures, can be remedied by the market, except to the extent that transactions cost money or the structure itself creates some impediments to bargaining.[5]

It may be that this welfare economics analogue to Say's law has always been quite obvious to economists, although if it has its relevance has too frequently been ignored. In any event, lawyers who use economics have in virtually every case been hopelessly confused on the subject.[6] For this reason, if no other, it is worthwhile elaborating on the practical implications of the proposition.

The primary implication is that problems of misallocation of resources and externalities are not theoretical but empirical ones. The resource allocation aim is to approximate, both closely and cheaply, the result the market would bring about if bargaining actually were costless.[7] The question then becomes: Is this accomplished most accurately

4. Any given individual qua individual might well be richer or poorer as a result of the liability rules in force at the beginning of the bargaining process. But this difference in distribution of wealth would *ex hypothesis* not be one which would affect total social product.

5. By transaction costs, I have in mind costs like those of getting large numbers of people together to bargain, and costs of excluding free loaders. But one may properly ask the question: in what way are these qualitatively different from the cost of establishing a bargain between two parties, i.e., the cost of walking over and dickering? And if they are not, then how are they different from common selling costs, which we all assume the market normally handles optimally? Perhaps the difference is a qualitative one which escapes me. If it is not, it may be that as to normal selling costs, or costs of one-for-one dickering we readily accept the probably justifiable empirical conclusion that no substitute for the market can achieve a similar result as cheaply. But see e.g., Calabresi, The Decision for Accidents, 78 Harv. L. Rev. 713, 725–729 (1965), and Calabresi, Fault, Accidents and the Wonderful World of Blum and Kalven, 75 Yale L.J. 216, 223–231 (1965) (suggesting that liability rules may occasionally be crucial even in these cases). With the kinds of costs which Coase seems to call transaction costs, that conclusion, though possibly often still valid, cannot be accepted without more data. See Coase, *supra* note 1.

6. One notable recent exception is Professor Frank I. Michaelman. See Michaelman, Property, Utility and Fairness: Comments on the Ethical Foundations of "Just Compensation" Law, 80 Harv. L. Rev. 1165, 1172–1176 (1967).

7. Professor Harold Demsetz in a very provocative article has recently suggested that the institution of private property and its protection by the law can be explained in these terms. Demsetz, Toward a Theory of Property Rights, 57 Am. Econ. Ass'n. Pap. & Proc. 347 (1967). Surprisingly, he does not suggest that one of the examples which he gives of externalities not internalized by private property, that of factory smoke pollution, might also be handled by a change in property rights, that is, by making the factory owner liable for the smoke, *id*. at 357. This might resolve the problem if transaction costs were significantly lower if the factory owner were initially liable than if homeowners bore the loss, a not unlikely hypothesis.

and most cheaply by structural rules (like anti-trust laws), by liability rules, by taxation and governmental spending, by letting the market have free play or by some combination of these? This question depends in large part on the relative *cost* of reaching the correct result by each of these means (an empirical problem which probably could be resolved, at least approximately, in most instances), and the relative *chances* of reaching a widely *wrong* result depending on the method used (also an empirical problem but one as to which it is hard to get other than "guess" type data). The resolution of these two problems and their interplay is *the* problem of accomplishing optimal resource allocations.

Two points are implied in the foregoing discussion. The first is that since transactions do cost money, and since substitutes for transactions, be they taxation, liability rules, or structural rules, are also not costless, the "optimal" result is not necessarily the same as if transactions were costless. Whatever device is used, the question must be asked: Are its costs worth the benefits in better resource allocations it brings about or have we instead approached a false optimum by a series of games which are not worth the candles used? This does not mean, though, that the actual optimum is necessarily the one an unaided market would reach. Further market improvements may well be prohibitive at a stage where laws and their enforcement are still a relatively cheap way of getting nearer the goal.[8]

The second point is that both the unreachable goal of "that point which would be reached if transactions were costless," and the gains which reaching nearer the goal would bring are not usually subject to precise definition or quantification. They are, in fact, largely defined by guesses. As a result, the question of whether a given law is worth its costs (in terms of better resource allocation) is rarely susceptible to empirical proof. This does not mean, of course, that the best we can do is adopt a laissez faire policy and let the market do the best it can. It is precisely the province of good government to make guesses as to what laws are likely to be worth their costs. Hopefully it will use what empirical information is available and seek to develop empirical information which is not currently available (how much information is worth *its* costs is also a question, however). But there is no reason to assume that in the absence of conclusive information no government action is better than some action. This is especially so if the guesses made take into account two factors. The first is: Action in an uncertain case is more likely to be justified if the market can correct an error resulting from the proposed action more cheaply than it could an error resulting from inaction. The second is: Action in an uncertain case is more likely to be justified if goals *other* than resource allocation (like proper income dis-

8. See Demsetz, *supra* note 7.

tribution) are served by the action. In effect the first factor says, in uncertainty increase the chances of correcting an error, while the second says, the achievement of other goals is accomplished very cheaply where the most that can be said about the resource allocation effect of a move is that we cannot be *sure* that it will be favorable.[9]

The relevance of the foregoing analysis may be seen in various areas of government intervention. I shall briefly mention three because they have brought forth different governmental responses: (1) the monopoly area, (2) highways or parks, (3) automobile accidents.

(1) Why should we have laws which attempt to control monopolies? Assuming no transaction costs, those who lose from the relative underproduction of monopolies could bribe monopolists to produce more. We know, however, that such market action is usually unrealistic—that is, it would be too expensive relative to the benefits it would bring. The problem of excluding free loaders would—absent any other problems—suffice to make it so. We believe that a series of structural rules, a series of laws in this case, are cheaper than market correction would be, and more important, we believe they are cheap enough to be worth having. This last belief involves certain guesses about what a "costless" market would do, and what the gains of approaching that goal by legislation are. Even if we assume that these guesses are in large terms fairly supported by empirical information, we certainly reach a point where the putative gains of further more stringent legislation or of more stringent enforcement are hard to justify in terms of the costs of such programs. In such a situation how far we go cannot help but be affected, and properly so, by what other goals—like dilution of power, or income redistribution—we think a further step will accomplish. Far from being irrelevant, these factors may be made an integral part of the law, if not the economics, of antitrust.[10]

9. Some may argue that other goals, like income, redistribution, are best achieved not through *ad hoc* decisions but rather as part of a general policy implemented through devices (like some forms of progressive taxation) chosen with the goal specifically in mind. Even if this position is accepted, it is possible that a general policy in favor of a particular distribution of income could be arrived at and its implementation be intentionally left to particular cases where no provably adverse resource allocation effects would come about. This would be especially attractive to those who accept the view that most forms of progressive taxation misallocate resources.

The mention of other goals may suggest that their benefits can be established with substantial certainty. That obviously is not so, but the process of deciding whether to accomplish these goals does not, in our society, seem to depend on the ability of social scientists to prove their desirability with substantial certainty.

10. Obviously there are also many cases where these other goals are pursued without regard to possible adverse resource allocation effects. My object in this comment is not to criticize or even to discuss such situations. Mine is a minimalist position and suggests only that taking such other goals into account is justified where doing so gives rise to no provable adverse resource allocation effects.

(2) The question of why we have public highways is not totally dissimilar. One can view the decision to have public highways as the product of certain assumptions about what people would do if there were no transaction costs. These assumptions are largely guess work—perhaps even more guess work than in the monopoly case. As in the monopoly situation, government intervention is believed to bring us nearer, and more cheaply, to what a costless market would establish than would the real market. The reason may be that exclusion of free loaders seems substantially more expensive than compulsion of payment by all putative gainers, including would-be free loaders. The result is taxation, in part, of those who supposedly gain from the presence of a better highway system. I say in part, because here again the guesses made are inevitably affected by other goals. "Free" parks can be analyzed in the same way. The validity of the assumptions, the availability of empirical information and the effect of other goals may be quite different. But the basic analysis remains the same.

(3) The case of automobile accidents is somewhat more complex. It presents in fairly typical fashion the problem of multiple cost causers. As such it raises quite clearly the issue of short and long run misallocations, and the issue of which governmental interventions are, if incorrect, more subject to market corrections.

Assume that the cheapest short run way of minimizing the sum of accident costs and of the costs of avoiding accidents involving pedestrians and cars is to have rubber bumpers rather than to have pedestrians wear fluorescent clothes. In a world of costless transactions rubber bumpers would become established in the market regardless of liability rules, regardless of whether cars or pedestrians bore the loss initially. Since transactions cost money, the short run effect would in fact be quite different, depending on who was held liable. Making the car owner liable would establish the proper number of rubber bumpers. This would be the desired short run resource allocation, unless, of course, the cost of establishing car owner liability were too great relative to the gains it brought about.[11] But making the car owner liable also has long run effects affecting the relative number of cars and pedestrians. Our assumption as to the best short run liability bearer does not carry with it any guarantee that the car owners are the best long run bearers. It might be that in a world of no transaction costs rubber bumpers would be established, but more cars relative to pedestrians would be desired than would come about if liability were placed on car owners.

Depending on how sure we are of our long and short run guesses, this problem can be handled by using different devices. For example,

11. This caveat has a long history in torts law and is in large part the basis of Holmes' famous justification of the fault system. See Holmes, The Common Law, 94–96 (1881). Needless to say, the caveat is perfectly valid even if it fails, as I believe it does, to justify the fault-liability system today.

the short run allocation could be accomplished by a car-owner liability rule, while the long run hypothetical and misallocation could be corrected by a subsidy to car makers raised from taxes on pedestrians. But the sureness necessary to justify this subsidy, in the absence of nonresource allocation goals which might support it, seems very hard to come by.

The automobile accident situation also raises the point that different devices for accomplishing seemingly "optimal" resource allocation vary in desirability depending on their relative costs and on the relative likelihood of error in our guess work. Returning to the rubber bumper example, if we are perfectly sure that rubber bumpers are always the cheapest way of minimizing the sum of car-pedestrian accident costs and the costs of avoiding such accidents, it seems likely that the cheapest way of getting rubber bumpers is by a law that requires them, rather than by liability rules. It is quite a different thing if the "cheapest way" is more complex and involves some rubber bumpers, and some more careful driving by owners without rubber bumpers.

This in turn suggests another factor in the decision. Suppose we are not sure whether rubber bumpers or wearing fluorescent clothing is the "cheapest way" of handling the car-pedestrian accident problem. In this case it may become necessary to consider the following question: Is an erroneous placing of liability on car owners or an erroneous placing of liability on pedestrians more likely to be corrected in the market? Whether car owners (or car makers) can bribe pedestrians more cheaply than pedestrians can bribe car owners or makers, becomes the relevant issue. Similarly, it becomes crucial to decide whether an error brought about by a "liability rule" is more subject to market correction than an error resulting from a law requiring a particular type of rubber bumper.

Clearly this sketchy description of the automobile accident problem (like those of monopoly and highways) can only indicate the range and complexity of the issues involved in deciding whom to hold liable, and what safety devices to require. A full analysis of the resource allocation issue in auto accidents has, to my knowledge, not yet been attempted. A fully adequate decision would clearly require immense amounts of empirical data. But here (as in the monopoly and highway cases) the lawyer cannot wait for near certainty. He must propose solutions which seem to be the best on the basis of data and impressions currently available. And here too he will be aided in making practical proposals by the fact that goals other than optimal resource allocations may give clear indications of the desirable course in situations where resource allocations policy gives only a hint. One such goal may well be the often mentioned goal of "adequate" loss spreading (which is, in fact, closely analogous to the income distribution policy).

The conclusion is that Coase's analysis, read as a kind of Say's law of welfare economics, gives us an admirable tool for suggesting what kind of empirical data would be useful in making resource allocation decisions, and for indicating what kinds of guesses are likely to be justifiably made in the absence of convincing data. Some may take Coase's analysis to suggest that little or no government intervention is usually the best rule. My own conclusions are quite different. His analysis, combined with common intuitions or guesses as to the relative costs of transactions, taxation, structural rules and liability rules, can go far to explain various types of heretofore inadequately justified governmental actions. This is especially so if one considers the relevance of goals other than resource allocations to those situations where inadequate data make resource allocations an unsatisfactory guide. Perhaps more precise data will some day prove some of these interventions to be improper from the standpoint of resource allocation. Then we shall have to choose, as we often do, between the bigger pie and other aims. Coase's analysis certainly suggests situations where this has been done. Its principal importance lies, however, in helping to delineate those areas of uncertainty where more facts would help us make better resource allocation judgments, and where, at least in the absence of more facts, the lawyer must be guided by guess work as to what the facts are and by goals other than resource allocations in suggesting workable solutions for problems which cannot wait till all the facts are in.

# 13

MILTON FRIEDMAN

# *The Role of Government: Neighborhood Effects*

*Milton Friedman is Paul Snowden Russell Distinguished Service Professor at the University of Chicago.*

The role of government . . . is to do something that the market cannot do for itself, namely, to determine, arbitrate, and enforce the rules of the game. We may also want to do through government some things that might conceivably be done through the market but that technical or similar conditions render it difficult to do in that way. These all reduce to cases in which strictly voluntary exchange is either exceedingly costly or practically impossible. There are two general classes of such cases: monopoly and similar market imperfections, and neighborhood effects. . . .

The second general class of cases arises when actions of individuals have effects on other individuals for which it is not feasible to charge or recompense them. This is the problem of "neighborhood effects". An obvious example is the pollution of a stream. The man who pollutes a stream is in effect forcing others to exchange good water for bad. These others might be willing to make the exchange at a price. But it is not feasible for them, acting individually, to avoid the exchange or to enforce appropriate compensation.

A less obvious example is the provision of highways. In this case, it is technically possible to identify and hence charge individuals for their use of the roads and so to have private operation. However, for general access roads, involving many points of entry and exit, the costs of collection would be extremely high if a charge were to be made for the specific services received by each individual, because of the necessity of establishing toll booths or the equivalent at all entrances. The gasoline tax is a much cheaper method of charging individuals roughly in proportion to their use of the roads. This method, however, is one in which the particular payment cannot be identified closely with the particular use. Hence, it is hardly feasible to have private enterprise provide the service and collect the charge without establishing extensive private monopoly.

These considerations do not apply to long-distance turnpikes with high density of traffic and limited access. For these, the costs of collection are small and in many cases are now being paid, and there are often numerous alternatives, so that there is no serious monopoly problem. Hence, there is every reason why these should be privately owned and operated. If so owned and operated, the enterprise running the highway should receive the gasoline taxes paid on account of travel on it.

Parks are an interesting example because they illustrate the difference between cases that can and cases that cannot be justified by neighborhood effects, and because almost everyone at first sight regards the conduct of National Parks as obviously a valid function of government. In fact, however, neighborhood effects may justify a city park; they do not justify a national park, like Yellowstone National Park or the Grand Canyon. What is the fundamental difference between the two? For the city park, it is extremely difficult to identify the people who benefit from it and to charge them for the benefits which they receive. If there is a park in the middle of the city, the houses on all sides get the benefit of the open space, and people who walk through it or by it also benefit. To maintain toll collectors at the gates or to impose annual charges per window overlooking the park would be very expensive and difficult. The entrances to a national park like Yellowstone, on the other hand, are few; most of the people who come stay for a considerable period of time and it is perfectly feasible to set up toll gates and collect admission charges. This is indeed now done, though the charges do not cover the whole costs. If the public wants this kind of an activity enough to pay for it, private enterprises will have every incentive to provide such parks. And, of course, there are many private enterprises of this nature now in existence. I cannot myself conjure up any neighborhood effects or important monopoly effects that would justify governmental activity in this area.

Considerations like those I have treated under the heading of neighborhood effects have been used to rationalize almost every conceivable intervention. In many instances, however, this rationalization is special pleading rather than a legitimate application of the concept of neighborhood effects. Neighborhood effects cut both ways. They can be a reason for limiting the activities of government as well as for expanding them. Neighborhood effects impede voluntary exchange because it is difficult to identify the effects on third parties and to measure their magnitude; but this difficulty is present in governmental activity as well. It is hard to know when neighborhood effects are sufficiently large to justify particular costs in overcoming them and even harder to distribute the costs in an appropriate fashion. Consequently, when government engages in activities to overcome neighborhood effects, it will in part introduce an additional set of neighborhood effects by failing to charge or to compensate individuals properly. Whether the original or the new neighborhood effects are the more serious can only be judged by the facts of the individual case, and even then, only very approximately. Furthermore, the use of government to overcome neighborhood effects itself has an extremely important neighborhood effect which is unrelated to the particular occasion for government action. Every act of government intervention limits the area of individual freedom directly and threatens the preservation of freedom indirectly. . . .

Our principles offer no hard and fast line how far it is appropriate to use government to accomplish jointly what it is difficult or impossible for us to accomplish separately through strictly voluntary exchange. In any particular case of proposed intervention, we must make up a balance sheet, listing separately the advantages and disadvantages. Our principles tell us what items to put on the one side and what items on the other and they give us some basis for attaching importance to the different items. In particular, we shall always want to enter on the liability side of any proposed government intervention, its neighborhood effect in threatening freedom, and give this effect considerable weight. Just how much weight to give to it, as to other items, depends upon the circumstances. If, for example, existing government intervention is minor, we shall attach a smaller weight to the negative effects of additional government intervention. This is an important reason why many earlier liberals, like Henry Simons, writing at a time when government was small by today's standards, were willing to have government undertake activities that today's liberals would not accept now that government has become so overgrown.

14

ROBERT DORFMAN AND
HENRY D. JACOBY

# A Public-Decision Model Applied to a Local Pollution Problem

*Robert Dorfman is David A. Wells Professor of Political Economy and Henry D. Jacoby is an Assistant Professor in the John F. Kennedy School of Government, both at Harvard University.*

Merely corroborative detail, intended to give artistic verisimilitude to an otherwise bald and unconvincing narrative.

W. S. Gilbert

## Introduction

Governmental decisions may be approached from either a normative or a descriptive point of view. The normative approach accepts well-defined objectives for governmental undertakings and recommends specific policies and actions for attaining them. It will not be followed here. The descriptive approach accepts the facts of life, including the nature of governmental agencies and the purposes of diverse interest groups in the community, and attempts to provide insight into what will happen under the circumstances. That will be our approach.

*"A Public-Decision Model Applied to a Local Pollution Problem," by Robert Dorfman and Henry D. Jacoby, based on* Models for Water Quality Management *by Robert Dorfman,* et al. *(Harvard University Press, forthcoming).*

Our method will be to construct a mathematical model of a political-decision problem. The model will contain room for a great many data, ranging from the technological features of the problem that technical experts have to take into account, to the political objectives and pressures that responsible officials have to evaluate. One of the advantages that we claim for the model, indeed, is that it provides a systematic framework for assembling such diverse data.

From these data the model will produce, mechanically, some predictions about the outcome of the political-decision process. These will not be unambiguous predictions like an astronomer's prediction of the moment at which an eclipse will take place. Rather, they will take the form of stating a range of likely outcomes of the process—perhaps a fairly wide range, but still much more limited than all the decisions that might be conceived of in advance. The power of a scientific theory, it has been said, is measured not by what it asserts but by what it precludes. This theory will preclude a great deal. Within the range of likely outcomes, the theory (or model) will provide some valuable information. It will highlight the political alignments that make some of the outcomes more likely than others and will indicate, rather specifically, the changes in the configuration of political influence that will tend to shift the decision from one outcome to another. Furthermore, it will express vividly just how interests oppose and how different decisions affect the welfares of different participants.

Concede, for the moment, that such a model is possible. Whether this is so or not is for the sequel to determine. Then, surely, the practice of politics should take it into account. (Note that we are being normative temporarily.) Any political decision should be made in the light of a realistic assessment of its consequences, which requires a prediction of how things, people, and political bodies will react to it. In the case of water-pollution control, for example, the federal government's policies are implemented by the states and by river-basin commissions; the states and river-basin commissions work through local agencies and authorities, individuals and business firms. Any decision at the federal level has to be based on a prediction about how the states and basin commissions will respond to it; the states and basin authorities must similarly predict the reactions of subordinate units and of the public at large. Many other programs of federal and state agencies are similarly affected by the responses of other governmental bodies. All legislation that concerns the powers and composition of government agencies is influenced by predictions of how those provisions will influence the behavior of the agency. In short, a predictive model of political behavior can help improve political decisions.

Constructing a predictive model of political decision processes is an ambitious, indeed presumptuous, undertaking. In fact, there is such good reason to believe that it cannot be done that we have construed our main task to be to convince ourselves and the reader that it is possible. Our method of proof is what mathematicians call "constructive." That is, rather than arguing the matter in the abstract, we have considered a political-decision problem taken from the field of water-pollution control (not a real problem, but one that catches the essence of real problems) and have constructed a theoretical model that predicts the outcome of the political-decision process in that instance. This construction shows that such a model is a possibility and exhibits its main features. It does not show that this model is a practical tool of political analysis. That showing would be a major research undertaking. In order to show that this political model or any political model is empirically valid, we should have to apply it to several real political decisions and compare its predictions with the observed political behavior. This would be a laborious and expensive task, for it would necessitate ascertaining all the physical, technical, economic, and political data that the model requires and then performing elaborate computations. We have not undertaken this mammoth empirical enterprise (and not merely out of laziness). This large task is not worth undertaking unless the analytic model to be tested shows at least fair promise of success. A necessary preliminary to serious empirical testing is trial experimentation under favorable circumstances to see whether the model can be implemented in principle and whether it behaves sensibly. If the model passes the preliminary screening, then it pays to go further with it. This chapter reports on such a preliminary testing of our conceptual model of political decision. To carry out this test we have conceived an artificial basin with a pollution problem, known as the Bow River Valley. It is small, it is simple; but it could exist in the sense that it does not violate any known principles of hydrology, sanitary engineering, economics, or politics. We have populated this valley with a large industrial source of pollution, two moderate-sized cities, and a recreation area, and placed it under the jurisdiction of a pollution-control commission organized under the Clean Water Restoration Act of 1966. We have provided to everyone concerned, the data and information he might actually have under the circumstances, and no more.[1] We have simplified the problem by suppressing much detail that would obscure the essential conflicts and issues likely to arise. For example, we have reduced the specification of water quality to a single dimension, namely, dissolved oxygen concentration (DO), and we have limited our description of the waste content of the various municipal and industrial effluents to the number of pounds of biochemical oxygen-demanding material (BOD) that they carry.[2] In addition, we have limited the powers of the regulatory au-

1. Or, anyway, not much more.

2. The description of the problem necessarily involves some technical concepts, such as these. Most of them will be explained below.

thority essentially to a single decision, namely, regulation of level of treatment by each polluter. We have ignored hydrologic uncertainty and other probabilistic complications and have made any number of simplifying approximations to facilitate computation. In short, we have loaded the dice heavily in favor of the model, as is perhaps appropriate for a preliminary test. The question is: Can this problem be expressed by a formal analytic model, and if so, does the model provide sensible and useful insights?

## The Test Problem

The situation we shall use to try out our conceptual scheme is sketched in Figure 1.[3] The Bow River flows generally from north to south. It is a respectable stream with a flow of 800 cfs during the summer months. But it is not a very high-quality stream. Because of the residual waste from upstream cities, the river, as it passes under the Gordon Bridge, has a dissolved-oxygen concentration of only 6.8 milligrams per liter (mg/1). Without the influence of effluent discharges upstream, one could expect the level of oxygen in the water to be near saturation (8.5 mg/1 at summer water temperatures).

The northernmost installation in the region under consideration is the Pierce-Hall Cannery. This is a large but somewhat outmoded vegetable cannery with an annual production of slightly over seven million equivalent cases a year, concentrated in the summer and autumn months. The cannery adds an ultimate BOD load of about one pound per case to the river, after primary treatment, to which it is already committed. The cannery is not very profitable. It employs about 800 workers, many of whom live in Bowville (pop. 250,000), ten miles downstream.

Bowville and the other riparian city, Plympton (pop. 200,000), are both fairly large centers supported by varied light manufacturing and commercial establishments serving the surrounding agricultural region. Both have waterfront parks of which they are proud, and Plympton, in particular, has some aspirations to being a tourist center because of its proximity to Robin State Park. Both cities discharge their wastes into the river after primary treatment. For simplicity, we shall assume that neither city anticipates its population or its waste load to grow significantly in the foreseeable future. This simplification will save us from having to forecast growth rates and from having to consider the possibility of "building ahead of demand." In fact, it enables us to neglect all dynamic considerations.

3. Professor Harold A. Thomas, Jr., acted as consulting engineer for this study. We are indebted to him for formulating and analyzing the hydrological and engineering aspects of our sample basin.

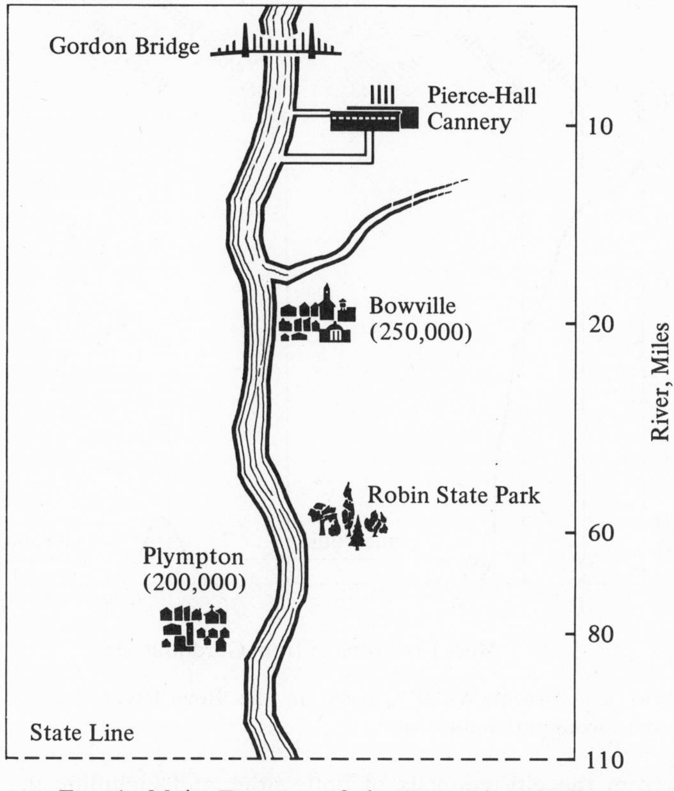


FIG. 1. Main Features of the Bow River Valley

Robin State Park, between Bowville and Plympton, has woodland recreational facilities. All concerned would like to develop its waterfront for boating, fishing, and—if at all possible—swimming. The quality of the water does not permit those uses at present. The park is used by the inhabitants of both cities, by the neighboring farm population, and by some tourists and day-trippers from outside the valley. Everyone is agreed that the quality of the water in the neighborhood of the park should be improved.

Thirty miles below Plympton, the river crosses a state line and flows out of our ken.

The current quality of the stream at critical points under low-flow conditions is shown in Figure 2. From just below Bowville down to the state line, water quality is very poor during summer droughts. For long stretches the river is anaerobic (i.e., the DO level falls to zero), and it is unfit for recreational or other use. In response to a generally felt need to improve the river, especially near the park, and in response to some pressure from the State Water Commission, the Bow Valley Water-Pollution Control Commission has been established, with membership

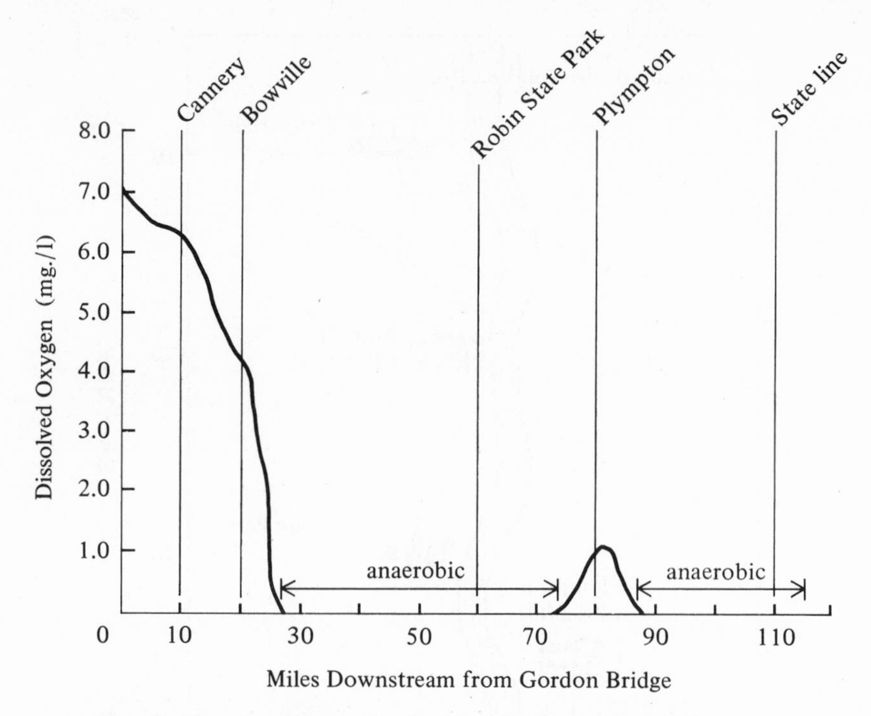


FIG. 2. Current Water Quality in the Bow River, Summer Drought Conditions

drawn from the city councils of both cities and including the Deputy State Commissioner of Parks and Recreation.

The commission faces two crucial problems. The first is to determine the quality classification of the river which, for political reasons, must be the same from the Gordon Bridge to the state line.[4] The second problem is to decide on the levels of treatment to be required of the three sources of waste within its jurisdiction. The cost of the improvement in quality to each of the polluters is simply the cost of the treatment required of him. Each polluter would therefore like to have a low treatment requirement for himself, but sufficiently high ones for the others to permit the achievement of the quality standards for the stream classification that has been adopted. The standards are expressed entirely in terms of dissolved-oxygen levels.

## *The Pierce-Hall Cannery*

The Pierce-Hall Cannery, ten miles upstream from Bowville, has already incorporated some internal process changes in order to reduce effluent volume and cut back on biological pollutants discharged into the

4. Quality classifications will be explained more fully below.

Bow River. They have also installed primary treatment facilities in the form of screening and sedimentation equipment. Even so, when the plant is in full operation it discharges a waste stream of approximately 30 million gallons per day (mgd), which carries an ultimate biochemical oxygen demand of 47,600 pounds per day.[5]

In order to reduce these waste flows further, Pierce-Hall would have to install additional treatment facilities. Preliminary estimates of the cost of different degrees of improvement in effluent quality are shown in Table 1. The firm's primary treatment plant removes 30 percent of the BOD in the Pierce-Hall effluent. To accomplish higher levels of removal, the waste stream would be passed through a tank where biological degradation—which in the absence of treatment would occur in the stream itself—can take place under controlled conditions. The degree of purification can be varied over a wide range by proper design of the plant. High degrees of BOD removal are naturally more expensive than low.

From the wide range of possible choices, the engineering consultants to Pierce-Hall have provided three alternative treatment-plant designs. Their estimates are summarized in Table 1. A "low-efficiency secondary" plant would bring the total removal up to 80 percent of the original load. A "high-efficiency secondary" design would remove 90 percent but would cost nearly four times as much. If still further treatment

TABLE 1. Cost of Additional Waste Treatment by Pierce-Hall Cannery

| Type of treatment | (1) Percent[a] of BOD removed | (2) Additional annual cost (*$/yr*) | (3) Additional net cost to Pierce-Hall (*$/yr*) | (4) Additional cost to Pierce-Hall per % removed (*$/yr/%*) | (5) Profit after taxes (*$/yr*) |
|---|---|---|---|---|---|
| Primary (now in place) | 30 | 0 | 0 | 0 | 375,000 |
| Primary *plus* low-efficiency secondary | 80 | 13,000 | 8,000 | 160 | 367,000 |
| Primary *plus* high-efficiency secondary | 90 | 59,000 | 35,000 | 2,700 | 340,000 |
| Primary *plus* high-efficiency secondary *plus* tertiary | 95 | 159,000 | 95,000 | 12,000 | 280,000 |

[a] The figures shown are percentages of the *gross* waste load.

5. Good engineering practice distinguishes between first-stage (or carbonaceous) and second-stage (or nitrogenous) BOD. For our purposes however, nothing is lost by omitting the additional technical discussion which this more complex specification would require.

is necessary, "tertiary treatment" may be used to reduce the waste contained in the effluent from the secondary unit, by such diverse and expensive methods as holding in stabilization ponds and adding special chemicals. Tertiary treatment costs two and a half times as much as secondary processes.

The plants shown in Table 1 are only three of an infinite number of alternative designs. The curvilinear cost function in Figure 3 indicates the actual range of alternatives, with additional treatment cost stated as a function of the percentage removal of BOD from the gross waste load. Because the cannery already has primary-treatment facilities, the zero point on the cost curve is at the 30-percent removal level. Since the data are so limited, we interpolate linearly between the data points provided, to estimate the cost of levels of removal other than 80, 90, and 95 percent. The resulting piecewise linear cost function is shown as a broken line in the figure. For example, Figure 3 shows that a plant that brings total BOD removal up to 60 percent of the gross waste load would cost $7,800 per year.[6]

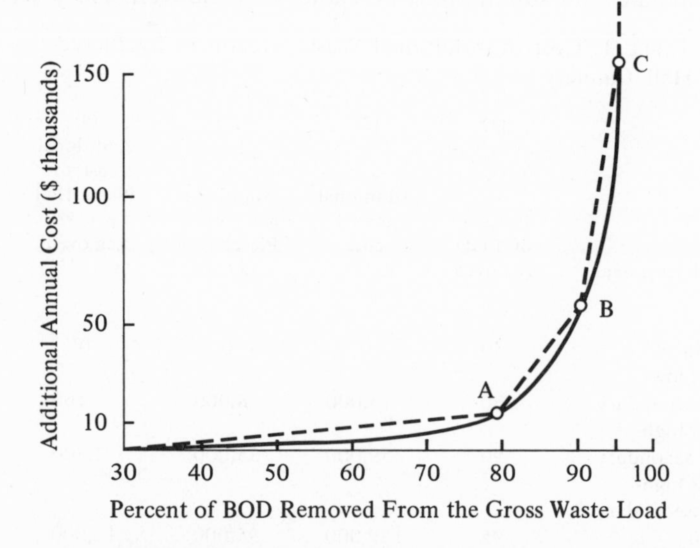


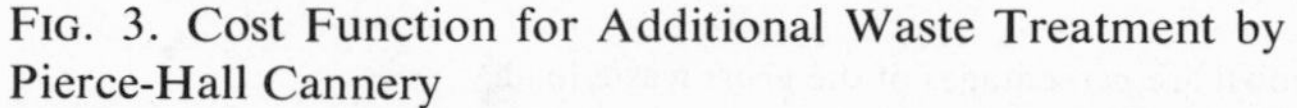
FIG. 3. Cost Function for Additional Waste Treatment by Pierce-Hall Cannery

Because of corporation income taxes, the net cost to the firm would only be $6/10$ of the total cost, and this latter figure is shown in column (3) of Table 1. Column (4) presents the same information, expressed in terms of the net cost of each additional percent of waste removed

6. Or from the table: $\frac{60-30}{80-30}(13{,}000) = \$7{,}800$.

from the plant's effluent. Thus, the figures in column (4) are the slopes of the piecewise linear cost function viewed by Pierce-Hall management.

The firm's net operating revenues, after income taxes, are 1.5 percent of gross sales when primary treatment only is employed, or approximately $375,000 a year, which is equivalent to 7.5 percent of the stockholders' equity. The firm is not a price leader and does not anticipate that it will be able to raise its prices appreciably, even if a large increase in treatment costs is imposed. Neither does it know of any changes in its methods of processing that would enable it to reduce its waste load at the current scale of operations. Therefore, any increase in treatment costs would have to come out of net profits. The estimated impact of different levels of treatment on net profits is also shown in Table 1. Notice that the effect is appreciable; the highest level of treatment would reduce annual profits by over 25 percent.

On the other hand, the management of Pierce-Hall is not adamantly opposed to making some contribution to improving the quality of the Bow River, since they recognize some responsibility to users and inhabitants farther downstream. But they are also concerned to maintain a return on equity capital of at least 6 percent.

## *Bowville*

The city of Bowville is the second major source of pollution in the valley. The city receives better-quality water than its downstream neighbors. According to the data in Figure 2, Bowville escapes the real pollution problems even during severe summer droughts. But the 250,000 inhabitants plus assorted light industries dump a heavy load of wastes into the river. Even after primary treatment, Bowville discharges 123,000 pounds of ultimate BOD on an average summer day. The total volume of effluent discharged is about 51 mgd. This load, added to the cannery wastes, renders the river unsuitable for recreational use farther downstream.

The public works department has prepared estimates of what it would cost the city to install additional treatment facilities. Like the cannery, cost estimates were made for three different treatment plant designs, as shown in Table 2.

Bowville would not have to bear the total cost of additional treatment. Under the provisions of the Federal Water Pollution Control Act, it could count on a federal grant of 50 percent of the construction cost of these facilities, and since capital cost is about half the total cost of waste treatment in this case, the citizens and local industries would have to bear only about 75 percent of the total outlay. The adjusted costs are shown in column (3) of Table 2. All are based on the assumption of a twenty-year life of facilities and a 5-percent interest rate. As was true

TABLE 2. Cost of Additional Waste Treatment by Bowville

| Type of treatment | (1) Percent[a] of BOD removed | (2) Additional annual cost ($/yr) | (3) Additional cost to city ($/yr) | (4) Additional cost to city per % removed ($/yr/%) | (5) Addition to Property Tax Rate ($/thousand) |
|---|---|---|---|---|---|
| Primary (now in place) | 30 | 0 | 0 | 0 | 0 |
| Primary *plus* low-efficiency secondary | 80 | 650,000 | 490,000 | 9,800 | 1.17 |
| Primary *plus* high-efficiency secondary | 90 | 880,000 | 660,000 | 17,000 | 1.58 |
| Primary *plus* high-efficiency secondary *plus* tertiary | 95 | 2,520,000 | 1,890,000 | 246,000 | 4.52 |

[a]The figures shown are percentages of the *gross* waste load.

for the cannery, both the total costs of different degrees of waste removal and the net costs to the city can be approximated by segmented linear cost curves of the type shown in Figure 3. Table 2 shows also the estimated effect of the different levels of treatment on the property-tax rate, which is now $63.50 per thousand assessed valuation.[7]

Bowville would gain only moderately from improvement in the quality of the Bow River. Its waterfront park is already a fine facility, although under extreme drought conditions the beach must be closed to swimming. If the river were cleaned up, these incidents could be avoided. In addition, the entire valley would be made more attractive to tourists and vacationers and, if the improvements were sufficient to permit the development of water-based recreation at Robin State Park, the park would become far more useful to inhabitants of the city. This latter consideration is important, since Bowville Waterfront Park is already overcrowded.

### *Plympton*

The difference between Plympton and Bowville is the difference between upstream and downstream riparian residents everywhere. Bowville has good water in all but the worst drought years; Plympton

7. There are many ways for a city to finance waste treatment facilities other than by the property tax as we assume here. Alternative financing methods differ not only in the impact of a particular facility on the city budget and in the distribution of costs among households and business firms; they also affect the total amount of waste to be handled. For example, certain types of sewer fees offer an incentive for industries and commercial establishments to cut back on total waste by means of internal process changes, while financing by the property tax offers no such incentive.

experiences poor river quality conditions almost every summer. Bowville's wastes degrade the river for most of the distance that concerns us, including the waterfront at Plympton; Plympton's wastes discommode no one, affecting only the quality checkpoint at the southern outlet of the valley. Bowville can improve the quality of the river by subjecting its wastes to a higher level of treatment; Plympton is virtually helpless—it cannot even protect its own waterfront. The stage is set for the classic conflict between upstream and downstream users.

Plympton is slightly smaller than Bowville and generally less affluent. It also has a primary-treatment plant. Its effluent volume is 43 mgd, containing after primary treatment 92,400 pounds of ultimate BOD.

Although Plympton has, in fact, little effect on the quality of the river in the region, it must expect to bear its share of responsibility for cleaning it up. Indeed, since its inhabitants will be the major direct beneficiaries of improved quality, it is eager to contribute what it can and to put pressure on the more consequential users upstream. The cost data for treatment at Plympton are shown in Table 3, and these also can be expressed as a cost function of the type shown in Figure 3. The dollars-and-cents cost of each level of treatment are lower for Plympton than for Bowville, because it is a smaller city, but the effect on the tax rate is greater, since Plympton is a poorer city and the value of taxable property per capita is lower there.

Nevertheless, Plympton is more eager to participate in a program of river-quality improvement than Bowville is. They would like to develop recreational facilities on their own waterfront, which does not now conform to sanitary standards and is occasionally beset by riverine odors (i.e., it stinks). They are more dependent than Bowville on tourism and are closer to the park.

TABLE 3. Cost of Additional Waste Treatment by Plympton

| Type of treatment | (1) Percent[a] of BOD removed | (2) Additional annual cost (*$/yr*) | (3) Additional cost to city (*$/yr*) | (4) Additional cost to city per % removed (*$/yr/%*) | (5) Addition to property tax rate (*$/thousand*) |
|---|---|---|---|---|---|
| Primary (now in place) | 30 | 0 | 0 | 0 | 0 |
| Primary *plus* low-efficiency secondary | 80 | 550,000 | 410,000 | 8,200 | 1.37 |
| Primary *plus* high-efficiency secondary | 90 | 730,000 | 550,000 | 14,000 | 1.83 |
| Primary *plus* high-efficiency secondary *plus* tertiary | 95 | 2,110,000 | 1,580,000 | 206,000 | 5.27 |

[a] The figures shown are percentages of the *gross* waste load.

### *Other Interested Parties*

The cannery and the two cities are the principal defilers of the river and represent the interests of most of the people who are directly concerned. But there are other interests that have to be taken into account and are likely to have considerable influence on any decisions about the river.

First, there is the federal government, which is interested in protecting and improving the quality of all interstate waters. This national interest is implemented by the Federal Water Pollution Control Administration. The FWPCA administers the grant and incentive programs established under federal water-quality legislation, and can be expected to contribute to the costs of increased levels of waste treatment undertaken by the two municipalities. It also has some enforcement powers.

The FWPCA's interest in the decision taken by the commission is twofold. First, it is responsible for protecting the interests of all users farther downstream, so that it feels compelled, and is empowered, to insist on a reasonable quality of water at the southern outlet from the valley. Specifically, the FWPCA has required that the river should meet Class C standards (as described below) at the state line. The FWPCA is, in fact, the only participant with a direct concern for the quality of the water so far south. Second, the FWPCA has a generalized interest in the quality of American streams, and thus will share the desires of the inhabitants of Bowville and Plympton for the quality of their waters and for the waterfront potential of the state park. It will, however, be less worried about the effect of increased treatment costs on local tax rates and more concerned with the national economic costs that different plans will imply. On balance, the FWPCA will attempt to induce the commission to undertake a combination of stream classification and treatment requirements that will yield the maximum benefit-cost ratio from a national point of view.

The state government is also concerned, particularly through the State Water and Sanitation Commission, and the State Department of Parks and Recreation. The State Industrial Commission may take a hand, however, if the financial health of the cannery should be jeopardized by any proposal. On balance, the influence of the state agencies can be expected to be similar to that of the FWPCA, except that the state will attach more importance to the quality of the water at Robin State Park and less to the quality at the state line.

There are, in addition, a variety of conservation and special-interest groups which make their influence felt through public communications media, hearings or investigative procedures of the Water Pollution Control Commission, participation in municipal politics, and even through direct representation on the commission.

All the groups mentioned in this subsection share a more keenly felt concern for the quality and usability of the river than for the cost of achieving high quality. For the purposes of our discussion, we shall lump them together and consider them to be represented adequately by the FWPCA.

## *Water Quality Standards*

The first task before the Bow Valley Water-Pollution Control Commission is to adopt a stream-use classification for the river between Bowville and the southern outlet at the state line. The State Water and Sanitation Commission has promulgated a set of standards that prescribe the quality of water to be used for different purposes. Once the commission has adopted a use classification, all its subsequent regulations must conform to it.

In actual practice, stream standards cover many characteristics—dissolved oxygen, floating solids, color and turbidity, coliform bacteria, taste and odor, temperature, pH, radioactivity, and others. For simplicity, however, we pretend that the state standard specifies only one dimension of stream quality, namely, instream dissolved oxygen.[8]

The state water standards are accordingly assumed to divide streams into five use classes: A, B, C, D, and U. Class A waters are very nearly in their pristine state—almost unaffected by man. Waters classified U are essentially uninhabitable for fish life, unsuitable for most recreational activities, and offensive to the sight, taste, and smell. The state standards [9] are specified in Table 4.

These standards must be met during the seven-day period with the lowest average daily flow that can be expected once in ten years: for the Bow River, this is 800 cfs. Under these low-flow conditions, the entire river below Bowville currently is below quality for Class D waters: indeed, as shown in Figure 2, in many parts of the valley the stream becomes anaerobic and is occasionally rather unpleasant to be near.

The imposition of defined standards of quality, which is almost inevitable in framing administrative regulations,[10] transforms the decision problem in a fundamental way. In the absence of codified classifications, a DO concentration of 5.1 mg / 1 will be recognized as only imperceptibly safer and more pleasant for swimming than one of 4.9 mg / 1. But once the higher concentration qualifies the river for a higher

8. Other pollutants that would be especially important in the sort of situation we are discussing include coliform bacteria (from municipal waste) and nutrients. We assume that the former is taken care of by chlorination (which is relatively cheap). Nutrients will not be discussed explicitly but deserve mention here because they are a source of increasing concern to water-quality managers.

9. Adapted from the Massachusetts standards.

10. Examples range from the definition of Grade A fresh eggs to the occupancy and safety regulations in building and zoning codes.

TABLE 4. State Water Quality Standards

| Use class | Minimum oxygen concentration * (mg/l) | Description |
|---|---|---|
| A Public water supplies | 7 | Character uniformly excellent. |
| B Bathing and contact-recreational use | 5 | Suitable for water-contact sports. Acceptable for public water supply with appropriate treatment. Suitable for agriculture, and certain industrial cooling and process uses; excellent fish and wildlife habitat; excellent aesthetic value. |
| C Boating and non-contact recreation | 3.5 | Habitat for wildlife and common food and game fishes indigenous to the region; certain industrial cooling and process uses; under some conditions acceptable for public water supply with appropriate treatment. Suitable for irrigation of crops used for consumption after cooking. Good aesthetic value. |
| D Minimum acceptable | 2 | Not objectionable; suitable for power, navigation, and certain industrial cooling and process uses. |
| U Unacceptable | <2 | Below Class D standards. Likely to be offensive. |

* Standards to be met under the minimum seven-day consecutive flow to be expected once in ten years.

use-classification, there is all the difference in the world between them—one permits the river to be developed legally for water-contact recreation; the other does not. A fundamental discontinuity is thereby introduced into decisions that impinge on water quality. This will have important consequences for our analysis.

### *Waste Discharge and Stream Quality*

Improvement in dissolved oxygen concentrations in the river can be obtained in a number of ways—e.g., artificial aerators or flow augmentation—but this example assumes the removal of oxygen demanding material at the waste source to be the only management method available.

In our model, we shall use the simplest formulation of the relationship between waste loads and stream quality, based on the work of Streeter and Phelps.[11] According to the Streeter-Phelps model of stream quality, the effect of discharging BOD into a stream at any point is to reduce the dissolved oxygen in the water at all points downstream by amounts that are directly proportional to the amount of BOD inserted and that depend in a complicated way on the distance from the point of waste discharge to the downstream points and on the hydrology of the stream. The factor of proportionality between waste discharge and quality response will be denoted by $d$. The value of $d$ depends on where the waste is discharged and where the water quality is measured. These data will be indicated by subscripts: $i$ for the point of BOD discharge and $j$ for the quality control point. Specifically, $d_{ij}$ will denote the decrease in dissolved oxygen at point $j$ (in milligrams per liter) caused by the increase of one lb/day in the amount of BOD discharged into point $i$. The full set of $d_{ij}$ values for the Bow River is given in Table 5. Furthermore, the effects of different waste sources on downstream quality are additive. For example, if both the cannery and Bowville reduce their waste discharges, then the impact of the two reductions on water

TABLE 5. Dissolved Oxygen Transfer Coefficients, $d_{ij}$ (Increase in DO at $j$ (mg l) resulting from decrease of 1000 lbs/day in BOD discharge at $i$)

| Discharge point ($i$) | Check point ($j$) (1) *Cannery* | (2) *Bowville* | (3) *Robin Park* | (4) *Plympton* | (5) *State Line* |
|---|---|---|---|---|---|
| 1 Cannery | 0.0 | 0.0461 | 0.0502 | 0.0333 | 0.0159 |
| 2 Bowville | — | 0.0 | 0.0585 | 0.0414 | 0.0206 |
| 4 Plympton | — | — | — | 0.0 | 0.0641 |
| Miles from Gordon Bridge | 10 | 20 | 60 | 80 | 110 |

11. Only the most casual treatment of the technical aspects of water quality is offered here. A brief introduction to the topic is provided also by Kneese and Bower [8, pp. 13–29]; for details of wastewater engineering see Fair, Geyer, and Okun [6]. References in brackets may be found at the end of the chapter.

quality at the park can be calculated by simply adding the effects of the two individual reductions, using appropriate values of $d_{ij}$.

## Model of the Bow Valley Water-Pollution Control Commission

The tedious recital of data in the last section will be recognized as a small-scale replica of the docket of any proceeding concerned with the use and control of public waters. Data and considerations of the sort that we have described are amassed in various forms, and the task of the commission is to digest, assimilate, and ultimately evaluate them. We are now concerned with what can be said about the upshot of this task.

One consideration that we have not yet mentioned is precisely what the commission is trying to achieve. In the preamble to their charter they are instructed

> so to regulate and control the use of the Bow River between the Gordon Bridge and the state line, and the discharge of liquid and solid matter of any form whatsoever thereinto, as to assure the highest practicable quality of water between and including the aforementioned points and to conform to all applicable state and Federal laws and regulations.

These vague and high-minded words, as is the nature of enabling legislation, need a great deal of interpretation. For example, what is meant by "the highest practicable quality of water?" The soundest interpretation, and the one that we shall adopt, seems to be that the commission is directed to enforce the highest quality of water that can be obtained without imposing undue burdens on anyone. But this is still too vague really to control specific decisions.

First, what is the "highest quality of water?" Water quality is a multidimensional concept. In a genuine instance, where many characteristics of water are taken into account, there are likely to be measures that will increase the dissolved oxygen concentration at the expense of increasing the amount of phosphates or other plant nutrients in the water. Would such a measure increase or reduce the quality of the water? The Bow River commission is spared this source of perplexity, since we have assumed that it pays attention only to dissolved oxygen, but sufficient ambiguity remains. For different improvement measures will have differential effects on the concentration of dissolved oxygen at different points in the river; and it is difficult to say which of two measures—one that makes a greater improvement at Plympton and one that makes its major improvement at the state park—improves the quality of the river more. Such questions must be decided somehow by the commission, and no single answer is likely to command the wholehearted assent of all interested parties.

Second, what are "undue burdens?" Clearly, if a decision requires Pierce-Hall to shut down or to operate at an abnormally low rate of profit, it implies an undue burden. Or if it requires Bowville to raise its tax to a level that is markedly out of line with taxes in Plympton or other competing towns, an undue burden has been imposed. Whether a burden is undue is a matter of judgment, and inherently vague. But the restriction is nevertheless a genuine limitation on the powers of the commission.

In practice the commission's range of discretion is circumscribed even more narrowly. If it is to be effective, it must secure the willing cooperation of the cities and industries under its jurisdiction; it cannot govern by blunt coercion. This means that it must issue regulations that all concerned will regard as reasonable and fair, and will submit to without recourse to law, higher political authority, or other forms of resistance. In our specific context, this means that the regulations must not endanger the current administrations of Bowville and Plympton if they agree to it; nor can any decision undermine the profitability of the cannery.

We shall refer to all these considerations as "political constraints." In the early stages it will not be clear whether these constraints impose loose or stringent restrictions on the commission's scope for decision; all interested parties will endeavor to make them appear very stringent. At any rate, the first order of business before the commission is to find some policy that attains a minimum acceptable improvement in the river while respecting all these requirements.

It is highly unlikely that there will be only one such policy. Then a choice must be made, and, as we have noted, the charter gives only a vague indication of how it is to be made. Within the limits set by the political constraints, therefore, the decision will depend largely on the judgments of the individual commissioners about where the interests of their constituents lie. These judgments are not likely to coincide. Therefore, the second order of business before the commission is one of pulling, hauling, and compromising to ascertain the decision that best reconciles the interests of all concerned.

Finally, of course, the decision must be technically feasible. If the commission decides to classify the river for Class C use, then it must require all polluters to take whatever measures are needed to bring the river up to this standard.

These, in general terms, are the outlines of the decision problem faced by the commission. This problem must now be formulated more exactly for purposes of further analysis.

## *Decision Variables, Costs, and Technological Constraints*

The first step in formulating the commission's problem precisely is to express the decisions open to it in numerical form. The first decision is the use classification of the river. It will be recalled from the description of the water-quality standards that the effect of adopting any use class is to prescribe the minimum permissible amount of dissolved oxygen in the river. For example, if Class C use is adopted, there must be at least 3.5 mg / l of dissolved oxygen in the river at all points. Let us then denote by $Q$ the minimum permissible concentration of dissolved oxygen. Then the commission must, in effect, choose a value of $Q$, either 7, 5, 3.5, or 2, depending on the use class selected. In order to avoid having to consider the effects of the commission's decisions on users farther downstream, we impose that the dissolved-oxygen concentration at the state line shall be 3.5 mg / l. This specification is regarded as outside the purview of the commission and will not be considered further.

The only other decisions facing the commission concern the level of treatment to be required of each polluter. These, too, can be expressed numerically. Each polluter has a range of possible treatment-plant designs. As illustrated in Figure 3, the more nearly complete the waste removal, the higher the cost. The commission has to prescribe the degree of removal which each discharger must attain, or in more precise language, the percentage of BOD which must be removed from each of the polluters' waste outflows. These decisions can be expressed numerically as follows. First, assign identifying numbers to the three polluters: 1 will designate the Pierce-Hall Cannery, 2 Bowville, and 4 Plympton (3 denotes the state park, and no wastes are generated there). Then, introduce a set of variables, called $x_i$, where $x_i$ denotes the percentage of BOD removed by polluter $i$ from his effluent. The commission must decide on $x_1$, $x_2$, and $x_4$.

The decision on the $x_i$ simultaneously determines the treatment cost that each polluter has to bear. The relationship is shown clearly in Figure 3. If $x_1 = 85$, the cannery will have to build the plant that lies half way between points $A$ and $B$ in that figure, which will cost it $21,500 a year—i.e., ½ (8,000 + 35,000).

An important property of all three of the cost curves is that the cost of removing an additional unit of BOD increases as the percentage already removed increases.

The commission's decision problem has now been reduced to the selection of four numbers: $Q$, which determines the use class of the river, and the three $x_i$ ($i = 1, 2, 4$), which specify the levels of treatment required of all polluters. The relationship between the $x_i$ and the concentration of dissolved oxygen at points downstream from polluter $i$

was discussed above. That analysis led to a set of coefficients, $d_{ij}$, that give the effect of a unit reduction in the oxygen demanding discharge of polluter $i$ on the dissolved-oxygen content of the water at any downstream point $j$. The $x_i$ have to be chosen so that the dissolved oxygen at all points is at least as great as $Q$, except that at the state line it must be at least 3.5 mg/l irrespective of $Q$. This requirement is expressed by constraint equations, one for each of the quality checkpoints in the river, designated by $j=2$ for Bowville, $j=3$ for Robin State Park, $j=4$ for Plympton, and $j=5$ for the state line. A typical equation of this group can be represented in the form: [12]

$$\sum_i d_{ij}L_i\,(x_i/100 - 0.3) \geqslant Q - \overline{q}_j.$$

In this equation $L_i$ is the gross BOD load (i.e., before primary treatment) generated by polluter $i$; $d_{ij}$ is the transport coefficient expressing the effect of his load on point $j$ downstream, and $x_i$ is his percentage removal of BOD from his effluent. The summation on the left-hand side expresses the increase in dissolved-oxygen concentration at point $j$ resulting from the waste treatment in excess of primary employed by all polluters.

On the right-hand side, $\overline{q}_j$ is the dissolved-oxygen concentration at point $j$ when only primary treatment is used by all polluters. These are the numbers plotted in Figure 2. The right-hand side, then, is the increase in dissolved oxygen at point $j$ required by the use classification corresponding to $Q$, and the entire inequality asserts that the $x_i$ must be chosen large enough to achieve the required improvement at point $j$. There is, as we said, one such equation for each quality checkpoint. These equations typify the technological constraints confronted by the commission.

## *Political Constraints*

Once the $x_i$ have been chosen, the increases in treatment costs imposed on the polluters follow, as we have seen. The formulas expressing the relationship of removal level to cost (as sensed by the individual polluters) are part of the model. In the sequel we shall frequently denote the cost imposed on polluter $i$ by $g_i$.

The costs imposed on the FWPCA are somewhat special. In this model the FWPCA plays the role of the custodian of the overall national interest. Therefore, the cost of any pollution-abatement plan, as perceived by the FWPCA, is the full economic cost of the additional treatment, without any allowance for the effects of tax advantages and

12. The constant 0.3 is subtracted from each $x_i$ since this is the BOD reduction achieved under the basal condition of primary treatment by all polluters.

federal grants-in-aid. In other words, the FWPCA is taken to react to economic-resource costs rather than to budgetary costs.

There are limits on the treatment costs that the commission can realistically impose on the polluters. As was mentioned above, the commission cannot impair Pierce-Hall's earning abilities excessively. This consideration can be expressed in the mathematical model by including a condition such as $g_1 \leq \$50{,}000$, which would prohibit any decision that cost Pierce-Hall more than $50,000 a year.

A different kind of political constraint is illustrated by the role of the FWPCA. The FWPCA, together with associated state agencies, is charged with maintaining the quality of the nation's waterways. It is empowered to enforce minimum standards and, in the case of the Bow River, will not permit it to continue in its present Class U condition. This means that the commission's field of choice of use class is confined to Class D or better, or, in numerical terms to $Q \geq 2$ mg/1.

In this section and the preceding one we have introduced the political and technological constraints that the commission's decision must satisfy. Within the limits set by those constraints, the commission will search for the best decision available to it. We now turn to the formulation of that search.

## *Pareto Admissibility*

We now envisage the commission in the process of choosing among decisions that meet the technological and political requirements discussed above. This is largely a process of compromise: each commissioner balances in his mind the advantages and drawbacks to his constituents of the decisions that seem within the range of possibility, and argues for the one that he deems most favorable. When the various positions have been aired, he finally agrees to a decision if he feels that it will be acceptable to his constituents and that no decision more favorable to them is obtainable.

Even this very simple and obvious characterization of the deliberations contains important information, for it entails that no decision will be adopted if some alternative decision is just as satisfactory to all interested parties and more satisfactory to some. Following the terminology of welfare economics, we call a decision *Pareto admissible* if it is not ruled out by the foregoing criterion—that is, if there exists no feasible alternative that some interested parties regard as superior and none regard as inferior. The commission's decision will almost surely be Pareto admissible, for if an inadmissible decision were proposed, someone would point out an alternative that he preferred and that no one would object to. We must therefore consider how the commissioners' preferences among various possible decisions are determined.

Every decision confers benefits on the participants by prescribing the quality of the water in the river, and imposes costs by requiring certain levels of waste treatment. Each commissioner bases his attitude toward a decision on a mental comparison of the benefits and costs the decision involves for the participants with whom he is most concerned, his constituents. The costs, of course, are a matter of dollars and cents. The benefits are not: they accrue largely in the form of amenities and facilities, and the eradication of distasteful conditions. To render the two terms of this balance comparable, we (and the commissioners, for that matter) have to assign monetary magnitudes to the benefits. The natural and most useful measure of benefits is "willingness to pay": that is to say, the value of the benefits of any decision to any participant is the greatest amount that he would be willing to pay to obtain them. We shall call the greatest amount that any participant, say participant *i*, would be willing to pay to obtain any decision his gross benefit from that decision, to be abbreviated $GB_i$.

This conversion of benefits to monetary magnitudes may seem a little strained. Yet it is done, and has to be done—usually very informally—all the time in the course of arriving at governmental decisions. For example, it would not be at all exceptional to have an exchange like the following at a public hearing on pollution abatement. The Bowville Commissioner of Water and Sanitation is on the stand:

> *Q*. How much would it cost you to reduce your daily waste load by an additional two percent?
>
> *A*. About $27,000 a year, I think.
>
> *Q*. If you did that, we could assure swimmable water at Robin State Park. How would you folks feel about that?
>
> *A*. I can't answer for the City Council, but I would certainly support that kind of proposal myself.

The commissioner from Bowville has made the monetary comparison and has concluded that swimming facilities at the state park are worth at least $27,000 a year to his city. Further questioning could elicit a more precise estimate.

These gross-benefit estimates are part of the basic data on which the commissioner's judgments and the commission's decisions rest. We shall have much more to say about them below. The costs of a decision to participant *i* have already been denoted by $g_i$. The difference between gross benefits and costs, $GB_i - g_i$, will be called net benefits, or $NB_i$ for short. Net benefits of any decision may be positive, zero, or negative, depending on how the actual costs imposed by the decision compare with the amount that the participant is willing to pay for the corresponding benefit. The net benefits are the crucial quantities that determine how the commissioners, individually and collectively, regard various possible decisions.

These concepts enable us to express some of the considerations we have encountered more formally and quantitatively. First, we have noted that political considerations place a limit on how disadvantageous a decision can be from the point of view of any participant. The net advantages of a decision to participant $i$ are measured by the corresponding value of $NB_i$. Thus we can express some of the political constraints by requiring that a decision satisfy

$$NB_i \geqslant b_i$$

for all participants $i$, where $b_i$ (presumably negative) is an estimate of the smallest net benefit that can be imposed upon $i$ without violating the guarantees of due process, or endangering the careers of $i$'s representatives, or otherwise provoking vigorous refusal to cooperate.

Second, we can express Pareto admissibility in quantitative terms. Call any particular decision $X$ and denote by $NB_i(X)$ the net benefits of that decision to participant $i$. $X$ is then a vector with four components, i.e., $X = [Q, x_1, x_2, x_4]$, specifying the use class of the river and the treatment levels required of the polluters. In this notation, decision $X$ is Pareto admissible if there does not exist any alternative permissible decision, say $Y$, for which $NB_i(Y) \geqslant NB_i(X)$ for all participants $i$ with strict inequality holding for some.

We have now formulated the main part of the commission's decision problem mathematically. Suppose, for the moment, that we had estimates of the $NB_i(X)$ for all decisions $X$ within the realm of consideration. It is a purely mathematical task to determine whether any particular decision satisfies the technological constraints typified by the equation on p. 223. It is also, now, simply a matter of mathematics to ascertain whether the decision satisfies the political constraints. And finally, as we shall see below, straightforward mathematics can be used to determine whether a decision is Pareto admissible. In short, given the net-benefit data and the technological properties of the system, we can compute mathematically the range of decisions that are likely candidates for adoption.

Three tasks remain: (1) to discuss the estimation of benefits, (2) to indicate how to compute Pareto-admissible decisions from these estimates, and (3) to consider how to evaluate the relative likelihoods of adoption of the different Pareto-admissible decisions.

## *Valuation of Benefits*

One way to estimate the values of the benefits anticipated from different decisions is simply to ask representatives of interested parties. This is impracticable: there are too many questions to ask too many people, and not enough reason to expect thoughtful and candid answers. A better way is to infer from past behavior and other evidence how much the participants have shown themselves to be willing to pay for the advantages that would be offered by the decisions.

Every decision is a package deal, resulting in a bundle of consequences to each participant. To estimate the value of the package we separate it into its components, estimate the value of each component, and total up these values. The conditions under which this procedure is valid will be discussed below.

The task of estimating the value of a specific improvement to a restricted group of citizens is arduous but not novel. Several methods are well established. One method, especially useful for recreational facilities, is the "user-days" approach. Following this method, to find the value to Bowville residents of an improvement at Robin State Park, one begins by estimating the additional use of the park that Bowvillers would make if the improvement were installed. Then, one estimates the value of each day's use of the park by a Bowville resident, and multiplies the two estimates.

Another method is the "alternative-cost" procedure. This is applicable when the improvement meets a need that will be satisfied by other means if the improvement is not undertaken. In such a case the value of the improvement is the saving it affords by rendering the alternative expenditure unnecessary.

Estimates of the value of improved water quality from the viewpoints of Bowville and Plympton, made by these and other methods, are recorded in Table 6. Each city places a value on improvements at its own waterfront and also on improvements at the park. In addition, the value of an improvement, wherever it occurs, depends on whether the water has already attained a quality of 5 mg / l of dissolved oxygen, which meets the state standard for water-contact use.

The table also contains estimates of the value of unit improvements from the point of view of the FWPCA, the custodian of the public interest. Its valuation of the improvement of water quality at Bowville and Plympton is simply a reflection of the values placed by the inhabitants on the improvement of their own water. The FWPCA's valuation of improvements at the park is the sum of the valuations of Bowville and Plympton, with 50 percent added to allow for the social value of use of the park by local residents who do not live in the two cities, and by outsiders.

Table 6 enables us to calculate the benefits of any decision from the point of view of each participant, provided that it is valid simply to add up the benefits of the different components of the improvement package. For example, if decision *X* increases the DO concentration at Bowville (where it is now 4 mg / l) by 2 mg / l, and at the park (where it is now virtually zero during summer droughts) by 3 mg / l, then its gross benefits to Bowville are:

$$GB_2(X) = 100{,}000 + 50{,}000 + 3 \times 33{,}000 = \$249{,}000.$$

If, in addition, the decision imposes on Bowville treatment costs of $390,000 a year, the net benefits are:

$$NB_2(X) = 249{,}000 - 390{,}000 = -\$141{,}000.$$

The calculation captures the major measurable benefits of improvement, but there are others. Most importantly, lifting the Bow River out of its current obnoxious Class U condition confers a gain on each participant that cannot be evaluated quantitatively. Fortunately, we do not have to assign a numerical magnitude to these benefits, since they will be enjoyed under all the decisions within the range of possibility, and do not form any basis for choosing among decisions.

TABLE 6. Value of Unit Improvements in Water Quality (*$/yr per mg/l*) *

| Place improved | Current water quality (mg/l) | Value perceived by *Bowville* | *Plympton* | *FWPCA* |
|---|---|---|---|---|
| Bowville waterfront | <5 | 100,000 | | 100,000 |
| | ≥5 | 50,000 | | 50,000 |
| Plympton waterfront | <5 | | 75,000 | 75,000 |
| | ≥5 | | 25,000 | 25,000 |
| Park | <5 | 33,000 | 30,000 | 94,500 |
| | ≥5 | 17,000 | 10,000 | 40,500 |

* Values valid only for qualities meeting Class D standards or higher.

Are we justified in breaking up the improvement package into its components, attaching a value to each component, adding up, and regarding the total as the value of the entire package? This is an old and classic question in economics. The answer is yes, provided that (1) the components are neither substitutes for nor complementary goods of each other, and (2) the total value of the package, computed in this way, is not a significant proportion of the budget of the participant. We should consider these provisos in order.

For simple summation of the values of improvements at different points to be valid, the benefits should not be interrelated. They would be interrelated if, for example, there were two state parks accessible to Plympton. For then, if a water-based recreational facility were created at one of them, the citizens of Plympton would be less willing to contribute to the cost of developing the other. In that case, the value of improvements to one of the parks would depend upon the extent of development of the other; those improvements would be substitutable goods as far as Plympton is concerned. Another possibility would be for the quality of fishing at the park to be affected by the quality of water both at Bowville and at the park. Then, if the water were improved at Bowville, it would be more valuable than otherwise to improve the water at the park, because the impact on fishing would be greater. This is an instance of complementarity: improvements of water quality at Bowville would enhance the value of each unit of improvement at the park. Substitutability and complementarity invalidate, in opposite ways, the simple addition of the values of improvement at different points along the river. Whether or not such interactions occur in any instance depends on the circumstances, and particularly on the use that would be made by the participants of the improved water at different points. In the case of the Bow River it appears reasonable to neglect such interactions.

Moreover, if the value of the improvement package is an appreciable proportion of the annual budget of a participant, simple addition of the values of components will not be valid. This is true because if a participant's budget has already been strained by paying for some environmental improvements, he will not be willing to pay as much for additional improvements as he otherwise would be.[13] This is a serious complication in principle. In our case, since expenditures on water-quality improvement are only a minor proportion of the total expenditures of any of the participants, we can, as a first approximation, ignore it. But we shall have to stay alert to the possibilities for error that this approximation may lead to.

On these grounds, we feel justified in adding up the values of the components of an improvement package, with due caution, in order to estimate the value of the entire package to any participant, as we have done.

## *Finding the Likely Decisions*

Matters now stand thus: we have expressed the commission's decision numerically (in fact, as a vector with four components) and have found the consequences of any decision to be fairly simple mathematical functions of the decision variables. Now we use these formulations to ascertain the Pareto-admissible decisions, for the commission's decision will surely be one of those.

Pareto-admissible decisions are found by creating and solving certain auxiliary problems. Suppose we choose any set of positive numbers, to be called $w_i$, one for each participant in the decision, and use them to form the weighted sum of net benefits,

13. Note the similarity to the difficulties caused by the income effect of expenditure in the conventional theory of consumer behavior.

$$\sum_i w_i NB_i(X).$$

Then if we set ourselves the problem of finding the decision $X$ that makes this sum as large as possible while conforming to all the technological and political constraints, we shall assuredly discover a Pareto-admissible decision. For if the solution to this problem, say Decision $X$, were not Pareto-admissible then there would be another decision, say $Y$, for which $NB_i(Y) \geqslant NB_i(X)$ for all participants $i$, with strict inequality for some $i$, so that

$$\sum_i w_i NB_i(Y) > \sum_i w_i NB_i(X)$$

contradicting the assumption that Decision $X$ solved the auxiliary problem. If the net benefit functions and the constraints are all piecewise-linear or linear, as in this example, this problem can be solved by linear programming. Otherwise more elaborate mathematics will be required.

Every time a set of weights $w_i$ is selected and the corresponding auxiliary problem is solved, a Pareto-admissible decision is discovered. A different set of weights is likely to lead to a different decision, more favorable to the participants whose relative weights have been increased. By trying out a number of different sets of weights, we can map the range of Pareto-admissible decisions. We can also see how much the relative weights have to be changed in order to move the solution from one decision to another.

We predict, of course, that the ultimate decision will be one or another of the solutions discovered by solving the auxiliary problems with judiciously chosen weights. Just which of these Pareto-admissible decisions will be chosen, we have no sure way of telling—even the members of the commission don't know that until their deliberations are finished—but we can narrow the range of uncertainty further by proper interpretation of the artificial weights. For they are not actually artificial or arbitrary; they are really measures of political weight or influence.

To see this, suppose that the commission had adopted Decision $X$, which is a solution to the auxiliary problem with weights $w_i$. Then the commission has revealed that it values benefits to participants 1 and 2 (for instance) in something like the ratio $w_1{:}w_2$. That is, the commission has shown itself to be willing to reduce the net benefits to participant 1 by \$1 if by so doing it can increase the benefits to participant 2 by $\$w_1 / w_2$ or more.

Some of the Pareto-admissible decisions will correspond to auxiliary problems with weights that diverge widely from any reasonable estimate of relative political influence. Those decisions can be ruled out; they are too favorable to participants with relatively little political influence, and not favorable enough to powerful participants. The remaining Pareto-admissible decisions, those that correspond to auxiliary problems with plausible political weights, are likely candidates for adoption. The theory provides no way for narrowing the prediction beyond this point.

## Prediction of Likely Decisions

The methods and devices thus far explained enable us to express quantitatively all the data, with one exception, that enter into a wide class of environmental decisions. The quantifiable data include the interests and preferences of all political groups concerned with the decision, along with the usual technological considerations. The one exception to this exercise in quantification was the "political weights," the measures of the relative influence of the various parties of interest. They are in principle unascertainable, and call for a different strategy.

Though we cannot know the relative weights, we can always explore the implications of any set of weights that might apply. For example, we can consider the implications of having the relative influences of the cannery, Bowville, Plympton, and the FWPCA stand in the ratios of 1: 6 : 2 : 1, or any other set of ratios. For any such set of weights, the decision problem is completely quantified. The task of finding the decision that satisfies everyone concerned as well as possible when, say, a dollar's worth of satisfaction to Bowville is considered to be six times as important as a dollar's worth to the cannery, etc., is so strictly routine that it can be turned over to an electronic computer.

Furthermore, such a computation is not an empty exercise, even though the weights are unknowable. Any sensible decision must correspond to some set of weights: we proved above that any decision of which that were not true could be altered so as to increase some party's satisfaction without detracting from anyone else's. So if we could solve the problem for all possible, or plausible, sets of weights, we should have a set of decisions that would include all those that might reasonably be expected to emerge. We could inspect this set of plausible decisions to detect any regularities, and thereby hope to gain insight into the points at issue and their likely resolution.

In essence, this is what we shall do. Formulating and solving the decision problem for all possible sets of weights is, of course, too vast a task, but we shall consider seventeen different sets of weights, covering a wide range of possibilities. The seventeen sets of weights are shown in Table 7.

The first set of weights used, designated (0,0,0,1), presumes that the FWPCA dominates the decision completely, i.e., controls 100 percent of the influence on the decision. Since, in this model, the FWPCA is presumed to be preoccupied with conventional national income benefits, or the maximization of the present value of net economic benefits, this

TABLE 7. Weight Allocations Used to Explore Possible Decisions

| Relative weight assigned to | | | |
|---|---|---|---|
| *Pierce-Hall* $w_1$ | *Bowville* $w_2$ | *Plympton* $w_4$ | *FWPCA* $w_5$ |
| 0 | 0 | 0 | 1 |
| 3 | 1 | 3 | 3 |
| 3 | 3 | 1 | 3 |
| 3 | 3 | 3 | 1 |
| 1 | 3 | 3 | 3 |
| 4 | 1 | 4 | 1 |
| 1 | 4 | 1 | 4 |
| 4 | 4 | 1 | 1 |
| 1 | 1 | 4 | 4 |
| 1 | 4 | 4 | 1 |
| 1 | 5 | 3 | 1 |
| 1 | 3 | 5 | 1 |
| 1 | 6 | 2 | 1 |
| 1 | 2 | 6 | 1 |
| 1 | 1 | 7 | 1 |
| 1 | 7 | 1 | 1 |
| 7 | 1 | 1 | 1 |

set of weights corresponds to deciding in accordance with the standard benefit-cost criterion. This set of weights, like each of the others, was applied three times: once with an assumed external requirement that the river attain use Class D, once where it was required to meet the standards for use Class C, and finally where the minimum use class permitted was Class B. The results of these computations, and similar computations for the other weighting schemes, are shown in Tables 9, 10, and 11. Since all fifty-one computations resulted in only fourteen different decisions, the decisions are listed and defined in Table 8 and will be referred to by number.

The weighting scheme (0,0,0,1) yielded decision No. 10 when the minimum use class permitted was either C or D. This accordingly is the "best" decision from the national point of view—or, in other words, when costs and benefits accruing to the cannery, Bowville, Plympton, and even outside taxpayers and users of the river are all accorded equal weight, as they are in the internal evaluations of the FWPCA. When the required use class was raised to B, however, decision No. 10 failed to meet the standard, and decision No. 11 had to be used, with a consequent reduction of $8,000 a year in net benefits. That is an estimate of the cost of the more stringent quality requirement.

Under both quality constraints for which it sufficed, decision No. 10 emerged as best under a wide variety of other sets of weights—indeed,

TABLE 8. Decisions Recommended by the Analysis

| Decision number | Quality class achieved | Treatment level in terms of percentage BOD removal by | | |
|---|---|---|---|---|
| | | *Cannery* | *Bowville* | *Plympton* |
| 1 | D | 95 | 50 | 67 |
| 2 | D | 90 | 52 | 67 |
| 3 | D | 80 | 55 | 67 |
| 4 | C | 95 | 65 | 61 |
| 5 | D | 90 | 66 | 61 |
| 6 | C | 90 | 67 | 61 |
| 7 | C | 80 | 70 | 60 |
| 8 | B | 95 | 80 | 54 |
| 9 | B | 94 | 80 | 54 |
| 10 | C | 90 | 80 | 55 |
| 11 | B | 90 | 81 | 54 |
| 12 | C | 80 | 80 | 56 |
| 13 | B | 80 | 85 | 54 |
| 14 | B | 90 | 90 | 51 |

under all weighting schemes except those in which the upstream users (the cannery and Bowville) or the downstream user (Plympton) was accorded a marked preponderance of the influence. Thus a balance of influence between the upstream and downstream users, as in schemes (1,4,4,1) and (1,3,5,1) led to a decision that coincided with the national interest as represented by the FWPCA. A preponderance of influence upstream (e.g., [1,7,1,1] or [1,6,2,1]) led to less costly decisions and lower use classifications. When Plympton was given a larger vote, the treatment levels and costs were higher. The three basic tables show all the details. Figure 4 compares the hydrologic consequences of two decisions: decision No. 5, a low-quality decision that emerges when the preponderance of influence is upstream, and decision No. 14, a high-quality decision that wins only when Plympton's interests are allowed to dominate.

A special computation labeled "L.C." for "least cost" was made to determine the cheapest plan for attaining use classes C and D. Notice, in the tables, that L.C. does not coincide with the results of any of the sets of weights; it is always worthwhile, in this example, to spend more than is absolutely necessary and thereby to increase the benefits to one or more of the interested parties. Since it does not reflect the interests of any of the parties, the L.C. solution is used as a neutral standard for comparing the effects of the other decisions on the welfares for the four interested parties. For example, Tables 9 and 10 show that the net benefits received by Pierce-Hall as a result of decision No. 10 would be $27,000 a year less than those resulting from the L.C. decision.

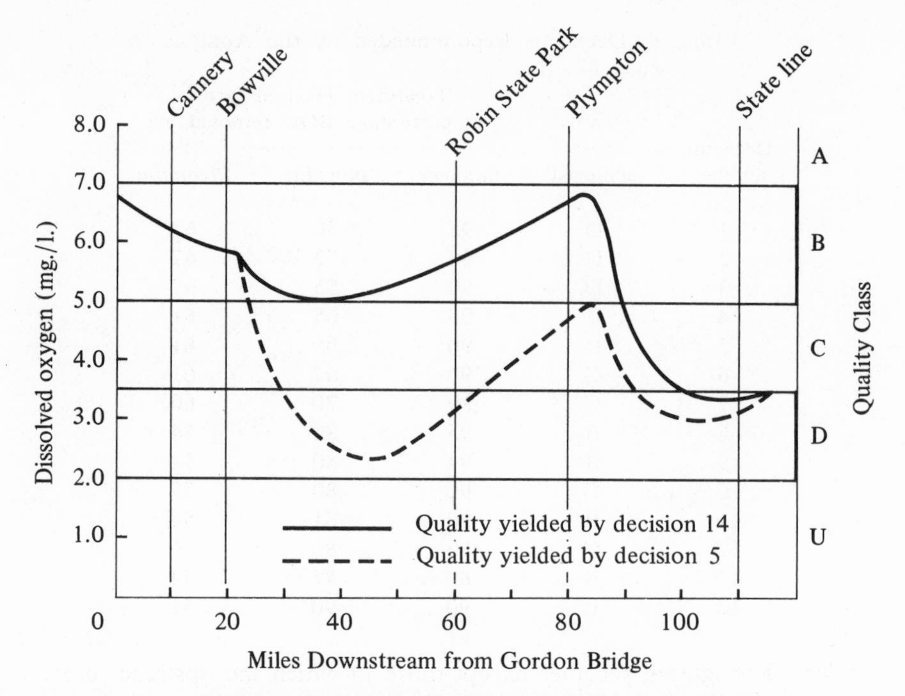


FIG. 4. Water Quality in the Bow River Under Alternative Decisions, Summer Drought Conditions

Figures 5 and 6 are what the economist would call the utility-possibility frontiers for the three use classification limits. In Figure 5, for example, the net benefits to Plympton and Bowville resulting from each of the decisions consistent with a minimum classification of D are plotted. Decisions No. 2, 5, 10, 11, and 14 all correspond to cannery net benefits of −$27,000. Notice as you move along the curve from right to left how Plympton gains at Bowville's expense. Notice also, from Table 9, how this movement corresponds to increasing the political influence that Plympton is assumed to have.

These remarks, and many other features of the tables and charts, show how the application of the model has not only quantified the expression of the problem; it has also led to a sharp, quantitative depiction of the conflicting results desired by the participants and of the compromises that can emerge from their joint deliberations. There are some persistent features. The cannery, for example, always removes at least 80 percent of its wastes, while, regardless of the weights, Plympton never removes more than 67 percent. Those characteristics of the decision can be predicted with confidence. They could not have been foretold without the analysis. The model has therefore been fruitful to the extent of placing some bounds on the outcome of the decision that could not have been predicted without it.

By exercising some political judgment we can go even further, though not with complete confidence. Notice, from Table 9, that the winning decision is either No. 5 or No. 10, except when the distribution of political influence is heavily skewed either upstream or downstream. From our political narrative at the outset, it appears that the cannery and Bowville together are almost certain to have enough influence to prevent their interests from being sacrificed to Plympton's. This consideration renders decisions No. 11, 12, 13, 14 highly implausible. On the other hand, the computations show that they would have to have, in effect, a 7 to 3 majority to force decisions No. 1 or 2 on Plympton or the FWPCA. (Notice that [1,5,3,1] or a 6 : 4 majority leads to decision No. 5.) If that extent of preponderance is implausible, and our narrative makes it seem so, then the victorious decision is almost sure to be No. 5 or No. 10. In either of those the cannery is required to treat its wastes for 90 percent of BOD removal, a fairly narrow band is established for Plympton, and a somewhat wider one for Bowville. In short, the range

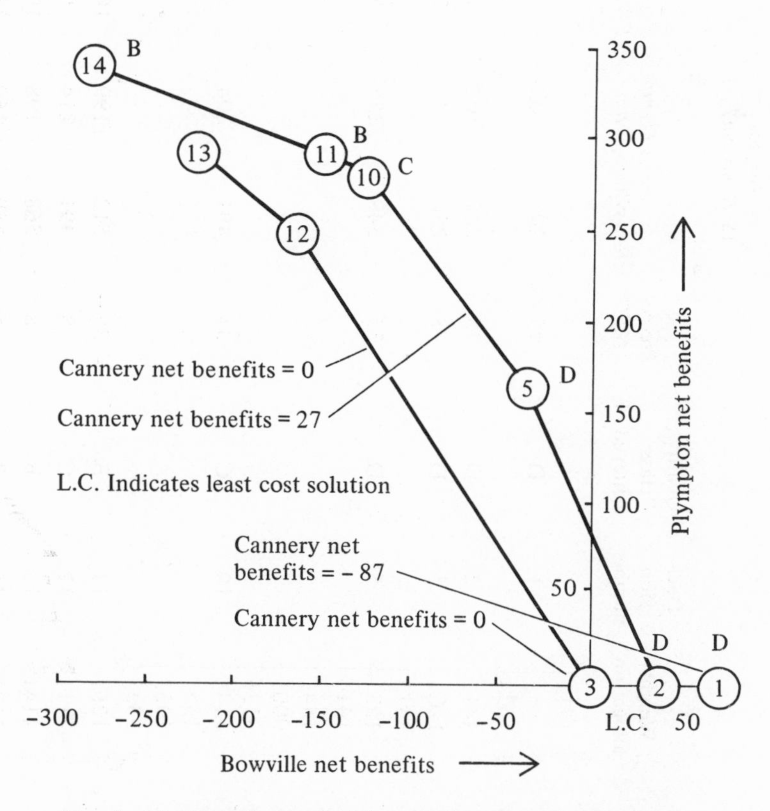


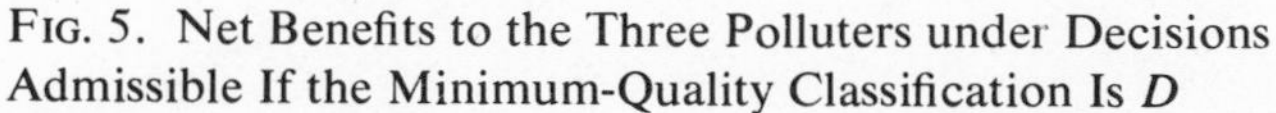
FIG. 5. Net Benefits to the Three Polluters under Decisions Admissible If the Minimum-Quality Classification Is *D*

TABLE 9. Decisions Corresponding to Specified Weight Allocations under a Minimum Quality Classification of D

| Weight allocation | Decision number | Quality class achieved | Net cost ($ thousand/yr) Pierce-Hall | Net cost Bowville | Net cost Plympton | Net cost FWPCA | Net benefits ($ thousand/yr) Pierce-Hall | Net benefits Bowville | Net benefits Plympton | Net benefits FWPCA |
|---|---|---|---|---|---|---|---|---|---|---|
| 1711, 1621 | 1 | D | 95 | 201 | 302 | 829 | −87 | 72 | −3 | −61 |
| 4411 | 2 | D | 35 | 217 | 302 | 750 | −27 | 48 | −2 | 10 |
| L.C. | 3 | D | 8 | 250 | 301 | 746 | 0 | 0 | 0 | 0 |
| 3313, 1531, 1414 | 5 | D | 35 | 349 | 255 | 863 | −27 | −38 | 160 | 101 |
| 1441, 0001, 3133, 1144, 1351, 1333, 3331 | 10 | C | 35 | 491 | 204 | 985 | −27 | −131 | 282 | 147 |
| 1261 | 11 | B | 35 | 512 | 199 | 1007 | −27 | −148 | 292 | 139 |
| 7111 | 12 | C | 8 | 491 | 214 | 953 | 0 | −158 | 255 | 125 |
| 4141 | 13 | B | 8 | 568 | 198 | 1034 | 0 | −219 | 294 | 97 |
| 1171 | 14 | B | 35 | 659 | 169 | 1162 | −27 | −279 | 348 | 36 |

"L.C." denotes least-cost plan for attaining specified quality.

TABLE 10. Decisions Corresponding to Specified Weight Allocations under a Minimum Quality Classification of C

| Weight allocation | Decision number | Quality class achieved | Net cost ($ thousand/yr) | | | | Net benefits ($ thousand/yr) | | | |
|---|---|---|---|---|---|---|---|---|---|---|
| | | | *Pierce-Hall* | *Bowville* | *Plympton* | *FWPCA* | *Pierce-Hall* | *Bowville* | *Plympton* | *FWPCA* |
| 1711, 1621 | 4 | C | 95 | 344 | 251 | 952 | −87 | −21 | 169 | 34 |
| 4411, 3313, 1414, 1531 | 6 | C | 35 | 360 | 251 | 873 | −27 | −45 | 170 | 105 |
| L.C. | 7 | C | 8 | 393 | 249 | 869 | 0 | −94 | 171 | 94 |
| 1441, 0001, 3133, 1144, 1351, 1333, 3331 | 10 | C | 35 | 491 | 204 | 985 | −27 | −131 | 282 | 147 |
| 1261 | 11 | B | 35 | 512 | 199 | 1007 | −27 | −148 | 292 | 139 |
| 7111 | 12 | C | 8 | 491 | 214 | 953 | 0 | −158 | 255 | 125 |
| 4141 | 13 | B | 8 | 568 | 198 | 1034 | 0 | −219 | 294 | 97 |
| 1171 | 14 | B | 35 | 659 | 169 | 1162 | −27 | −279 | 348 | 36 |

TABLE 11. Decisions Corresponding to Specified Weight Allocations under a Minimum Quality Classification of B

| | | | Net cost ($ thousand/yr) | | | | Net benefits ($ thousand/yr) | | | |
|---|---|---|---|---|---|---|---|---|---|---|
| Weight allocation | Decision number | Quality class achieved | *Pierce-Hall* | *Bowville* | *Plympton* | *FWPCA* | *Pierce-Hall* | *Bowville* | *Plympton* | *FWPCA* |
| 1711<br>1621 | 8 | B | 95 | 487 | 200 | 1074 | −87 | −115 | 292 | 80 |
| 1441<br>1531 | 9 | B | 81 | 491 | 200 | 1056 | −73 | −121 | 292 | 96 |
| 4411<br>3313<br>0001<br>3133<br>1144<br>1351<br>1333<br>3331<br>1261 | 11 | B | 35 | 512 | 199 | 1007 | −27 | −148 | 292 | 139 |
| 7111<br>4141 | 13 | B | 8 | 568 | 198 | 1034 | 0 | −219 | 294 | 97 |
| 1171 | 14 | B | 35 | 659 | 169 | 1162 | −27 | −279 | 348 | 36 |

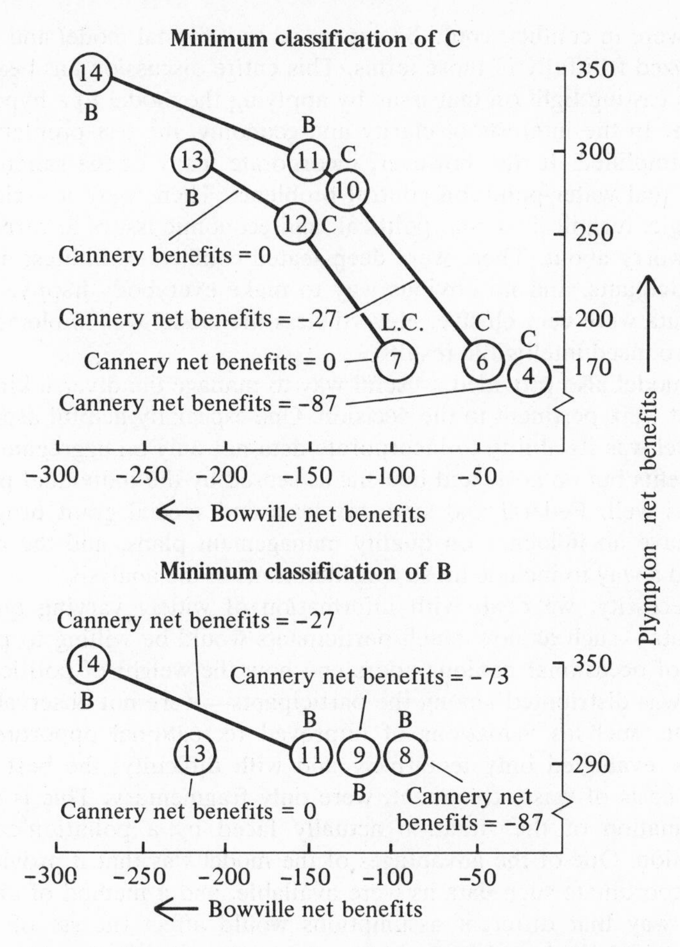


FIG. 6. Net Benefits to the Three Polluters under Decisions Admissible If the Minimum-Quality Classification Is *C* or *B*

of possibilities for this decision has been very substantially narrowed by the application of the model followed by a modicum of political judgment. And that is what the model was intended to achieve.

# Conclusions

## *Appraisal of the Test*

At the outset we raised the question of whether a complicated problem of governmental decision-making, in which the interests of influential groups were in conflict, could be expressed as a formal model and could be analyzed fruitfully in those terms. This entire discussion has been devoted to casting light on that issue by applying the model to a hypothetical case. In the interests of clarity and economy, the test problem was highly simplified. It did, however, incorporate many of the salient features of real water-pollution control problems. There were a variety of hydrologic, technical, social, political, and economic issues to formulate and to worry about. There were deep-seated conflicts of interest among the participants, and no obvious way to make everybody happy. Some of the data were very elusive. Nevertheless the model was implemented, and it produced intelligible results.

The model also provided a useful way to manage the diverse kinds of data that were pertinent to the decision. One especially helpful aspect of the model was its ability to incorporate data not only on aggregate costs and benefits but on costs and benefits as sensed by the individual participants as well. Federal and state tax laws and special grant programs surely have an influence on quality management plans, and the model provided a way to include those considerations in the analysis.

Of necessity, we dealt with information of widely varying quality. Some data—such as how much participants would be willing to pay to be free of occasional noxious odors and how the weight of political influence was distributed among the participants—were not observable at all; some, such as valuations of improved recreational opportunities, could be evaluated only tentatively and with difficulty; the best data, such as costs of waste treatment, were only fragmentary. This is a fair approximation of the situation actually faced by a pollution-control commission. One of the advantages of the model was that it provided a way to coordinate such data as were available, and a method of analyzing the way that different assumptions would affect the set of plans likely to be adopted by the commission.

On this last point, our sample results proved highly informative. Even our limited selection of weight allocations allowed us to explore the set of plans that would be adopted under extremes of imbalance in the consideration given to the different participants. We found, of course, that there was conflict among the participants and that, in general, the greater the political weight the individual had, the greater were his net benefits. This is no profound discovery; it is only common sense. But the analysis has quantified and sharpened this vague and obvious perception, and has shown concretely how the response to a change in influence is likely to be implemented and how it will affect the welfares of the individual participants.

Our results also provided a framework within which we could use reasonable political judgments to predict the outcome of the decision process. Without the analysis, we should not have been able to guess what the different decisions would look like or to foresee which set of political circumstances would favor the adoption of one or another of the Pareto-admissible plans. It is not that the analysis can provide any new knowledge of the political situation; but it allows us to combine what judgments we may possess with data on the physical alternatives and on the relative importance of different aspects of the problem, as viewed by the individual participants, to learn what outcomes are more or less likely than others.

The analysis also taught us something about the commission itself. We have seen how its actions are likely to be influenced if more or less stringent quality requirements are imposed from the outside. In a similar way we could have analyzed the effects of other changes in external conditions. For example, one of the assumptions in our example was that the commission had no authority to tax or to offer grants-in-aid. Our results show, however, that there are circumstances under which such powers would be very useful. Consider the situation where Bowville and the FWPCA carry little influence in the commission so that the (4,1,4,1) weights turn out to be appropriate. Table 9 indicates that decision No. 13 will be taken under these circumstances. It happens, however, that a shift from 13 to decision No. 11 would maintain the same quality class and leave Plympton essentially unaffected, yet the change would save Bowville over $2 for every dollar that it cost the cannery. If fiscal expedients were available, the wasteful decision No. 13 would never be adopted: the commission could tax Bowville and use the proceeds to reduce the financial burden on the cannery. In this way the technically efficient decision No. 11 could be implemented to everybody's benefit and without imposing any financial hardship. The value of such authority, though it might have been suspected, could not have been established without invoking a model capable of bringing out the economic consequences of the technical peculiarities of the river that make the cannery the strategic site for treating waste.

This finding illustrates that our concept of Pareto admissibility is relative. It depends heavily on the range of decisions that are assumed to be technically and legally practicable. Change those assumptions, and the Pareto-admissible set changes correspondingly. Thus the model can be used to estimate the value of changing either the technical possibilities available (by research) or the legal possibilities permissible (by legislation), by comparing the Pareto-admissible sets that correspond to alternative assumptions. Among the legal possibilities that merit consideration are the power to tax and make grants, the authority to impose effluent charges, and the authority to operate regional treatment facilities. All of these are issues now being debated earnestly in the several water-pollution control commissions that are currently operative.

## *Applicability of the Model*

Like any model, this one depends on numerous assumptions and can be applied only to circumstances that correspond fairly well to the assumptions. The fundamental assumption is that a water-pollution control commission, like any other government agency, is responsive to the wishes of its constituency. This seems highly reasonable. By being responsive, an agency reduces its exposure to complaints, litigation, and animosity; builds its reputation for efficiency and fair-mindedness, accumulates political support and influence, and fosters cooperative attitudes among the people with whom it must deal. In short, it gains the consent of the governed, which is an essential prerequisite of effective government. Any governmental agency or commission that acts according to this simple principle will arrive at one or another of the decisions that we have called Pareto admissible.

The kind of analysis that follows from this assumption contrasts with the approach used in much of the literature, which rests on the postulate that government agencies endeavor to achieve some overarching goals typically called "the general welfare" or "the national interest." Such an agency would be unresponsive to the sectional or special interests of its constituents. There is much empirical evidence that agencies in fact behave predominantly as we have assumed. See, for example, the work of Matthew Holden [7].

Furthermore, the institutional arrangements that are employed by water pollution control agencies in order to be responsive to their constituents are frequently similar to those used by the Bow Valley Water-Pollution Control Commission. This is no coincidence. We have modeled our commission in the image of some established pollution-control commissions. The Delaware River Basin Commission (DRBC) is made up of the governors of the four states through which the Delaware flows, plus a representative of the federal government, just as the Bow Valley Water-Pollution Control Commission is made up of representatives from the municipalities and industries located along the river. The Ohio River Valley Water Sanitation Commission (ORSANCO) has a similar makeup.

The formal composition of the commission is not decisive, however; an anxious concern for the interests and responses of the people affected by its decisions is. Direct representation is one expedient for assuring this concern. More fundamental is the maintenance of intimate contact with the constituency, by informal consultations, formal conferences, and public hearings, before any important decision is made. Both the DRBC and the ORSANCO follow such procedures, as does the Bow Valley Water-Pollution Control Commission. Indeed, the federal legislation that provides the charter for all the newer commissions requires elaborate consultations. The requirement to conduct public hearings at which all affected "interstate agencies, states, municipalities and industries involved" can present their cases is stressed repeatedly in the federal laws. Operating practice, so far as we have been able to discern it, sincerely implements these requirements. In these respects the Bow Valley Water-Pollution Control Commission appears to be a fair replica of the pollution-control agencies now in existence.

A subsidiary assumption that we made was that the constituency of the Bow Valley Commission consists entirely of local residents except for a generalized public interest represented by the FWPCA. This characteristic is by no means universal, even in the field of water-pollution control. For example, the Water Quality Act of 1965 gave primary responsibility for regulating the use of interstate waters to "State water pollution control agencies," which are simply departments of the state governments. Now, the range of interest groups represented on the State Water Commission is quite different from those that we have envisaged on the Bow Valley Commission. Therefore, if the problems of the Bow Valley were brought before the State Water Commission instead of before the Bow Valley Commission, the decision might well be affected. It would remain true that in any proceeding before the State commission concerning the regulation of the Bow River, the parties with predominant interest and influence would be precisely the ones that we have considered. But the political-influence weights would be likely to differ: purely local and sectoral interests would have smaller leverage, while statewide and national interests would gain greater recognition. One useful application of the model is to assist in forming judgments about the consequences of changes in the constitution of decision-making authorities.

Next, a number of specialized assumptions were made purely to keep computations simple and tractable. They concern the policy instruments considered, the technical aspects of waste management and water quality, the question of uncertainty, and the problem of growth in waste loads. We can review each one of these four aspects in turn and give a brief indication of some of the limitations of the sample case we have used.

First and most important was the narrow range of discretion permitted the Bow Valley Commission. Many actual commissions have a scarcely wider range of authority but, as we noted above, additional powers and privileges can permit superior decisions (in the Pareto-admissible sense). Additional policy instruments can be incorporated into the model, but only at the cost of some increase in complexity.

Second, we vastly simplified the technological and hydrologic aspects of the problem. In particular, we utilized a highly simplified description of the biochemical composition of municipal and industrial sewage and of the technology of waste treatment. We ignored altogether such waste-management methods as manufacturing-process change and waste-water recirculation. In addition, we reduced water-quality standards to a single dimension, dissolved oxygen, and built our analysis on the simplest available formulation of waste transport and decay. The model can be revised to incorporate more adequate assumptions about these aspects of pollution control, but then solution and interpretation of the problem will be more difficult than in our case. For example, in the reformulation it may not be possible to maintain certain convenient mathematical properties (such as linearity) that were built into the Bow Valley example.

A third simplification was to ignore the unpredictability and variability of streamflows, gross waste loads, treatment-plant performance, and all of the other aspects of uncertainty that are vexatious parts of most real decision problems. To introduce uncertainty in its manifold aspects, we should have to modify and complicate our model along well-established lines.

In our fourth area of simplification, we evaded dynamic considerations by assuming that neither the towns nor the cannery were expected to grow, and that no new industries were likely to be established along the river. Although the possibility of growth presents serious analytic difficulties of its own, a satisfactory method for dealing with them could be built into our model.

Finally, we ignored two important features of the political-decision process. One is the influence of logrolling, pressure politics, and side-payments of all sorts. There are pressures, threats, and inducements that Bowville can use to persuade the cannery to acquiesce in a decision that would otherwise be unacceptable, and vice versa. Even Bowville and Plympton, though their interests in the treatment of the river are almost diametrically opposed, can bargain a bit out in the corridor. Such bargaining can influence where in the range of likely outcomes the ultimate decision will fall. It reinforces the finding that our prediction is a range rather than a point. But such side bargaining will not push the decision outside the Pareto-admissible range.

Our other political simplification is, in principle, more fundamental. We assumed that each of the participants had a firm and immutable evaluation of the consequences of every decision for his own welfare. This assumption was contained in our estimation of the values attached by the participants to increments in water quality at different points in the river, and in our treatment of those evaluations as unchanged throughout the decision process. In fact, a significant feature of any group decision process is the attempt by each participant to persuade the others to alter their psychological evaluations so that they are more in line with his. The representatives of Plympton and conservation interests will emphasize the benefits to everyone of high quality water throughout the river and, in fact, will urge it as a moral imperative. The representatives of the cannery will point out that the prosperity of the entire basin depends on the economical use of the waste-absorption potential of the river. To the extent that this rhetoric is not in vain it will succeed in causing some of the participants to change their subjective valuations of the importance of improvements in water quality. It will make our Table 6 invalid.

How serious is this difficulty? This is a significant and open question, fundamental to the understanding of the political process. Perusal of the histories of ORSANCO and the DRBC suggests, however, that when specific decisions are being debated most of the rhetoric is ineffectual, that people want at the end much what they wanted at the beginning, and that the operative aspect of the bargaining process is a reconciliation of the pressures that the different interest groups have been able to mobilize.[14]

## *Comparison with Current Practice*

The prevalent method for analyzing public policy decisions of the sort dealt with here is benefit-cost analysis. Superficially, benefit-cost analysis applies to the government sector the calculus of profit and loss that is used in business decisions. Its popularity is at least partly a response to the businesslike ethic that prevails in our culture. But its contrast with the approach here advocated is, in fact, more profound than the question of whether the government should follow businesslike practices. It is really a reflection of the most ancient cleavage in the tradition of political philosophy.

One great school of political philosophy views a government as the leader of its people, responsible for defining the goals of its citizens and formulating their social standards, preferably under the wise guidance of a philosopher-king or benevolent despot. The other great school sees the government as the corporate embodiment of its people, serving their communal interests and carrying out their wishes, preferably as expressed in direct (nowadays, participatory) democracy.

Traditionally these two views have been advanced as norms, as expressions of what governments ought to be. But they deserve also to be taken seriously as expressions of how governments actually behave. Examples of both governmental leadership and responsiveness are easy to find. ORSANCO, which provides the best documented experience in the water pollution area thanks to Cleary, was established in response to public demand (i.e., the demand of a few leading private citizens who mobilized widespread support) and went on to exercise a good deal of leadership and initiative of its own. It did so, however, less by pursuing its own goals than by undertaking a series of educational activities that

14. For an instructive account of the issues faced by ORSANCO and their resolution see Cleary [3].

increased its constituents' awareness of the importance of abating the pollution in the Ohio.

We cannot pursue here the rather ill-defined issue of leadership versus responsiveness. In practice, government agencies appear to mix the two in varying proportions, with responsiveness proponderating except for transitory episodes, usually at the highest levels of government.[15]

Benefit-cost analysis is, however, an expression of the leadership role of government. A benefit-cost formula is a tool for evaluating the desirability of different undertakings by the government's standards. For example, one of the chronic condundrums in benefit-cost analysis is the choice of the rate of discount to be used in evaluating deferred benefits and costs. This rate, selected by agency officials or expert consultants, represents the official evaluation of the relative importance of consequences that emerge at different dates. Similarly, the Flood Control Act of 1936 instructs agencies to compute benefit-cost ratios by adding up the benefits and costs of a project "to whomsoever they may accrue." The intent is to maintain neutrality, but the effect is to impose a judgment about the relative social importance of effects upon upstream users and downstream users, rich and poor, farmers and urbanites, and so on. All governmental undertakings redistribute income in some manner.[16] Any evaluative formula must incorporate some appraisal of dedistributive consequences, either implicitly or explicitly.

A third example of the governmental evaluations built into benefit-cost analyses is provided by the problem of aggregating benefits, and costs of different kinds. When the beneficial results are priced on economic markets, as is the case with irrigation water and hydroelectric power, market prices are used. Otherwise prices representing social evaluations have to be found. Outdoor recreation benefits, for example, are frequently valued at $1.50 per user-day, but some authorities insist that use by comparatively deprived urban dwellers should be assigned higher "merit" values.[17] Clearly, some values have to be used, and any values represent the judgment of the agency that adopts them. Finally, some consequences of government undertakings which are deemed excessively difficult to evaluate are simply omitted from benefit-cost calculations. This, too, represents an implicit governmental judgment, and one that has drawn much criticism.[18] In sum, there is no way out of it:

15. Truman [12] presents a full-dress analysis of governmental responsiveness.

16. This problem is discussed more completely by Marglin [11, pp. 67ff.]. Marglin recommends that different values be assigned to different consequences, dependent on the beneficiary. For a full treatment of the theoretical significance of income redistribution see Little [9].

17. Mack and Myers [10] is a careful analysis of the problem of evaluating recreational benefits from a social point of view. There is much additional literature, some cited by them.

18. These criticisms are reviewed by Dorfman [4] and elsewhere in the volume in which it appears.

a benefit-cost formula incorporates many judgments, implicit and explicit, of the relative importance of the consequences of an undertaking. These judgments must be those of the agency doing the evaluation or its superiors in the governmental hierarchy.

In fundamental contrast to the benefit-cost approach, the analysis used here invokes no other evaluations than those of the people affected. It is explicitly noncommittal with respect to the relative importance or influence of the different participants; that is why it does not lead to an unambiguous prediction. It assembles the data from which those people and the agency concerned derive their decisions, but it does not presume that anyone has a formula for global social evaluation. Therefore it does not purport to recommend what should be done, but only to describe how actors in a political process interact to produce a decision.

This analysis cannot be regarded as an alternative, even in principle, to the decision-making methods actually used. But it can be of assistance in understanding and facilitating those methods. Its advantage lies in its ability to sketch quickly and cheaply the range of alternative decisions that is worth considering. In our test case, with the data and a moderately fast computer at hand, between two and three minutes were required to determine the corresponding admissible decision. By varying the assumed data artfully, the main outlines of the entire range of admissible decisions were mapped out with about sixty repetitions of this quick computation.

Of course, the computations that would be required in a practical instance are of an entirely different order of magnitude from those encountered in this simple test. Contemplate any actual river basin. It would contain a half-dozen or more cities and towns, several dozen factories or other points of waste discharge, a mainstream, and a number of tributaries with complicated hydrology. In addition, many among our impressive list of simplifying assumptions would be inappropriate. Policy instruments besides the enforcement of a certain percentage waste removal at the source would be relevant, and a rich variety of alternative technologies of waste management would have to be considered. Our uni-dimensional measure of water quality would be considered inadequate, as would our simple specification of the relationship between waste discharge and stream condition. Provision for future growth in wastes would need to be introduced, along with some appropriate method of dealing with uncertainty. If all these complications were taken into account, the calculations would not only be much larger than those that we have encountered, but would be beyond the capacity of any computer now extant or envisaged.

These complexities, however, do not render mathematical analysis inapplicable. They do necessitate a good deal of simplification of the full richness of reality. Simplifying assumptions would have to be made about waste management options, though probably not as severe as the simplifications that we have indulged in. A water quality index of one or two dimensions would have to be introduced in place of the multidimensional specifications set forth in water quality standards documents. The hydrology would have to be simplified. The number of points of pollution would have to be reduced by consolidating groups of nearby installations into a single synthetic polluter. All these and other simplifications would have to be carried to the point where the calculation became manageable.

The result would be an inevitable loss in accuracy. But this necessity does not invalidate the method, for the relevant standard of accuracy is not some unattainable ideal but the level of accuracy that is attainable by alternative procedures. The truth is that the economic-hydrologic-biologic-political ecology of a live river basin in the full majesty of its intricacy far transcends the capacity of any method of analysis of decision-making available to man. All methods of decision-making require severe simplifications, as a perusal of the dockets of any water-control authority will establish. And, there is reason to believe, the simplifications required for mathematical analysis are less disabling than the simplifications that are conventional in more informal procedures. Only hard experience can determine how practical and helpful mathematical analyses will be in actual instances, but the fact that they must invoke some serious simplifications is not *ipso facto* decisive.

## References

[1] Bator, F. M., "The Simple Analytics of Welfare Maximization," *American Economic Review, 47* (March 1957), pp. 22–59.

[2] Baumol, W. J. and R. C. Bushnell, "Error Produced by Linearization in Mathematical Programming," *Econometrica, 35* (July–October 1967), pp. 447–471.

[3] Cleary, Edward J., *The ORSANCO Story—Water Quality Management in the Ohio Valley under an Interstate Compact*. Baltimore: Johns Hopkins Press, 1967.

[4] Dorfman, R., "Introduction" in *Measuring Benefits of Government Investments,* R. Dorfman, ed. Washington: The Brookings Institution, 1965, pp. 1–11.

[5] Dorfman R. and H. D. Jacoby. "A Model of Public Decisions Illustrated by a Water Pollution Policy Problem" in *The Analysis and Evaluation of Public Expenditures: The PPB System,* submitted to the Subcommittee on Economy in Government, U.S. Congress, Joint Economic Committee, May 29, 1969.

[6] Fair, G. M., J. C. Geyer, and D. A. Okun, *Water and Wastewater Engineering,* Volume II. New York: John Wiley and Sons, 1968.

[7] Holden, Matthew, Jr., "Political Control as a Bargaining Process: An Essay on Regulatory Decision-Making," pub. no. 9, Cornell University Water Resources Center, October 1966.
[8] Kneese, A. V. and B. T. Bower, *Managing Water Quality: Economics, Technology, Institutions*. Baltimore: Johns Hopkins Press, 1968.
[9] Little, I. M. D., *A Critique of Welfare Economics,* 2d ed. London: Oxford University Press, 1957.
[10] Mack, R. P. and S. Myers, "Outdoor Recreation" in *Measuring Benefits of Government Investments,* R. Dorfman, ed. Washington: The Brookings Institution, 1965, pp. 71–101.
[11] Marglin, S. A., "Objectives of Water Resource Development: A General Statement," *Design of Water-Resource Systems,* Arthur Maass, et al. Cambridge: Harvard University Press, 1962, Chapter 2.
[12] Truman, David B., *The Governmental Process: Political Interests and Public Opinion*. New York: Alfred A. Knopf, 1951.

15

HAROLD A. THOMAS, JR.

# *The Animal Farm: A Mathematical Model for the Discussion of Social Standards for Control of the Environment*

*Harold A. Thomas, Jr., is Gordon McKay Professor of Civil and Sanitary Engineering at Harvard University.*

1. Old MacDonald has a farm; and on his farm he has a herd of $CN$ ferthings. Each year he sells $N$ mature animals and each year $N$ young ferthings join the herd. Ferthings are worth most when they are $C$-years old. The profit from each animal sold is $U$ groats.

2. Lately, however, because of a water-borne disease the farmer's profit has been reduced. It is less than $NU$. He must make a decision. He would like to have more groats.

3. The farmer's daughter, Honoria, who is taking a course in epidemiology, explains the etiology of the disease. Ferthing pathologists call it *hyfertitus*. It is carried by a small pathogen that lives in water. The

*"The Animal Farm: A Mathematical Model for the Discussion of Social Standards for Control of the Environment," by Harold A. Thomas, Jr., from* The Quarterly Journal of Economics, *February 1963.* 

animals contract the disease only from drinking the water. They do not get it from each other. Every pathogen ingested has the same likelihood of causing death, and this likelihood does not depend upon how many other pathogens have been ingested previously. The mortality : morbidity ratio is one. *Hyfertitus* is a very serious disease.

4. Honoria, who is also taking a course in probability, writes on a piece of paper as follows:

Let $Q(X)$ = the probability of any ferthing not dying from *hyfertitus* during the $C$-year period before sale, the pathogen density being $X$ organisms in the water volume ingested per animal in this period. Then in accordance with the foresaid specifications as to the mechanism of the pathogenicity the probability of surviving if the density of pathogens were increased by a small amount $\delta X$ would be the probability of surviving the first $X$ micro-organisms, multiplied by one minus the probability of fatal infection by the incremental micro-organisms, or

$$Q(X + \delta X) = Q(X)[1 - K\delta X],$$

where $K$ is an infectivity parameter measuring the virulence of the pathogens. Forming the differential quotient and letting $\delta X \to 0$ we obtain $dQ / dX = -KQ$. Integrating and adjusting the constant of integration so that $Q(0) = 1$ gives.

$$Q(X) = e^{-KX}. \qquad (1)$$

The larger $X$ for a given $K$ the smaller the proportion of surviving animals. If, for example, $X$ happens to be equal to $0.69 / K$ then only one-half of the ferthings will live. Honoria is on the honor roll in her course in probability.

5. A salesman named Young Sam comes to the farm. He is an honest and candid sales representative of an equipment company. Young Sam says that his company makes a remarkable water treatment unit called a "Disinfeclarminator" that will eliminate the disease 100 percent. It is easily installed and is completely effective in killing the pathogens of *hyfertitus* over all ranges of density $X$. It is simple to operate. There is nothing to do. And best of all, a treatment unit for a herd of size $N$ costs only $V$ groats!

6. MacDonald moves toward decision. To treat or not to treat. That is the question. First he makes a preliminary decision that he will try to maximize economic efficiency. This is his objective function.

(1) If the treatment unit is not installed his annual income or gain will be $NUQ(X)$. The present value of the future time stream of income will be

$$G_1 = \beta NUQ(X), \qquad (2)$$

where

$$\beta = \frac{1-(1+r)^{-T}}{r}, \tag{3}$$

$r$ = the discount rate for capital (opportunity cost of capital), and
$T$ = the economic time horizon.

(2) If the treatment unit is installed, the disease will be eliminated and the gain will be

$$G_2 = \beta NU - V. \tag{4}$$

The break-even point in terms of pathogen density, $X_c$, is obtained by setting

$$G_1 = G_2. \tag{5}$$

Solving equation (5) for the quality criterion $X_c$ by means of equations (1), (2) and (4), and using the approximate relation, $Q(X) = 1 - KX$, which is valid when $KX$ is small,

$$X_c = \frac{V}{K\beta NU}. \tag{6}$$

7. MacDonald decides. He has numbers for all the factors on the right-hand side of equation (6). He calculates $X_c$ and finds it to be smaller than the actual $X$ in his water supply so that $G_2$ is larger than $G_1$. Therefore he installs the unit and obtains a greater income. Now he will have enough groats to pay for his daughter's courses in probability and epidemiology. Young Sam makes a sale. Everybody gains! This is a nonzero sum model. Honoria takes equation (6) to her professor.

8. The professor is brave; he wishes to apply the model to people. He would like to dispel the mystique that often enshrouds the setting of quality standards for control of the environment. But he is not foolhardy and he speaks in the subjunctive mood.

9. The professor makes some inferences from equation (6). They are interesting if they pertain to ferthings; they are surprising if the model is considered as possibly applicable to human beings. It would appear that in a general way the setting of all criteria, standards, or rules for administration of man's environment might follow the rationale underlying equation (6).

(1) The tolerance level of the pathogen is not zero. A calculated risk is taken.

(2) The factor $V$ appears in the numerator of equation (6). The critical concentration depends on the cost of treatment. If a technical innovation occurs that reduces the cost of treatment, in due course of time the tolerance level should be reduced. The stringency of a standard ought to be proportioned to cost.

(3) If as normally would be the case the per capita cost of treatment, $V/N$, decreases as $N$ increases because of economies of scale, then big herds of animals should drink purer water than small herds. Equation (6) says this is true for a noncontagious disease. For a contagious disease the argument would have even greater force. It is pertinent to remark that in human populations the water quality as measured by coliform organism density is often found to be better in the large cities of a region than in the smaller towns. There are sizable economies of scale in rapid sand filtration plants.

(4) The tolerance concentration depends in part upon the economic parameters $r$ and $T$. The equation (6) shows that if the interest rate, $r$, is large, $X_c$ should be large; and if the economic life, $T$, is small, $X_c$ should be large. A large $r$-value in a capital market indicates a high discounting of the future in favor of the present. Accordingly, if MacDonald's farm were in Brazil he would use a higher tolerance level in his decision process than he would if his farm were in New Zealand. Underdeveloped countries tend to have high capital discount rates relative to those in developed countries. Should people in Brazil drink less pure water than people in New Zealand? If the answer is not yes, and Brazilians are constrained to drink overpure water, then it must be asked whether such constraint does not deprive them of resources that might better go into highways, schools or into other sectors of the public health budget where the prevailing discount rate is considered.

(5) In equation (6) it is evident that the tolerance level, $X_c$, in the ferthing problem depends upon two physical parameters ($N$ and $K$) and four economic factors ($r$, $T$, $U$ and $V$). MacDonald had numbers for all of these and could compute $X_c$.

If we now consider potable water for human beings (or air, or food, or comfort and safety) we must ask which, if any, of these parameters becomes fuzzy? The answer, of course, is loud and quick—we do not usually assign a utility to a human being in groats or dollars. The parameter $U$ is not known numerically. The remaining parameters do not become fuzzy with the switch.

(6) But for many years decisions have been made as to quality criteria and standards. In the United States drinking water should have an average coliform concentration not greater than about one organism per 100 milliliters. In British practice Class 3 water (coliform density 3 to 10 per 100 milliliters, presumptive) is rated as suspicious. The upper limit for concentration of boron in irrigation water classed as "good" is 0.67 milligram per liter if sensitive plants are to be grown. The World

Health Organization specifies that cyanide (as CN) has a maximum allowable concentration of 0.01 milligram per liter. The Standards for Protection Against Radiation adopted by the U.S. Atomic Energy Commission require that the gross quantity of radioactive material released into the waste water disposal system at any licensed plant or institution be less than one curie per year. There are recommended limiting concentrations in the atmosphere of such pollutants as ozone, lead and hydrogen sulfide. All civilized countries have legislated on the question of preservatives in food. At the First Opium Convention at the Hague in 1911 regulatory criteria were promulgated. The maximum permissible speed on the New York Thruway is 65 miles per hour.

These criteria and many others like them represent socioeconomic decisions and reflect in each case the inherent tension between human desires and human capability. The criteria of quality represent a balance of costs and benefits. In equation (6) it is seen that the tolerance level, $X_c$, is proportional to a cost : benefit ratio. In a general way using more elaborate and realistic mathematical models it may be shown that without exception *every quality criterion or rule whether it pertains to health, to aesthetics or to property damage is always equal to a function of a cost: benefit ratio.*

(7) The setting of any quality criterion or standard relating to health and well-being inevitably entails making an implicit estimate of a cost : benefit ratio based on whatever data or other factors are available for judgment. In situations where the costs are known within reasonable limits, the establishment of a standard amounts to assigning a certain utility to human life, health or well-being. For example, if in the simple decision model for MacDonald's farm we are able to set down numerical values for the factors of cost, infectivity, discount rate, and economic life ($V$, $K$, $r$, and $T$), then an assignment of a numerical value for the water quality criterion is tantamount to assignment of a numerical value for the benefit parameter $U$. The setting of a standard is in effect an imputation of a certain finite level of utility or benefit accruing to life, to health, or to well-being.

(8) A close connection always obtains between quality criteria for water, air, food, safety, and aesthetics and a relevant, implicit, estimated cost : benefit ratio. In the mathematical model of the farm that has been discussed, the relation is one of simple proportion. The constant of proportionality is equal to the infectivity parameter $K$. More elaborate models are needed to take into account more complicated features of reality such as physical and economic uncertainty, errors in data, measurement difficulties, more complex patterns of epidemiology, meteorology, hydrology, technology, human psychology and economic behavior that enter into the decision process.

The modern large electronic digital computer has made it possible to construct elaborate models and to apply them in simulating the problems of environmental science so as to elucidate the effect of various factors and interactions that are germane to the establishment of quality criteria. But despite the greater complexity of these models, the principle stated in section (7) retains its central thrust: the fixing or setting of quality criteria for control of the environment always involves a value judgment; and this value judgment always has the form of a cost : benefit ratio. *To set a criterion is to impute a cost: benefit ratio.*

(9) The professor proposes an arduous, long-range program to strengthen the rational base of decision-making in the management of the environment. He says that the following steps should be taken:

(i) Identify and classify the problems of environmental control on the basis of (a) the mathematical structure; and (b) the type of utility or disutility pertaining to people such as longevity, health, aesthetics, well-being and safety. This classification would differ markedly from previous problem-classifying schemes that have been used, such as, for example, the classification based upon the dominant branch of science involved (fluid mechanics, microbiology, etc.).

(ii) For each class construct an appropriate generalized mathematical model that relates the quality criterion with the corresponding estimated cost : benefit ratio. The relation should make explicit the effect and importance of all relevant physical and economic factors.

(iii) Apply the generalized model for each class to those areas of environmental management where decisions have already been made and quality criteria have been established. Compute from the model that value of the utility parameter (benefit vector) which has been implicitly assigned in the assignment of a definite numerical value (or range of values) for the quality criterion. In the decision model for MacDonald's farm, equation (6) would be solved for *U*, all other factors being known.

(iv) Compare these utilities within classes and establish norms and ranges. Within each class a wide range of results may be expected. Large inconsistencies will appear. Some classes will yield more consistent and reliable estimates than others. Particularly useful perhaps would be an analysis of the technological function relating highway speed and highway death (or other disutility) with existing highway speed control legislation and enforcement. A large segment of the public has active participation with officialdom in the balancing of costs and benefits and the setting of norms in this activity. In other enterprises the direct feedback of public response may be impeded or obscured by technical complexity. In these cases bureaucratic bias and institutional constraints may develop that hinder or retard the equilibrating process.

(v) Use these norms and ranges as rational guidelines in setting new quality criteria and in the re-evaluation of old standards. In this way the entire structure of regulatory codes for control of the environment can be systemized. Internal consistency will beget external viability.

(10) The professor concludes. Not with a Q.E.D. but with an E.I.E.I.O.

# IV

# *THE ROOTS OF ENVIRONMENTAL DEGRADATION*

We have found general agreement on the proposition that, so long as resources are treated as unrestricted common property, there will be a tendency for individuals to use them excessively from the point of view of society at large. No such consensus exists on the question of why the condition of the environment has become so critical in recent times. In this section the roots of the problem are sought variously in the cultural, political, and technological conditions of modern society. Most contributors concede that industrial growth lies somewhere close to the heart of the matter, but each has a different interpretation of its role, and none regards it as sufficient cause of the present state of environmental decay.

The tenor of most articles suggests that the questions addressed here remain fertile fields for investigation. Most views are fairly conjectural. Only Barry Commoner produces solid documentation, and in doing so must be credited with having raised a body of issues that invite further economic research.

Commoner, an ecologist, examines the growth rates of a large number of economic activities and places the blame for the current state of the environment on post-World War II developments in technology which have promoted a variety of products and processes whose environmental impacts are singularly noxious, as substitutes for older ones whose influence on the environment was relatively benign.

Lewis W. Moncrief takes a broad and somewhat eclectic view of the problem in confronting historian Lynn White's popular thesis that responsibility for the West's neglect of the environment is to be found in the Judeo-Christian tradition with its axiom that nature exists to serve man. He argues, on the contrary, that religion can be implicated only to the extent that it underlay the rise of industrialization and of democratic institutions, two great revolutions of the nineteenth century which vastly increased man's means of production and redistributed them more widely. The result was a surge in output and its by-product, waste. According to his thesis, population growth, capital-intensive production, and the American frontier philosophy all played supporting roles.

As something of a counterpoise, Marshall I. Goldman points out that environmental degradation is no stranger to the leading communist country in the world today. It is as serious in the Soviet Union as anywhere else. His findings might set the reader to pondering about the meaning of externalities and the conditions that give rise to them in a centrally planned economy where, according to the theories expounded earlier, we should expect all social costs to be internalized.

J. H. Dales focuses on the need for constant adaptation at what he calls the "interface" between law, which defines a society's system of property rights, and economics, which is concerned with the purchase and sale of such rights. As circumstances change, the boundary between the two may shift, giving rise to new sets of social problems. Thus, as land, air, and water become increasingly scarce, the laws which originally defined them as common property become less and less appropriate.

Both E. J. Mishan and John Kenneth Galbraith find the fault in modern society's system of values. Mishan believes that the fetish of economic growth has led to a costly contempt for those qualities of life that cannot be measured in dollars and cents. Roger Weiss, disagreeing, argues that the great majority of mankind are quite prepared to make some sacrifice of environmental amenities in order to enjoy the benefits, to them, of economic growth. This position is a partial echo of Thomas's paper in the preceding section. The conflicting views of Mishan and Weiss illustrate the inconclusiveness of economic arguments that rest heavily on personal value-judgments or that weigh the contending claims of different segments of the community. Weiss makes the additional point that it is often economic growth itself which provides the means for coping with environmental problems.

Galbraith asserts that esthetic values are relegated to inferior status by the industrial system, which creates for itself the social goals that reflect its own needs—in particular, goals commensurate with rapid economic growth and all its trappings.

One possible reason, not emphasized here, for society's failure to come to grips with environmental degradation is the sheer complexity of the task. As both the preceding and subsequent sections in this volume demonstrate, the problems of disentangling conflicting interests and agreeing on optimal environmental policies are formidable ones. Within the framework of existing social institutions, there is no automatic tendency toward optimal use of the environment. The extent and complexity of the problem must first be recognized and the weight of inertia overcome before deliberate policies for dealing with it can be formulated and put into effect.

# 16

BARRY COMMONER

# *The Environmental Costs of Economic Growth*

*Barry Commoner is Professor of Plant Physiology at Washington University in Saint Louis and Director of the Center for the Biology of Natural Systems.*

## Introduction

This paper is concerned with an evaluation of the environmental costs of economic growth in the United States. This is a complex issue which has appeared rather suddenly on the horizon of public affairs; it therefore suffers somewhat from a high ratio of concern to fact. In addition, the issue is one which happens not to coincide with the domain of any established academic discipline. For environmental costs have been, until rather recently, so far removed from the concerns of orthodox economics as to have been nearly banished from that realm under the term "externalities." And, for its part, the discipline of ecology has, also until very recently, maintained a position of lofty disdain for such mundane matters as the price of ecological purity.

*"The Environmental Costs of Economic Growth," by Barry Commoner, based on a paper prepared for presentation at a Resources for the Future Forum on "Energy, Economic Growth, and the Environment," Washington, D.C., April 20, 1971. Reprinted by permission of the author.*

Let us begin with a brief summary of the ecological background of the issue.

The environment is defined as a system comprising the earth's living things and the thin global skin of air, water, and soil which is their habitat. This system, the ecosphere, is the product of the joint, interdigitated evolution of living things and of the physical and chemical constituents of the earth's surface. On the time scale of human life the evolutionary development of the ecosphere is irreplaceable; if the system should be destroyed, it could never be reconstituted or replaced either by natural processes or by human effort.

The basic functional element of the ecosphere is the ecological cycle, in which each separate element influences the behavior of the rest of the cycle, and is in turn influenced by it. For example, in surface water, fish excrete organic waste which is converted by bacteria to inorganic products; the latter are in turn nutrients for algal growth; the algae are eaten by the fish, and the cycle is complete. Such a cyclical process accomplishes the self-purification of the environmental system in that wastes produced in one step in the cycle become the necessary raw materials for the next. Such cycles are cybernetically self-governed, dynamically maintaining a steady-state condition of indefinite duration. However, if stressed by an external agency, such a cycle may exceed the limits of its self-governing processes and eventually collapse. Thus, if the water cycle is overloaded with organic animal waste, the amount of oxygen needed to support waste decomposition by the bacteria of decay may be greater than the oxygen available in the water. The oxygen level is then reduced to zero; lacking the needed oxygen, the bacteria die, and this phase of the cycle stops, halting the cycle as a whole. It is evident that there is an inherent limit to the turnover rate of local ecosystems and of the global ecosystem as a whole.

Human beings are dependent on the ecosphere not only for their biological requirements (oxygen, water, food) but also for resources which are essential to all their productive activities. These resources, together with underground minerals, are the irreplaceable and essential foundation of all human activities.

If we regard economic processes as the means by which human society uses and disposes of the resources available to it, then it is evident that the continued availability of the resources derived from the ecosphere (i.e., nonmineral resources), and therefore the stability of the ecosystem, are essential to the functioning of any economic system. More bluntly, any economic system that hopes to survive must be compatible with the continued operation of the ecosystem.

Because the turnover rate of an ecosystem is inherently limited, there is a corresponding limit to the rate of production of any of its constituents. Different segments of the global ecosystem (e.g., soil, fresh water, marine ecosystems) operate at different intrinsic-turnover rates and therefore differ in the limits of their productivity. On purely theoretical grounds it is self-evident that any economic system which is impelled to grow by constantly increasing the rate at which it extracts wealth from the ecosystem must eventually drive the ecosystem to a state of collapse. Computations of the limits of the rate of growth of the global ecosystem or of any major part of it are, as yet, in a rather primitive state. Such a limit may be reached ever more rapidly if the growth of the economic system is dependent on productive activities which are especially destructive of the stability of the ecosystem.

Unlike all other forms of life, human beings are capable of exerting environmental effects which extend, both quantitatively and qualitatively, far beyond their influence as biological organisms. Human activities have also introduced into the environment not only intense stresses due to natural agents (such as bodily wastes), but also wholly new substances not encountered in natural environmental processes: artificial radioisotopes, detergents, pesticides, plastics, a variety of toxic metals and gases, and a host of man-made, synthetic substances. These human intrusions on the natural environment have thrown major segments of the ecosystem out of balance. Environmental pollution is the symptom of the resultant breakdown of the environmental cycles.

## The Problem

In order to evaluate the cost of economic growth in terms of environmental deterioration, it is necessary to define both terms—if possible, in quantitative dimensions that permit a description of their relationship. The common definition of economic growth is applicable here: the increase in output of goods and services generated by economic activity. Environmental deterioration is a more elusive concept. On the basis of the foregoing discussion it may be defined as degradative changes in the ecosystems which are the habitat of all life on the planet. The problem is, then, to describe such changes in terms that can be related, quantitatively if possible, to the processes of economic growth—i.e., to increased production of economic goods.

To begin with, we must recognize the self-governing nature of the ecosystem. It is this basic property which ensures its stability and continued activity. It helps to define both the process of ecological degradation and the nature of the agencies that can induce it. We can define ecological, or environmental, degradation as a process which so stresses an ecosystem as to reduce its capability for self-adjustment, and which, therefore, if continued can impose an irreversible stress on the system and cause it to collapse.

An agency which is capable of exerting such an effect on an ecosystem must arise from *outside* that system. This results from the cyclical nature of the ecosystem, which brings about, automatically, the system's readjustment to any *internal* change in the number or activity of any of its normal biological constituents. For what characterizes the behavior of a constituent of an ecological cycle is that it both influences and is influenced by the remainder of the cycle. For example, organic waste produced by fish in a closed aquatic ecosystem, such as a balanced aquarium, cannot degrade the system because the waste is converted to algal nutrients and simply moves through the ecological cycle back to fish. In contrast, if organic waste intrudes upon this same ecosystem from without, it is certain to speed up the cycle's turnover rate and, if sufficiently intense, to consume all of the available oxygen and bring the cycle to a halt.

The internal changes in an ecosystem which occur in response to an external stress are complex, nonlinear processes and not readily reduced to simple quantitative indices. The aquatic ecosystem is one of the relatively few instances in which this goal can, to some degree, be approached because oxygen tension is a sensitive internal indicator of the system's approach to instability. However, in most cases, such internal measures of the state of an ecosystem have not yet been elucidated. Hence, as a practical expedient, we must fall back on a measure of the *impact* on the ecosystem of an external degradative agency as an index of environmental quality. This expedient has the virtue of making possible the quantitative comparison of the effects of ecological impacts of different origins, which is of particular importance in considering their relation to economic processes. Such data can be translated to the resultant internal changes when the necessary ecological information becomes available.

In what follows, then, the environmental cost of a given economic process will be represented by its *environmental impact,* a term which represents the amount of an agency external to the ecosystem which, by intruding upon it, tends to degrade its capacity for self-adjustment.

Turning to the possible environmental impacts that may result from *human* activity, we find the situation somewhat complicated by the special role of human beings on the earth. In one sense, human beings are simply another animal in the earth's ecosystem, consuming oxygen and organic foodstuff and producing carbon dioxide, organic wastes, heat, and more people. In this role the human being is a constituent part of an ecosystem and in terms of the previous definition exerts no impact on it. However, a human population has a zero environmental impact only as long as it is in fact part of an ecosystem, which is the case, for example, if food is acquired from soil which receives the population's organic waste. If a population is separated from this cycle by, for example, settling in a city, their wastes are intruded, with or without treatment, into surface waters. Now the population is no longer a part of the soil ecosystem, and the wastes are *external* to the *aquatic* system on which they intrude. An environmental impact is generated, leading to water pollution.

On this basis, people—viewed simply as biological organisms—generate an environmental impact only insofar as they become separated from the ecosystem to which terrestrial animals naturally belong. This is, of course, nearly universally true in the United States. The intensity of this environmental impact is generally proportional to the population size.

All other environmental impacts are generated not by human biological activities, but by human *productive activities,* and are governed by economic processes. They may be generated in several different ways:

1) Certain economic gains are derived from an ecosystem by exploiting its biological productivity. In these cases, a constituent of the ecosystem which has economic value—for example, an agricultural crop, timber, or fish—is withdrawn from the ecosystem. Insofar as neither the withdrawn substance or a suitable substitute returns as nutrient to the ecosystem from which it came, it constitutes a drain on the system which cannot continue indefinitely without causing its collapse. Destructive erosion of the soil following excessive exploitation or the incipient destruction of the whaling industry due to the extinction of whales are examples of such effects.

2) Environmental stress may also arise from an opposite sort of intrusion. The amount of some component of the ecosystem may be augmented from outside that system rather than withdrawn from it. This may be done either to dispose of waste or to accelerate the system's rate of turnover and increase its yield. Examples are the intrusion of sewage into surface water, and the intensive use of fertilizer nitrogen in agriculture. In the latter case, following a reduction in the nitrogen available from the soil's natural store of nutrient (its organic humus) due to over-exploitation through uncompensated crop withdrawal (i.e., a stress of the type described above), the nitrate level is artificially raised by adding fertilizer to the soil. Because of the low efficiency of nutrient uptake by the crop's roots (which is in turn a result of inadequate soil oxygen due to reduced porosity steming from the decreased humus content), a considerable portion of the fertilizer leaches from the soil into surface waters. There it becomes an external stress on the aquatic ecosystem, causing algal overgrowths and a breakdown of the self-purifying aquatic cycle.

3) Apart from stresses caused by externally altered concentrations of natural constituents, environmental impact may be due to the intrusion into an ecosystem of a substance wholly foreign to it. Thus, DDT has a powerful environmental impact in part because it readily upsets the naturally balanced ecological relations among insect pests, the plants they attack, and the insects which, in turn, prey on the pests. DDT-induced

outbreaks of insect pests often result. In general, there is a considerable risk of environmental pollution whenever productive activity introduces substances which are foreign to the natural environment.

We turn now to the practical problem of evaluating the environmental cost of economic growth. The most general theoretical aspect of this problem has already been alluded to. Since the global ecosystem is closed, and its integrity is essential to the continued operation of any conceivable economic system, it is evident that there must be an upper limit to the growth of productive activities on the earth.

However, such a theoretical statement is hardly an effective guide to practice. It fails to specify the time scale in which the ecological limitation on economic growth is likely to take effect. One can readily grant the truth of such an abstract theorem—for example, that economic growth will eventually be limited by the extinction of the sun—and disregard its practical consequences because of the rather long time-scale involved, in this case some billions of years. Accordingly, it would seem useful to make the problem more concrete by examining the relationship between economic growth and environmental impact in the real world. And, since growth is, of course, a time-dependent process, an historical approach is called for.

## The Origins of Environmental Impacts

In what follows, I shall report the results of an initial effort to describe the origins of environmental impacts in the United States.* Most pollution problems are of relatively recent origin. The post-war period, 1945–46, is a convenient benchmark, for a number of pollutants—man-made radioisotopes, detergents, plastics, synthetic pesticides and herbicides—are due to the emergence, after the war, of new technologies. The statistical data available for this period enable us to compare changes in the levels of various pollutants with the concurrent economic activities that might have significant environmental effects.

Although we lack sufficient data on the actual levels of most pollutants, some estimates of historical changes can be made from intermittent observations and from data on emissions of pollutants. Some of the data are summarized in Table 1, which indicates that since 1946, emissions of various pollutants have increased anywhere from 200 to 2000 per cent. In the case of phosphate, which is a pollutant of surface waters that enters mainly from municipal sewage, data on long-term trends show that in the 30-year period between 1910 and 1940, phosphorus

* This study has been carried out as part of the program of the American Association for the Advancement of Science Committee on Environmental Alterations in collaboration with Michael Corr and Paul J. Stamler.

output from municipal sewage increased gradually from about 17 million to about 40 million pounds per year. Thereafter the rate accelerated so rapidly that in the 30-year period from 1940 to 1970 phosphorus output rose to about 300 million pounds per year.

TABLE 1. Postwar Increases In Pollutant Emissions

| | Annual Production | | | | *Percent increase over indicated period* |
|---|---|---|---|---|---|
| *Pollutant* | *Year* | *Amount* | *Year* | *Amount* | |
| Inorganic Fertilizer Nitrogen | 1949 | $.91 \times 10^6$ tons | 1968 | $6.8 \times 10^6$ tons | 648 |
| Synthetic Organic Pesticides | 1950 | $286 \times 10^6$ lbs. | 1967 | $1050 \times 10^6$ lbs. | 267 |
| Detergent Phosphorus | 1946 | $11 \times 10^6$ lbs. | 1968 | $214 \times 10^6$ lbs. | 1,845 |
| Tetraethyl Lead[b] | 1946 | $.048 \times 10^6$ tons | 1967 | $.25 \times 10^6$ tons | 415 |
| Nitrogen Oxides[b] | 1946 | 10.6[a] | 1947 | 77.5[a] | 630 |
| Beer Bottles | 1950 | $6.5 \times 10^6$ gross | 1967 | $45.5 \times 10^6$ gross | 595 |

[a] Dimension = NOx (ppm) × gasoline consumption (gals $\times 10^{-6}$); estimated from product of passenger vehicle gasoline consumption and ppm of NOx emitted by engines of average compression ratio 5.9 (1946) and 9.5 (1967) under running conditions, at 15 in. manifold pressure. NOx emitted: 500 ppm in 1946; 1200 ppm in 1967 (Ref.)
[b] Automotive emissions

It should be noted that these data record the *emissions* of pollutants, which do not necessarily correspond to their concentrations in the environment or to their ultimate effects on the ecosystems or human health. Numerous, complex, and interrelated processes intervene between the entry of a pollutant into the ecosystem and its biological effect. Moreover, two or more pollutants may interact synergistically to intensify the separate effects. Most of these processes are still too poorly understood for us to convert the amount of a pollutant entering an ecosystem to a quantitative estimate of its destructive effects. Nevertheless, it is evident that these effects (such as the incidence of respiratory disease due to air pollutants, or of algal overgrowths due to phosphate and nitrate) have increased sharply, with the rapid rise of pollutant levels since 1946. Since pollutant emission is a direct measure of the activity of the source, it is a useful way to estimate the contributions of different sources to the overall degradation of the environment.

Let us call the amount of a given pollutant introduced annually into the environment the *environmental impact, I.* We can relate this value to the effects of three major factors that influence it by means of the identity:

$$I = \text{Population} \cdot \frac{\text{Economic Good}}{\text{Population}} \cdot \frac{\text{Pollutant}}{\text{Economic Good}}.$$

*Population* refers to the size of the population in a given year. *Economic Good* is the amount of a designated good produced (or where appropriate, consumed) during the year, and *Pollutant* refers to the amount of a specific pollutant (defined as above) released into the environment as a result of the production (or consumption) of the good during the year. This relationship enables us to estimate the contributions of three factors to the total environmental impact: the size of the population, production (or consumption) per capita, and the amount of pollutant generated per unit of production (or consumption), which reflects the nature of technology.

Since we are concerned with identifying the sources of the sharp increases in the environmental impacts in the United States in the period from roughly 1946 to the present, it is of interest to examine the changes in the nation's productive activities that occurred during that period. Between 1946 and 1968 population increased, at an approximately constant rate, by about 42 percent; GNP (adjusted to 1958 dollars) increased by about 126 percent, and GNP per capita grew approximately exponentially by about 59 percent.

As a first approximation the contribution of population growth to the overall values of environmental impacts since 1946 is of the order of 40 percent. In most cases this represents a relatively small contribution to the total impact, since, as indicated in Table 1, these values increased by from 200 to 2000 percent in that period.

In order to evaluate the effects of the remaining factors it is useful to examine the growth rates of different sectors of the economy. A series of productive activities which are likely to contribute significantly to environmental impact and which are representative of the overall pattern of the economy were selected, and their annual percentage rates of change between 1947 and 1970 are presented in Table 2, from which it is possible to derive some useful generalizations about the pattern of economic growth in relation to environmental impacts.

Production and consumption of certain goods have increased at an annual rate about equal to that of the population. This group includes food, fabric and clothing, major household appliances, and certain basic metals and building materials, including steel and copper and brick. With respect to these basic items, *per capita* production (or consumption) has remained essentially unchanged from 1947 to 1970.

The annual production of certain other goods has decreased since 1947 or has increased at an annual rate below that of the population. Horse-power produced by work animals is the extreme case; it declined at an annual rate of about 8 percent. Other items in this category are cotton fiber, wool fiber, lumber, railroad horsepower, and railroad

TABLE 2. Percent Annual Increase of Production (Consumption) of Various Products in the United States, 1947–70.

| Item | % Annual increase |
|---|---|
| Mercury for Chloralkali * | 15.931 |
| No Return Beer Bottles * | 14.815 |
| Plastics | 12.864 |
| Fertilizer Nitrogen * | 10.242 |
| Surface Active Agents | 10.034 |
| Synthetic Organic Chemicals | 9.843 |
| Chlorine Gas | 8.112 |
| Aluminum | 7.907 |
| Automotive Horsepower | 7.837 |
| Electric Power | 7.609 |
| Total Horsepower | 7.576 |
| Synthetic Fibers | 7.087 |
| Synthetic Pesticides | 6.596 |
| Wood Pulp | 5.832 |
| Truck Freight | 4.795 |
| Consumer Electronics | 4.723 |
| Motorfuel * | 4.363 |
| Total fuel * | 2.731 |
| Newsprint | 2.715 |
| Total Freight | 2.438 |
| Steel | 2.088 |
| All Fibers * | 2.046 |
| New Copper | 2.014 |
| Bricks | 1.811 |
| Population | 1.643 |
| Hosiery | 1.609 |
| Shoes | 1.332 |
| Major Appliances * | 1.053 |
| Railroad Freight | .618 |
| Lumber | − .068 |
| Cotton Fiber | − .317 |
| Returnable Beer Bottles * | −1.785 |
| Wool Fiber * | −2.167 |
| Railroad Horsepower | −3.660 |
| Work Animal Horsepower | −7.948 |

* Consumption (all other figures: production)

freight. These are goods which have been significantly displaced in the pattern of production during the course of the overall growth of the economy. Cultivated farm acreage also declined in this period.

A third group of productive activities increased at an annual rate in excess of population. Among these, three different classes can be discerned. Certain rapidly increasing productive activities are substitutes for others that have declined relative to population. These generally represent technological displacement of an older process by a newer one, with the sum of goods produced remaining essentially constant, per capita, or increasing somewhat. These displacements include natural fibers (cotton and wool) by synthetic fibers; lumber by plastics; soap by detergents; steel by aluminum and concrete; railroad freight by truck freight; harvested acreage by fertilizer, and returnable by nonreturnable bottles. The second class of activities includes secondary consequences of displacements. The displacement of natural products by synthetic ones, for example, involves the use of a greatly increased amount of synthetic organic chemicals, so that this category has increased sharply. Moreover, since many organic syntheses require chlorine as a reagent, the rate of chlorine production has also increased rapidly. Finally, because chlorine is efficiently produced in a mercury electrolytic cell, the use of mercury for this purpose has increased at a very considerable rate. Similarly, the rapidly rising rate of power utilization is, in part, a secondary consequence of certain displacement processes, for a number of the new technologies are more power-consumptive than the technologies which they replace. Finally, some of the rapidly growing activities represent neither displacements of older technologies nor sequelae to such displacements but increments in per-capita availability of consumer goods such as consumer electronics (radios, television sets, sound equipment, etc.).

In sum, the pattern of growth in the United States economy since 1946–47 may be generalized as follows. The annual production of basic life necessities, representing perhaps one-third of the total GNP, grew at about the pace of the population, so that no significant overall change in *per capita* production took place. However, within these general categories—food, fiber, and clothing; freight haulage; household necessities—there was a pronounced displacement of natural products by synthetic ones, of power-conservative products by relatively power-consumptive ones, of reusable containers by "disposable" ones.

## The Environmental Impact of Economic Growth

We can now rephrase the original question: What are the relative costs, in intensity of environmental impact, of the several distinctive features of the growth of the United States economy from 1946 to the present? Reasonably complete quantitative answers to this question are, unfortunately, well beyond the present state of knowledge. At present it is possible in most cases to provide only an informal, qualitative description of the changes in environmental impact which have been induced by the postwar transformation of the economy. In some cases it is also possible to produce a quantitative evaluation in the form of an Environmental Impact Index. Such evidence leads to the general conclusion that in most of the technological displacements which have accompanied the growth of the economy since 1946, *the new technology has an appreciably greater environmental impact than the technology which it has displaced, and the postwar technological transformation of productive activities is the chief reason for the present environmental crisis.*

Let us now consider six economic activities whose environmental impacts are especially significant.

## *Agricultural Production*

Agricultural production in the United States, as measured by the U.S. Department of Agriculture Crop Index, has increased at about the same rate as the population since 1946. However, the technological methods for achieving agricultural production have changed significantly in that period. Although agricultural production per capita has increased slightly, harvested acreage has decreased, and the use of inorganic nitrogen fertilizer has risen sharply. This displacement process—i.e., of fertilizer for land—leads to a considerably increased environmental impact.

Briefly stated, the relevant ecological situation is as follows.[1] Nitrogen, an essential constituent of all living things, is available to plants in nature from organic nitrogen stored in the soil in the form of humus. Humus is broken down by bacteria to eventually release inorganic forms of nitrogen as nitrate. The latter is taken up by plant roots and reconverted to organic matter, such as the plant's protein. Finally, the plant may be eaten by a grazing animal, which returns the nitrogen not retained in its own body to the soil as bodily wastes.

Agriculture imposes a drain on this cycle: nitrogen is removed from the system in the form of the plant crop or of livestock produced from it. In ecologically sound husbandry, all of the organic nitrogen, other than the food itself, produced by the soil system—plant residues, manure, garbage—is returned to the soil, where it is converted by complex microbial processes to humus and thus helps to restore the soil's organic nitrogen content. The deficit, if it is not too large, can be made up by the process of nitrogen fixation in which bacteria, usually in close association with the roots of certain plants, take up nitrogen gas from the air and convert it into organic form. If the nitrogen cycle is not in bal-

1. See Commoner, Barry, "Threats to the Integrity of the Nitrogen Cycle: Nitrogen Compounds in Soil, Water, Atmosphere and Precipitation," *Global Effects of Environmental Pollution,* Symposium organized by American Association for the Advancement of Science, Dallas, Texas, December 1968, edited by S. Fred Singer, Reidel, Dordrecht-Holland (1970).

ance, agriculture "mines" the soil nitrogen, progressively depleting it. This process does more than reduce the store of organic nitrogen available to support plant growth, for humus is more than a nutrient store. Humus is also responsible for the porosity of the soil to air. And air is essential to the soil, not only as a source of nitrogen for fixation, but also because its oxygen is essential to the root's metabolic activity. For example, in Corn Belt soils about one-half of the original organic nitrogen has been lost since 1880. Naturally, other things being equal, such soil is relatively infertile and produces relatively poor crop yields. However, beginning after World War II a technological solution was intensively applied to this problem. Sharply increasing amounts of inorganic nitrogen were applied to the soil in the form of fertilizer. Annual nitrogen-fertilizer usage in the United States increased by an order of magnitude from 1946 to 1968.

In effect, then, nitrogen fertilizer can be regarded as a substitute for land. With the intensive use of fertilizer it becomes possible to accelerate the turnover rate of the soil ecosystem so that each acre of soil produces more food than before. The economic benefits of this new technology are appreciable and self-evident. However, this economic advantage may be counterbalanced by the increased impact on the environment because, in view of the reduced humus content of the soil, the plant's roots do not absorb the added fertilizer efficiently. As a result an appreciable part leaches from the soil as nitrate and enters surface waters, where it becomes a serious pollutant. Nitrate may encourage algal overgrowths which, on their inevitable death and decay, tend to break down the self-purifying aquatic cycle.

Excess nitrate from fertilizer drainage leads to still another environmental impact which may affect human health. While nitrate in food and drinking water appears to be relatively innocuous, *nitrite* is not, for it combines with hemoglobin in the blood, converting it to methemoglobin, which cannot carry oxygen. Unfortunately, nitrate can be converted to nitrite by the action of bacteria in the intestinal tract, especially in infants, causing asphyxiation and even death. On these grounds, the United States Public Health Service has established 10 ppm of nitrate nitrogen as the acceptable limit of nitrate in drinking water. In a number of agricultural areas in the United States nitrate levels in water supplies obtained from wells, and in some instances from surface waters, have exceeded this limit. Our own studies in the area of Decatur, Illinois, show quite directly that in the Spring of 1970, when the city's water supply, which is derived from an impoundment of the Sangamon River, recorded 9 ppm of nitrate nitrogen, a minimum of 60 percent of the nitrate was derived from inorganic fertilizer applied to the surrounding farmland.[2] It is possible to compute the environmen-

2. Kohl, D. H., Shearer, G. B., and Common, Barry, "Isotopic Analysis of the Movement of Fertilizer Nitrogen into Surface Water." In press 1971.

tal impact of the shift to intensive use of fertilizer from the formula given for *I*. During the period 1949–68 the total annual use of fertilizer nitrogen, i.e., the total environmental impact, increased by 648 percent. In other words, the amount used in 1968 was 7.48 times the 1949 level. Population in 1968 was 1.34 times and crop production per capita 1.11 times the 1949 amount. It follows that pollution generated per unit of output must have been 5.05 times the 1949 quantity ($1.34 \times 1.11 \times 5.05 = 7.48$), an increase of 405 percent compared with an increase of 47.7 percent in crop production. Clearly the change in technology was the dominant factor in the large increase in the total environmental impact of fertilizer. Specifically, in 1949 about 11,000 tons of fertilizer nitrogen were used *per unit of crop production,** while in 1968 about 57,000 tons were employed for the *same* output. This means that the efficiency with which fertilizer nitrogen contributes to crop yield *declined* by 80 percent. Obviously, an appreciable part of the added nitrogen does not enter the crop and must appear elsewhere in the ecosystem.

A similar situation exists in the case of pesticides. Between 1950 and 1967 there was a 168 percent increase in the amount of pesticides used *per unit of crop production,* as a national average. By killing off natural insect predators and parasites of the target pest, while the latter often becomes resistant to insecticides, the use of modern synthetic insecticides tends to exacerbate the pest problems that they were designed to control. As a result *increasing* amounts of insecticides must be used to maintain agricultural productivity. Insecticide usage is, so to speak, self-accelerating—resulting in a decreased efficiency and an increased environmental impact.

Another technological displacement in agriculture is the increased use of feedlots in preference to range feeding for the production of livestock. Range-fed cattle are integrated into the soil ecosystem; they graze the soil's grass crop and restore nutrient to the soil as manure. When the cattle are maintained instead in huge pens, where they are fed on corn and desposit their wastes intensively in the feedlot itself, the wastes do not return to the soil. Instead the waste drains into surface waters where it adds to the stresses due to fertilizer nitrogen and detergent phosphate. The magnitude of the effect is considerable. At the present time the organic waste produced in feedlots is more than the organic waste produced by all the cities of the United States. Again, the newer technology has a serious environmental impact, and in this case has displaced a technology with an essentially zero environmental impact.

* Editors' note: A "unit" of crop output is equal to one percent of total agricultural production in the year 1958.

## *Textiles*

While total fiber production per capita has remained constant since 1946, natural fibers (cotton and wool) have been significantly displaced by synthetic ones. This technological change considerably increased the environmental impact of fiber production and use.

One reason is that the energy required for the synthesis of the final product, a linear polymer (cellulose in the case of cotton, keratin in the case of wool, and polyamides in the case of nylon) is greater for the synthetic material. Nylon production involves as many as 10 steps of chemical synthesis, each requiring considerable energy in the form of heat and electric power to overcome the entropy associated with chemical mixtures and to operate the reaction apparatus. In contrast, energy required for the synthesis of cotton is derived, free, from an essentially inexhaustible source—the sun—and is transferred without combustion and resultant air pollution. Moreover, the raw material for cellulose synthesis is carbon dioxide and water, both freely available renewable resources, while the raw material for nylon synthesis is petroleum or a similar hydrocarbon—non-renewable resources. As a result the environmental stress due to the *production* of such an artificial fiber is probably well in excess of that due to the production of an equal weight of cotton. This is only an approximation, for we need far more detailed, quantitative estimates, in the form of the appropriate environmental impact indices, that also take into account the fuel and other materials used in the production of cotton.

Because a synthetic fiber such as nylon is unnatural, it also has a greater impact on the environment as a waste material than cotton or wool. The natural polymers in cotton and wool, cellulose and keratin, are important constituents of the soil ecosystem. Through the action of molds and decay bacteria, they contribute to the formation of humus. Cellulose and keratin are *not* "wastes" to nature, but nutrients for soil microorganisms.

For every polymer which is produced in nature by living things, there exist in some living things enzymes which have the specific capability of degrading that polymer. In the absence of such an enzyme the natural polymers are quite resistant to degradation, as is evident from the durability of fabrics which are protected from biological attack.

The contrast with synthetic fibers is striking. The structure of nylon and similar synthetic polymers is a human invention and does not occur in natural living things. Hence, unlike natural polymers, synthetic ones find no counterpart in the armamentarium of degradative enzymes in nature. Ecologically, synthetic polymers are literally indestructible, so that every bit of synthetic fiber or polymer that has been produced on the earth is either destroyed by burning—and thereby pollutes the air

—or accumulates as rubbish. One result is that microscopic fragments of plastic fibers, often red, blue or orange, have now become common in certain marine waters.[3] For technological displacement has been at work in this area too; in recent years natural fibers such as hemp and jute have been nearly totally replaced by synthetic fibers in fishing operations. A chief reason for this use of synthetic fibers is that they resist degradation by molds, which, as already indicated, readily attack cellulosic net materials such as hemp or jute. Thus, the property which enhances the economic value of the synthetic fiber over the natural one is precisely the property which increases the environmental impact of the synthetic material.

## *Detergents*

Synthetic detergents have largely replaced soap in the United States as domestic and industrial cleaners, with the total production of cleaners per capita remaining essentially unchanged. Soap is based on a natural organic substance, fat. Being a natural product, fat is extracted from an ecosystem (for example, a coconut palm plantation), and when released into an aquatic ecosystem after use, soap is readily degraded by the bacteria of decay. Since most municipal wastes in the United States are subjected to treatment which degrades organic waste to its inorganic products, in actual practice the fatty residue of soap wastes is degraded by bacterial action within the confines of a sewage-treatment plant. What is then emitted to surface waters is only carbon dioxide and water. Hence, there is little or no impact on the aquatic ecosystem due to biological oxygen demand (which accompanies bacterial degradation of organic wastes) arising from soap wastes. Nor is the product of soap degradation, carbon dioxide, usually an important ecological intrusion, since it is in plentiful supply from other environmental sources, and in any case is an essential nutrient for photosynthetic algae. The production of synthetic detergents is a more serious source of pollution than the production of soap.

Once used and released into the environment as waste, detergents, furthermore, generate a more intense environmental impact than a comparable amount of soap. Soap is wholly degradable to carbon dioxide. In contrast, even the newer detergents, which are regarded as degradable because the paraffin chain of the molecule (being unbranched, in contrast with the earlier non-degradable detergents) is broken down by bacterial action, nevertheless leave a residue of phenol which may not be degraded and may accumulate in surface waters. Phenol is a rather toxic substance, being foreign to the aquatic ecosystem.

Unlike soap, detergents are compounded with considerable amounts

3. See note in *Marine Pollution Bulletin 2*, p. 23, February 1971.

of phosphate in order to enhance their cleansing action and to soften water. Phosphate may readily induce water pollution by stimulating heavy overgrowths of algae which, on dying, release organic matter into the water and overburden the aqueous ecosystem. Nearly all of the increase in sewage phosphorus in the United States can be accounted for by the phosphorus content of detergents. Since soap, which has been displaced by detergents, is quite free of phosphate, the environmental impact due to phosphate is clearly a consequence of the technological change in the production of cleaners.

The environmental-impact index of phosphate in cleaners between 1946 and 1968 increased 1845 percent. The increase in the effect of population size was 42 percent; the effect of per-capita use of cleaners did not change; the technological factor increased about 1270 percent.

## *Secondary Environmental Effects of Technological Displacements*

Increased production of synthetic organic chemicals leads to intensified environmental impacts in several different ways. This industry has heavy power requirements which add to the rising levels of air pollutants emitted by power plants. In addition, organic synthesis releases into the environment a wide variety of reagents and intermediates, which are foreign to natural ecosystems and often toxic, generating important, often poorly understood, environmental impacts. Common examples are massive fish kills and plant damage resulting from release of organic wastes, insecticides, and herbicides to surface waters or the air.

Perhaps the most serious environmental impact attributable to the increased production of synthetic organic chemicals is the intrusion of mercury into surface waters. This effect is mediated by chlorine production. This substance is a vital reagent in many organic syntheses; about 80 percent of present chlorine production finds its end use in the synthetic organic chemical industry. Moreover, a considerable proportion of chlorine production is carried out in electrolytic mercury cells; until recent control measures were imposed on the industry, 0.2 to 0.5 pounds of mercury were released to the environment per ton of chlorine manufactured in mercury electrolytic cells. This means, for example, that the substitution of nylon for cotton has generated an intensified environmental impact due to mercury, for nylon production (unlike cotton production) involves the use of chlorinated intermediates, therefore chlorine, and hence the release of mercury into the environment.

Similarly the displacement of steel and lumber by aluminum adds to the burden of air pollutants, for aluminum production is extremely power-consumptive. The production of chemicals, aluminum, and cement account for about 28 percent of the total industrial use of electricity in the United States.

## *Packaging*

The displacement of older forms of packaging by "disposable" containers, such as nonreturnable bottles, is another example of the intensification of environmental impact due to the postwar pattern of U.S. economic growth. There has been a striking increase in environmental impact due to beer bottles, which are not assimilated by ecological systems and are, in their manufacture, quite power-consumptive. The major factor in this intensified environmental impact is the new technology—the use of nonreturnable bottles to contain beer—rather than an increase in the per-capita consumption of beer, or increased population. At the same time a recent study shows that the total expenditure of energy (for bottle manufacture, processing, shipping, etc.) required to deliver equal amounts of fluid in nonreturnable bottles is 4.7 times that for returnable ones.[4]

## *Automotive Vehicles*

Finally, there is the environmental impact of changes in patterns of passenger travel and freight traffic since 1946. Particularly important has been the increased use of automobiles, busses, and trucks.

The environmental impact of the internal-combustion engine is due to the emission of nitrogen oxides, carbon monoxide, waste fuel, and lead. The intensities of these impacts, as measured by the levels of these pollutants in the environment, is a function not only of the vehicle-miles traveled, but also of the nature of the engine itself—i.e., technological factors are relevant as well.

The technological changes in automotive engines since World War II have worsened environmental impact. For passenger automobiles, overall mileage per gallon of fuel declined from 14.97 in 1949 to 14.08 in 1967, largely because average horsepower increased from 100 to 240. Another important technological change was in the average compression ratio, which increased from about 5.9 to 9.5 between 1949 and 1968. This engineering change has had two important effects on the environmental impact of the gasoline engine. First, increasing amounts of tetraethyl lead are needed as a gasoline additive in order to suppress the engine knock that occurs at high compression ratios. The annual use of tetraethyl lead increased by over 400 percent from 1946 to 1968. Essentially all of this lead is emitted from the engine exhaust and is disseminated into the environment. Since lead is not a functional element in any biological organism, and is in fact toxic, it represents an external intrusion on the ecosystem and generates an appreciable environmental effect.

4. Bruce Hannon (University of Illinois, Urbana), personal communication.

A second consequence of the increase in engine-compression ratio has been a rise in the concentration of nitrogen oxides emitted in engine exhaust. This has occurred because the engine temperature increases with the compression ratio. The combination of nitrogen and oxygen, present in the air taken into the engine cylinder, to form nitrogen oxides is enhanced at elevated temperatures. Nitrogen oxide is the key ingredient in the formation of photochemical smog. Through a series of light-activated reactions involving waste fuel, nitrogen oxides induce the formation of peroxyacetyl nitrate, the noxious ingredient of photochemical smog. Smog of this type was first detected in Los Angeles during 1942 and 1943; it was unknown in most other United States cities until the late 1950s and 1960s, but is now a nearly universal urban pollutant. Peroxyacetyl nitrate is a toxic agent to man, agricultural crops, and trees. Introduction of this agent probably increased by about an order of magnitude from 1946 to 1968.

The total environmental impact of nitrogen oxides increased by about 630 percent between 1946 and 1967. The technological factor (the amount of nitrogen oxides emitted per vehicle-mile) increased by 158 percent, vehicle-miles traveled per capita increased by about 100 percent, and the population factor by about 41 percent. In the case of tetraethyl lead, the largest increase in impact is in vehicle-miles traveled per capita (100 percent), followed by the technological factor (83 percent) and the population factor (41 percent). It is evident that the major influences on automotive air pollution are increased per-capita mileage (in part because of changes in work-residence distribution due to the expansion of the suburbs) and the increased environmental impact per mile traveled, due to technological changes in the gasoline engine.

A similar situation obtains with respect to overland shipments of inter-city freight. Here truck freight has tended to displace railroad freight. And, again, the displacing technology has a more severe environmental impact than does the displaced technology. This is evident from the energy required to transport freight by rail and truck: 624 BTU per ton-mile by rail and 3462 BTU per ton-mile by truck. It should be noted as well that the steel and cement required to produce equal lengths of railroad and expressway (suitable for heavy truck traffic) differ in the amount of power required in the ratio of 1 to 3.6. This is due to the power-consumptive nature of cement production and to the fact that four highway lanes are required to accommodate heavy truck traffic. In addition, the divided roadway requires a 400-foot right-of-way, while a train roadbed needs only 100 feet. In all these ways the displacement of railroads by automotive vehicles, not only for freight, but also for passenger travel, has intensified the resultant environmental impact.

## Some Conclusions

The data presented here reveal a functional connection between economic growth—at least in the United States since 1946—and environmental impact. It is significant that the range of increase in the computed environmental impacts agrees fairly well with the independent measure of the actual levels of pollutants occurring in the environment. Thus, the increase in the environmental impact index for tetraethyl lead computed from gasoline-consumption data for 1946–67 is about 400 percent; a similar increase in environmental-lead levels has been recorded from analyses of layered ice in glaciers.[5] Similarly, the 648 percent increase in the 19-year period 1949–68 in the environmental-impact index computed for nitrogen fertilizer is in keeping with the few available large-scale field measurements. Thus field data show that nitrate entering the Missouri River as it traversed Nebraska in the 6-year period 1956–62 increased a little over 200 percent.[6] The environmental-impact indices computed for several aspects of automotive vehicle use are also in keeping with general field observations. It is widely recognized that the most striking increase among the several aspects of environmental deterioration due to automotive vehicles has occurred with respect to photochemical smog. It is significant, then, that this disparity between the observed increase in smog levels and the increase in vehicle use is accounted for by the environmental-impact index computed for nitrogen oxides, the agent which initiates the smog reaction, for that index increased by 630 percent from 1946 to 1967.

These agreements with field data support the conclusion that the environmental-impact indices provide useful approximations of the changes in environmental impact associated with the relevant features of the growth of the United States economy since 1946. We can therefore place some reliance on the subdivision of the total-impact index into the several factors: population size, per-capita production or consumption, and the technology of production and use.

It is of interest to make a direct comparison of the relative contributions of increases in population size, per-capita consumption, and changes in the technology of production, to the increases in total environmental impact since 1946. The ratio of the value of the most recent total index to the value of the 1946 index (or to the value for the earliest year for which the necessary data are available) is indicative of the change in the total impact over this period of time. The relative contri-

5. Patterson, C. C., *Environment 10,* p. 72 (1967).
6. Commoner, *op. cit.*

butions of the several factors to these total changes is given by the ratios of their respective partial indices. Figure 1 reports this comparison for the six productive activities evaluated. The population factor contributes only between 12 and 20 percent of the total changes in impact indices. For all but the automotive pollutants, the increase in per-capita consumption makes a rather small contribution—no more than 5 percent—to the total changes in impact index. For nitrogen oxides and tetraethyl lead (from automotive sources), this factor accounts for about 40 percent of the total effect, reflecting a considerable increase in the

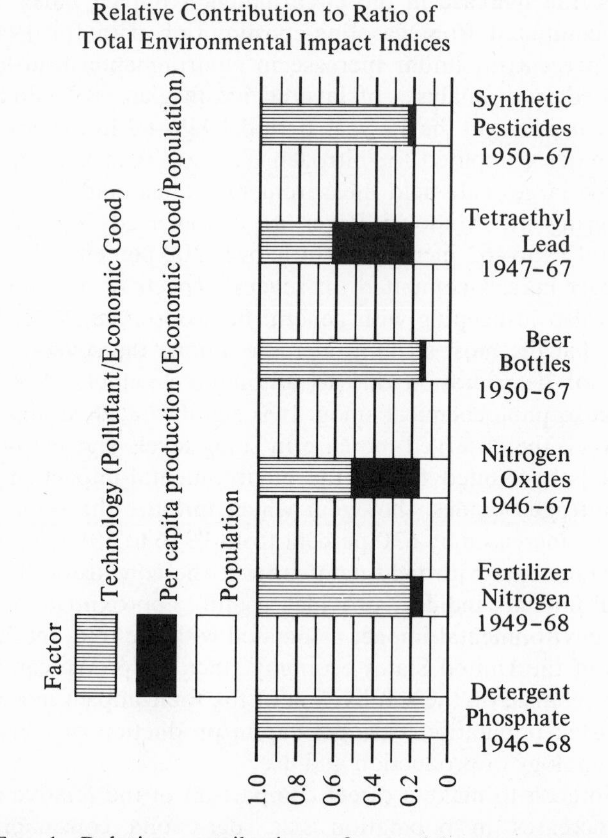


FIG. 1. Relative contributions of several factors to changes in environmental-impact indices. The contributions of population size, production per capita, and technological characteristics (amount of pollutant released per unit of production) to the total environmental-impact indices were computed as shown in the text. Each bar is subdivided to show the *relative* contribution, on a scale of 1.0, of each factor to the change in the environmental-impact index between the earlier year and the later year.

number of vehicle-miles traveled per capita since 1946. The technological changes in the processes which generate the various economic goods contribute from 40 to 90 percent of the total increase in impact.

In evaluating these results it should be noted that automotive travel is itself strongly affected by a kind of technological transformation: the rapid increase of suburban residences in the United States and the concomitant failure to provide adequate railroad and other mass transportation to accommodate to this change. That the overall increase in vehicle-miles traveled per capita since 1946 (about 100 percent) is related to increased residence-work travel incident upon this change is suggested by the results of a 1963 survey. It was found that 90 percent of all automobile trips, representing 30 percent of the total mileage traveled, are ten miles or less in length. The mean residence-work travel distance was about 5.5 miles. Thus, it is probably appropriate to regard the increase in per-capita vehicle-miles traveled by automobile as not totally attributable to increased "affluence," but rather as a response to new work-residence relationships which are costly in transportation.

Since 1946, pollution levels in the United States have increased sharply—generally by an order of magnitude or so. It seems evident from the data that most of this increase is due to the technology of production and that growth in both population and per capita consumption exert a much smaller influence. Thus the chief reason for the sharp increase in environmental stress in the United States is the sweeping transformation in production technology in the postwar period. *Productive activities with intense environmental impacts have displaced activities with less serious environmental impacts; the growth pattern has been counter-ecological.*

This conclusion is easily misconstrued to mean that technology is, *per se,* ecologically harmful. A few examples will make it clear that this interpretation is unwarranted.

Suppose that the sewage, instead of being introduced into surface waters as it is now, whether directly or following treatment, were transported from urban collection systems by pipeline to agricultural areas, where—after appropriate sterilization procedures—it was incorporated into the soil. Such a pipeline would literally reincorporate the urban population into the soil's ecological cycle, restoring the integrity of that cycle and, incidentally, removing the need for inorganic nitrogen fertilizer—which also stresses the aquatic cycle. The urban population would no longer be external to the soil cycle and would therefore be incapable of generating a negative biological stress upon it or exerting a positive ecological stress on the aquatic ecosystem. But note that this state of zero environmental impact is not achieved by a return to "primitive" conditions, but by an actual technological advance, the construction of a sewage-pipeline system.

Or consider the technological treatment of gold and other precious metals. Gold is, after all, subject to numerous technological manipulations which generate economic value. Yet we manage to accomplish all of this without intruding more than a small fraction of all the gold ever acquired by human beings into the ecosphere. Because we value it so highly very little gold is "lost" to the environment. In contrast, most of the mercury which has entered commerce in the last generation has been disseminated into the environment, with very unfortunate effects on the environment. Clearly, given adequate technology—and motivation—we could be as thrifty in our handling of mercury as we are of gold, thereby preventing the entry of this toxic material into the environment. Again, what is required is not necessarily the abandonment of mercury-based technology, but rather the improvement of that technology to the point of satisfactory compatibility with the ecosystem.[7]

Generally speaking, then, it is possible to reduce the environmental impact of human activities by developing alternatives to ecologically faulty activities. This can be accomplished, not by abandoning technology and the economic goods which it yields, but by developing *new* technologies which incorporate not only the knowledge of the physical sciences (as most do moderately well; the new machines do, after all, usually produce their intended goods), but ecological wisdom as well.

The deterioration of the environment, and its cost in money, social distress, and personal suffering, is chiefly the result of the ecologically faulty technology which has been employed to remake productive enterprises. The resulting environmental impacts stress the basic ecosystems which support the life of human beings, destroy the "biological capital" which is essential to the operation of industry and agriculture, and may, if unchecked, lead to the catastrophic collapse of these systems. The environmental impacts already generated are sufficient to threaten the continued development of the economic system—witness the current difficulties of the United States in siting new power plants at a time of severe power shortage, the recent curtailment of industrial innovation in the fields of detergents, chemical manufacturing, insecticides, herbicides, chlorine production, oil drilling, oil transport, supersonic aviation, nuclear-power generation, industrial uses of nuclear explosives, all resulting from public rejection of the concomitant environmental deterioration.

If we are to survive economically as well as biologically, much of the technological transformation of the United States economy since 1946

7. Editors' note: This is true, but if we were as meticulous in handling mercury as we are with gold, it might make mercury virtually as expensive as gold. Whether or not this would be a good thing, it does suggest that misplaced efforts to protect the environment can impair the quality of life in other respects. They may also have adverse effects on the environment, as in the following chain: expensive mercury—no fluorescent lights—more incandescent lights—more power consumption—more thermal pollution, sulphur oxides and the like.

will need to be, so to speak, redone in order to bring the nation's productive technology much more closely into harmony with the inescapable demands of the ecosystem. This will require the development of massive new technologies, including: systems to return sewage and garbage directly to the soil; the replacement of synthetic materials by natural ones; the reversal of the present trend to retire soil from agriculture and to elevate the yield per acre; the development of land transport that operates with maximal fuel efficiency at low combustion temperatures; the sharp curtailment of the use of biologically active synthetic organic agents. In effect, what is required is a new period of technological transformation of the economy which will reverse the counter-ecological trends developed since 1946. The cost of the new transformation might represent a capital investment of hundreds of billions of dollars. To this must be added the cost of repairing the ecological damage which has already been incurred, such as the eutrophication of Lake Erie, again a bill to be reckoned in the hundreds of billions of dollars.

The enormous size of these costs raises a final question: Is there some functional connection in the economy between the tendency of a given productive activity to inflict an intense impact on the environment (and the size of the resultant costs) and its role in economic growth? It is evident from even a cursory comparison of the productive activities which have rapidly expanded in the economy since 1946 with those which they have displaced, that the former are considerably more profitable than the latter. The correlation between profitability and rapid growth is one that is presumably accountable by economics. Is the additional linkage to intense environmental impact also functional, or only accidental?

It has been pointed out often enough that environmental pollution represents a long-unpaid debt to nature. Is it possible that the United States economy has grown since 1946 by deriving much of its new wealth through enlargement of that debt? If this should turn out to be the case, what strains will develop in the economy if, for the sake of survival of our society, that debt should now be called? How will these strains affect our ability to pay the debt—to survive?

17

LEWIS W. MONCRIEF

# *The Cultural Basis for Our Environmental Crisis*

*Lewis W. Moncrief is Associate Professor and Director of the Recreation Research and Planning Unit. He holds a joint appointment in the Departments of Park and Recreation Resources and Resource Development at Michigan State University, East Lansing.*

One hundred years ago at almost any location in the United States, potable water was no farther away than the closest brook or stream. Today there are hardly any streams in the United States, except in a few high mountainous reaches, that can safely satisfy human thirst without chemical treatment. An oft-mentioned satisfaction in the lives of urbanites in an earlier era was a leisurely stroll in late afternoon to get a breath of fresh air in a neighborhood park or along a quiet street. Today in many of our major metropolitan areas it is difficult to find a quiet, peaceful place to take a leisurely stroll and sometimes impossible to get a breath of fresh air. These contrasts point up the dramatic changes that have occurred in the quality of our environment.

It is not my intent in this article, however, to document the existence of an environmental crisis but rather to discuss the cultural basis for

*"The Cultural Basis for Our Environmental Crisis," by Lewis W. Moncrief, from* Science, *Vol. 170, October 30, 1970, pp. 508–512.* 

such a crisis. Particular attention will be given to the institutional structures as expressions of our culture.

## Social Organization

In her book entitled *Social Institutions,*[1] J. O. Hertzler classified all social institutions into nine functional categories: (i) economic and industrial, (ii) matrimonial and domestic, (iii) political, (iv) religious, (v) ethical, (vi) educational, (vii) communications, (viii) esthetic, and (ix) health. Institutions exist to carry on each of these functions in all cultures, regardless of their location or relative complexity. Thus, it is not surprising that one of the analytical criteria used by anthropologists in the study of various cultures is the comparison and contrast of the various social institutions as to form and relative importance.[2]

A number of attempts have been made to explain attitudes and behavior that are commonly associated with one institutional function as the result of influence from a presumably independent institutional factor. The classic example of such an analysis is *The Protestant Ethic and the Spirit of Capitalism* by Max Weber.[3] In this significant work Weber attributes much of the economic and industrial growth in Western Europe and North America to capitalism, which, he argued, was an economic form that developed as a result of the religious teachings of Calvin, particularly spiritual determinism.

Social scientists have been particularly active in attempting to assess the influence of religious teaching and practice and of economic motivation on other institutional forms and behavior and on each other. In this connection, L. White [4] suggested that the exploitative attitude that has prompted much of the environmental crisis in Western Europe and North America is a result of the teachings of the Judeo-Christian tradition, which conceives of man as superior to all other creation and of everything else as created for his use and enjoyment. He goes on to contend that the only way to reduce the ecologic crisis which we are now facing is to "reject the Christian axiom that nature has no reason for existence save to serve man." As with other ideas that appear to be new and novel, Professor White's observations have begun to be widely circulated and accepted in scholarly circles, as witness the article by religious writer E. B. Fiske in the *New York Times* early in 1970.[5] In this

1. J. O. Hertzler, *Social Institutions* (McGraw-Hill, New York, 1929), pp. 47–64.
2. L. A. White, *The Science of Culture* (Farrar, Straus & Young, New York, 1949), pp. 121–145.
3. M. Weber, *The Protestant Ethic and the Spirit of Capitalism,* translated by T. Parsons (Scribner's, New York, 1958).
4. L. White, Jr., *Science* 155, 1203 (1967).
5. E. B. Fiske, "The link between faith and ecology," *New York Times* (4 January 1970), section 4, p. 5.

article, note is taken of the fact that several prominent theologians and theological groups have accepted this basic premise that Judeo-Christian doctrine regarding man's relation to the rest of creation is at the root of the West's environmental crisis. I would suggest that the wide acceptance of such a simplistic explanation is at this point based more on fad than on fact.

Certainly, no fault can be found with White's statement that "Human ecology is deeply conditioned by beliefs about our nature and destiny—that is, by religion." However, to argue that it is the primary conditioner of human behavior toward the environment is much more than the data that he cites to support this proposition will bear. For example, White himself notes very early in his article that there is evidence for the idea that man has been dramatically altering his environment since antiquity. If this be true, and there is evidence that it is, then this mediates against the idea that the Judeo-Christian religion uniquely predisposes cultures within which it thrives to exploit their natural resources with indiscretion. White's own examples weaken his argument considerably. He points out that human intervention in the periodic flooding of the Nile River basin and the fire-drive method of hunting by prehistoric man have both probably wrought significant "unnatural" changes in man's environment. The absence of Judeo-Christian influence in these cases is obvious.

It seems tenable to affirm that the role played by religion in man-to-man and man-to-environment relationships is one of establishing a very broad system of allowable beliefs and behavior and of articulating and invoking a system of social and spiritual rewards for those who conform and of negative sanctions for individuals or groups who approach or cross the pale of the religiously unacceptable. In other words, it defines the ball park in which the game is played, and, by the very nature of the park, some types of games cannot be played. However, the kind of game that ultimately evolves is not itself defined by the ball park. For example, where animism is practiced, it is not likely that the believers will indiscriminately destroy objects of nature because such activity would incur the danger of spiritual and social sanctions. However, the fact that another culture does not associate spiritual beings with natural objects does not mean that such a culture will invariably ruthlessly exploit its resources. It simply means that there are fewer social and psychological constraints against such action.

In the remainder of this article, I present an alternative set of hypotheses based on cultural variables which, it seems to me, are more plausible and more defensible as an explanation of the environmental crisis that is now confronting us.

No culture has been able to completely screen out the egocentric tendencies of human beings. There also exists in all cultures a status hierarchy of positions and values, with certain groups partially or totally excluded from access to these normatively desirable goals. Historically, the differences in most cultures between the "rich" and the "poor" have been great. The many very poor have often produced the wealth for the few who controlled the means of production. There may have been no alternative where scarcity of supply and unsatiated demand were economic reality. Still, the desire for a "better life" is universal; that is, the desire for higher status positions and the achievement of culturally defined desirable goals is common to all societies.

## The Experience in the Western World

In the West two significant revolutions that occurred in the 18th and 19th centuries completely redirected its political, social, and economic destiny.[6] These two types of revolutions were unique to the West until very recently. The French revolution marked the beginnings of widespread democratization. In specific terms, this revolution involved a redistribution of the means of production and a reallocation of the natural and human resources that are an integral part of the production process. In effect new channels of social mobility were created, which theoretically made more wealth accessible to more people. Even though the revolution was partially perpetrated in the guise of overthrowing the control of presumably Christian institutions and of destroying the influence of God over the minds of men, still it would be superficial to argue that Christianity did not influence this revolution. After all, biblical teaching is one of the strongest of all pronouncements concerning human dignity and individual worth.

At about the same time but over a more extended period, another kind of revolution was taking place, primarily in England. As White points out very well, this phenomenon, which began with a number of technological innovations, eventually consummated a marriage with natural science and began to take on the character that it has retained until today.[7] With this revolution the productive capacity of each worker was

6. R. A. Nisbet, *The Sociological Tradition* (Basic Books, New York, 1966), pp. 21–44. Nisbet gives here a perceptive discourse on the social and political implications of the democratic and industrial revolutions to the Western world.

7. It should be noted that a slower and less dramatic process of democratization was evident in English history at a much earlier date than the French revolution. Thus, the concept of democracy was probably a much more pervasive influence in English than in French life. However, a rich body of philosophic literature regarding the rationale for democracy resulted from the French revolution. Its counterpart in English literature is much less conspicuous. It is an interesting aside to suggest that the industrial revolution would not have been possible except for the more broad-based ownership of the means of production that resulted from the long-standing process of democratization in England.

amplified by several times his potential prior to the revolution. It also became feasible to produce goods that were not previously producible on a commercial scale.

Later, with the integration of the democratic and the technological ideals, the increased wealth began to be distributed more equitably among the population. In addition, as the capital to land ratio increased in the production process and the demand grew for labor to work in the factories, large populations from the agrarian hinterlands began to concentrate in the emerging industrial cities. The stage was set for the development of the conditions that now exist in the Western world.

With growing affluence for an increasingly large segment of the population, there generally develops an increased demand for goods and services. The usual by-product of this affluence is waste from both the production and consumption processes. The disposal of that waste is further complicated by the high concentration of heavy waste producers in urban areas. Under these conditions the maxim that "Dilution is the solution to pollution" does not withstand the test of time, because the volume of such wastes is greater than the system can absorb and purify through natural means. With increasing population, increasing production, increasing urban concentrations, and increasing real median incomes for well over a hundred years, it is not surprising that our environment has taken a terrible beating in absorbing our filth and refuse.

## The American Situation

The North American colonies of England and France were quick to pick up the technical and social innovations that were taking place in their motherlands. Thus, it is not surprising that the inclination to develop an industrial and manufacturing base is observable rather early in the colonies. A strong trend toward democratization also evidenced itself very early in the struggle for nationhood. In fact, Thistlewaite notes the significance of the concept of democracy as embodied in French thought to the framers of constitutional government in the colonies.[8]

From the time of the dissolution of the Roman Empire, resource ownership in the Western world was vested primarily with the monarchy or the Roman Catholic Church, which in turn bestowed control of the land resources on vassals who pledged fealty to the sovereign. Very slowly the concept of private ownership developed during the Middle Ages in Europe, until it finally developed into the fee simple concept.

In America, however, national policy from the outset was designed to

8. F. Thistlewaite, *The Great Experiment* (Cambridge Univ. Press, London, 1955), pp. 33–34, 60.

convey ownership of the land and other natural resources into the hands of the citizenry. Thomas Jefferson was perhaps more influential in crystallizing this philosophy in the new nation than anyone else. It was his conviction that an agrarian society made up of small landowners would furnish the most stable foundation for building the nation.[9] This concept has received support up to the present and, against growing economic pressures in recent years, through government programs that have encouraged the conventional family farm. This point is clearly relevant to the subject of this article because it explains how the natural resources of the nation came to be controlled not by a few aristocrats but by many citizens. It explains how decisions that ultimately degrade the environment are made not only by corporation boards and city engineers but by millions of owners of our natural resources. This is democracy exemplified!

## Challenge of the Frontier

Perhaps the most significant interpretation of American history has been Fredrick Jackson Turner's much criticized thesis that the western frontier was the prime force in shaping our society.[10] In his own words,

> If one would understand why we are today one nation, rather than a collection of isolated states, he must study this economic and social consolidation of the country. . . . The effect of the Indian frontier as a consolidating agent in our history is important.

He further postulated that the nation experienced a series of frontier challenges that moved across the continent in waves. These included the explorers' and traders' frontier, the Indian frontier, the cattle frontier, and three distinct agrarian frontiers. His thesis can be extended to interpret the expansionist period of our history in Panama, in Cuba, and in the Philippines as a need for a continued frontier challenge.

Turner's insights furnish a starting point for suggesting a second variable in analyzing the cultural basis of the United States' environmental crisis. As the nation began to expand westward, the settlers faced many obstacles, including a primitive transportation system, hostile Indians, and the absence of physical and social security. To many frontiersmen, particularly small farmers, many of the natural resources that are now highly valued were originally perceived more as obstacles than as assets. Forests needed to be cleared to permit farming. Marshes needed to be

9. *Ibid*., pp. 59–68.
10. F. J. Turner, *The Frontier in American History* (Henry Holt, New York, 1920 and 1947).

drained. Rivers needed to be controlled. Wildlife often represented a competitive threat in addition to being a source of food. Sod was considered a nuisance—to be burned, plowed, or otherwise destroyed to permit "desirable" use of the land.

Undoubtedly, part of this attitude was the product of perceiving these resources as inexhaustible. After all, if a section of timber was put to the torch to clear it for farming, it made little difference because there was still plenty to be had very easily. It is no coincidence that the "First Conservation Movement" began to develop about 1890. At that point settlement of the frontier was almost complete. With the passing of the frontier era of American history, it began to dawn on people that our resources were indeed exhuastible. This realization ushered in a new philosophy of our national government toward natural resources management under the guidance of Theodore Roosevelt and Gifford Pinchot. Samuel Hays [11] has characterized this movement as the appearance of a new "Gospel of Efficiency" in the management and utilization of our natural resources.

## The Present American Scene

America is the archetype of what happens when democracy, technology, urbanization, capitalistic mission, and antagonism (or apathy) toward natural environment are blended together. The present situation is characterized by three dominant features that mediate against quick solution to this impending crisis: (i) an absence of personal moral direction concerning our treatment of our natural resources, (ii) an inability on the part of our social institutions to make adjustments to this stress, and (iii) an abiding faith in technology.

The first characteristic is the absence of personal moral direction. There is moral disparity when a corporation executive can receive a prison sentence for embezzlement but be congratulated for increasing profits by ignoring pollution abatement laws. That the absolute cost of society of the second act may be infinitely greater than the first is often not even considered.

The moral principle that we are to treat others as we would want to be treated seems as appropriate a guide as it ever has been. The rarity of such teaching and the even more uncommon instance of its being practiced help to explain how one municipality can, without scruple, dump its effluent into a stream even though it may do irreparable damage to the resource and add tremendously to the cost incurred by down-

11. S. P. Hays, *Conservation and the Gospel of Efficiency* (Harvard Univ. Press, Cambridge, Mass., 1959).

stream municipalities that use the same water. Such attitudes are not restricted to any one culture. There appears to be an almost universal tendency to maximize self-interests and a widespread willingness to shift production costs to society to promote individual ends.

Undoubtedly, much of this behavior is the result of ignorance. If our accounting systems were more efficient in computing the cost of such irresponsibility both to the present generation and to those who will inherit the environment we are creating, steps would undoubtedly be taken to enforce compliance with measures designed to conserve resources and protect the environment. And perhaps if the total costs were known, we might optimistically speculate that more voluntary compliance would result.

A second characteristic of our current situation involves institutional inadequacies. It has been said that "what belongs to everyone belongs to no one." This maxim seems particularly appropriate to the problem we are discussing. So much of our environment is so apparently abundant that it is considered a free commodity. Air and water are particularly good examples. Great liberties have been permitted in the use and abuse of these resources for at least two reasons. First, these resources have typically been considered of less economic value than other natural resources except when conditions of extreme scarcity impose limiting factors. Second, the right of use is more difficult to establish for resources that are not associated with a fixed location.

Government, as the institution representing the corporate interests of all its citizens, has responded to date with dozens of legislative acts and numerous court decisions which give it authority to regulate the use of natural resources. However, the decisiveness to act has thus far been generally lacking. This indecisiveness cannot be understood without noting that the simplistic models that depict the conflict as that of a few powerful special interests versus "The People" are altogether inadequate. A very large proportion of the total citizenry is implicated in environmental degradation; the responsibility ranges from that of the board and executives of a utility company who might wish to thermally pollute a river with impunity to that of the average citizen who votes against a bond issue to improve the efficiency of a municipal sanitation system in order to keep his taxes from being raised. The magnitude of irresponsibility among individuals and institutions might be characterized as falling along a continuum from highly irresponsible to indirectly responsible. With such a broad base of interests being threatened with every change in resource policy direction, it is not surprising, although regrettable, that government has been so indecisive.

A third characteristic of the present American scene is an abiding faith in technology. It is very evident that the idea that technology can overcome almost any problem is widespread in Western society. This optimism exists in the face of strong evidence that much of man's technology, when misused, has produced harmful results, particularly in the long run. The reasoning goes something like this: "After all, we have gone to the moon. All we need to do is allocate enough money and brainpower and we can solve any problem."

It is both interesting and alarming that many people view technology almost as something beyond human control. Rickover put it this way: [12]

> It troubles me that we are so easily pressured by purveyors of technology into permitting so-called "progress" to alter our lives without attempting to control it—as if technology were an irrepressible force of nature to which we must meekly submit.

He goes on to add:

> It is important to maintain a humanistic attitude toward technology; to recognize clearly that since it is the product of human effort, technology can have no legitimate purpose but to serve man—man in general, not merely some men; future generations, not merely those who currently wish to gain advantage for themselves; man in the totality of his humanity, encompassing all his manifold interests and needs, not merely some one particular concern of his. When viewed humanistically, technology is seen not as an end in itself but a means to an end, the end being determined by man himself in accordance with the laws prevailing in his society.

In short, it is one thing to appreciate the value of technology; it is something else entirely to view it as our environmental savior—which will save us in spite of ourselves.

## Conclusion

The forces of democracy, technology, urbanization, increasing individual wealth, and an aggressive attitude toward nature seem to be directly related to the environmental crisis now being confronted in the Western world. The Judeo-Christian tradition has probably influenced the character of each of these forces. However, to isolate religious tradition as a cultural component and to contend that it is the "historical root of our ecological crisis" is a bold affirmation for which there is little historical or scientific support.

To assert that the primary cultural condition that has created our environmental crisis is Judeo-Christian teaching avoids several hard questions. For example: Is there less tendency for those who control the resources in non-Christian cultures to live in extravagant affluence with attendent high levels of waste and inefficient consumption? If non-Ju-

12. H. G. Rickover, *Amer. Forests* 75, 13 (August 1969).

deo-Christian cultures had the same levels of economic productivity, urbanization, and high average household incomes, is there evidence to indicate that these cultures would not exploit or disregard nature as our culture does?

If our environmental crisis is a "religious problem," why are other parts of the world experiencing in various degrees the same environmental problems that we are so well acquainted with in the Western world? It is readily observable that the science and technology that developed on a large scale first in the West have been adopted elsewhere. Judeo-Christian tradition has not been adopted as a predecessor to science and technology on a comparable scale. Thus, all White can defensibly argue is that the West developed modern science and technology *first*. This says nothing about the origin or existence of a particular ethic toward our environment.

In essence, White has proposed this simple model:

| I | | II | | III |
|---|---|---|---|---|
| Judeo-Christian tradition | → | Science and technology | → | Environmental degradation |

I have suggested here that, at best, Judeo-Christian teaching has had only an indirect effect on the treatment of our environment. The model could be characterized as follows:

| I | | II | | III | | IV |
|---|---|---|---|---|---|---|
| Judeo-Christian tradition | → | 1) Capitalism (with the attendant development of science and technology)<br>2) Democratization | → | 1) Urbanization<br>2) Increased wealth<br>3) Increased population<br>4) Individual resource ownership | → | Environmental degradation |

Even here, the link between Judeo-Christian tradition and the proposed dependent variables certainly have the least empirical support. One need only look at the veritable mountain of criticism of Weber's conclusions in *The Protestant Ethic and the Spirit of Capitalism* to sense the tenuous nature of this link. The second and third phases of this model are common to many parts of the world. Phase I is not.

Jean Mayer,[13] the eminent food scientist, gave an appropriate conclusion about the cultural basis for our environmental crisis:

> It might be bad in China with 700 million poor people but 700 million rich Chinese would wreck China in no time. . . . It's the rich who wreck the environment . . . occupy much more space, consume more of each natural resource, disturb ecology more, litter the landscape . . . and create more pollution.

13. J. Mayer and T. G. Harris, *Psychol. Today* 3, 46 and 48 (January 1970).

MARSHALL I. GOLDMAN

# *The Convergence of Environmental Disruption*

*Marshall I. Goldman is Professor of Economics at Wellesley College and an Associate of the Russian Research Center, Harvard University.*

By now it is a familiar story: rivers that blaze with fire, smog that suffocates cities, streams that vomit dead fish, oil slicks that blacken seacoasts, prized beaches that vanish in the waves, and lakes that evaporate and die a slow smelly death. What makes it unfamiliar is that this is a description not only of the United States but also of the Soviet Union.

Most conservationists and social critics are unaware that the U.S.S.R. has environmental disruption that is as extensive and severe as ours. Most of us have been so distressed by our own environmental disruption that we lack the emotional energy to worry about anyone else's difficulties. Yet, before we can find a solution to the environmental disruption in our own country, it is necessary to explain why it is that a socialist or communist country like the U.S.S.R. finds itself abusing the environment in the same way, and to the same degree, that we abuse it. This is especially important for those who have come to believe as basic doctrine that it is capitalism and private greed that are the root cause of

*"The Convergence of Environmental Disruption," by Marshall I. Goldman, from* Science, *Vol. 170, October 2, 1970, pp. 37–42.* 

environmental disruption. Undoubtedly private enterprise and the profit motive account for a good portion of the environmental disruption that we encounter in this country. However, a study of pollution in the Soviet Union suggests that abolishing private property will not necessarily mean an end to environmental disruption. In some ways, state ownership of the country's productive resources may actually exacerbate rather than ameliorate the situation.

## The Public Good

That environmental disruption is a serious matter in the Soviet Union usually comes as a surprise not only to most radical critics of pollution in the West but also to many Russians. It has been assumed that, if all the factories in a society were state-owned, the state would insure that the broader interests of the general public would be protected. Each factory would be expected to bear the full costs and consequences of its operation. No factory would be allowed to take a particular action if it meant that the public would suffer or would have to bear the expense. In other words, the factory would not only have to pay for its *private costs,* such as expenses for labor and raw materials; it would also have to pay for its *social costs,* such as the cost of eliminating the air and water pollution it had caused. It was argued that, since the industry was state-run, including both types of costs would not be difficult. At least that was what was assumed.

Soviet officials continue today to make such assumptions. B. V. Petrovsky, the Soviet Minister of Public Health, finds environmental disruption in a capitalist society perfectly understandable: "the capitalist system by its very essence is incapable of taking radical measures to ensure the efficient conservation of nature." By implication he assumes that the Soviet Union can take such measures. Therefore it must be somewhat embarrassing for Nikolai Popov, an editor of *Soviet Life,* to have to ask, "Why, in a socialist country, whose constitution explicitly says the public interest may not be ignored with impunity, are industry executives permitted to break the laws protecting nature?"

Behind Popov's question is a chronicle of environmental disruption that is as serious as almost any that exists in the world. Of course in a country as large as the U.S.S.R. there are many places that have been spared man's disruptive incursions. But, as the population grows in numbers and mobility, such areas become fewer and fewer. Moreover, as in the United States, the most idyllic sites are the very ones that tend to attract the Soviet population.

Just because human beings intrude on an area, it does not necessarily follow that the area's resources will be abused. Certainly the presence of human beings means some alteration in the previous ecological balance, and in some cases there may be severe damage, but the change need not be a serious one. Nonetheless, many of the changes that have taken place in the Soviet Union have been major ones. As a result, the quality of the air, water, and land resources has been adversely affected.

## Water

Comparing pollution in the United States and in the U.S.S.R. is something like a game. Any depressing story that can be told about an incident in the United States can be matched by a horror story from the U.S.S.R. For example, there have been hundreds of fish-kill incidents in both countries. Rivers and lakes from Maine to California have had such incidents. In the U.S.S.R., effluent from the Chernorechensk Chemical Plant near Dzerzhinsk killed almost all the fish life in the Oka River in 1965 because of uncontrolled dumping. Factories along major rivers such as the Volga, Ob, Yenesei, Ural, and Northern Dvina have committed similar offenses, and these rivers are considered to be highly polluted. There is not one river in the Ukraine whose natural state has been preserved.[1] The Molognaia River in the Ukraine and many other rivers throughout the country are officially reported as dead. How dangerous this can be is illustrated by what happened in Sverdlovsk in 1965. A careless smoker threw his cigarette into the Iset River and, like the Cuyahoga in Cleveland, the Iset caught fire.

Sixty-five percent of all the factories in the largest Soviet republic, the Russian Soviet Federated Socialist Republic (RSFSR), discharge their waste without bothering to clean it up.[2] But factories are not the only ones responsible for the poor quality of the water. Mines, oil wells, and ships freely dump their waste and ballast into the nearest body of water. Added to this industrial waste is the sewage of many Russian cities. Large cities like Moscow and Leningrad are struggling valiantly, like New York and Chicago, to treat their waste, but many municipalities are hopelessly behind in their efforts to do the job properly. Only six out of the 20 main cities in Moldavia have a sewer system, and only two of those cities make any effort to treat their sewage.[3] Similarly, only 40 percent of the cities and suburbs in the RSFSR have any equipment for treating their sewage. For that matter, according to the last completed census, taken in 1960, only 35 percent of all the housing units in urban areas are served by a sewer system.[4]

1. *Rabochaia Gaz.* 1967, 4 (15 Dec. 1967).
2. *Ekon. Gaz.* 1967, No. 4, 37 (1967).
3. *Sovet. Moldaviia* 1969, 2 (1 June 1969).
4. V. G. Kriazhev, *Vnerabochee Vremia i Sphera Obslyzhivaniia* (Ekonomika, Moscow, 1966), p. 130.

Conditions are even more primitive in the countryside. Often this adversely affects the well-water and groundwater supplies, especially in areas of heavy population concentration. Under the circumstances it is not surprising to find that major cities like Vladimir, Orenburg, and Voronezh do not have adequate supplies of drinking water. In one instance reported in *Pravda,* a lead and zinc ore enriching plant was built in 1966 and allowed to dump its wastes in the Fragdon River, even though the river was the sole source of water for about 40 kilometers along its route. As a result the water became contaminated and many people were simply left without anything to drink.

Even when there are supplies of pure water, many homes throughout the country are not provided with running water. This was true of 62 percent of the urban residences in the U.S.S.R. in 1960.[5] The Russians often try to explain this by pointing to the devastation they suffered during World War II. Still it is something of a shock, 25 years after the war, to walk along one of the more fashionable streets in Kharkov, the fifth largest city in the U.S.S.R., and see many of the area's residents with a yoke across their shoulders, carrying two buckets of water. The scene can be duplicated in almost any other city in the U.S.S.R.

Again, the Soviet Union, like the United States, has had trouble not only with its rivers but with its larger bodies of water. As on Cape Cod and along the California coast, oil from slicks has coated the shores of the Baltic, Black, and Caspian seas. Refineries and tankers have been especially lax in their choice of oil-disposal procedures.

Occasionally it is not only the quality but the quantity of the water that causes concern. The Aral and Caspian seas have been gradually disappearing. Because both seas are in arid regions, large quantities of their water have been diverted for crop irrigation. Moreover, many dams and reservoirs have been built on the rivers that supply both seas for the generation of electric power. As a result of such activities, the Aral Sea began to disappear. From 1961 to 1969 its surface dropped 1 to 3 meters. Since the average depth of the sea is only about 20 to 30 meters, some Russian authorities fear that, at the current rate of shrinkage, by the turn of the century the sea will be nothing but a salt marsh.[6]

Similarly, during the past 20 years the level of the Caspian Sea has fallen almost 2½ meters. This has drastically affected the sea's fish population. Many of the best spawning areas have turned into dry land. For the sturgeon, one of the most important fish in the Caspian, this has meant the elimination of one-third of the spawning area. The combined effect of the oil on the sea and the smaller spawning area reduced the fish catch in the Caspian from 1,180,400 centners in 1942 to 586,300 centners in 1966. Food fanciers are worried not so much about the stur-

5. *Ibid.*
6. *Soviet News* 1970, 6 (7 Apr. 1970).

geon as about the caviar that the sturgeon produces. The output of caviar has fallen even more drastically than the sea level—a concern not only for the Russian consumers of caviar but for foreigners. Caviar had been a major earner of foreign exchange. Conditions have become so serious that the Russians have now begun to experiment with the production of artificial caviar.

The disruption of natural life in the Caspian Sea has had some serious ecological side effects. Near Ashkhabad at the mouth of the Volga a fish called the belyi amur also began to disappear. As a consequence, the mosquito population, which had been held in check by the belyi amur, grew in the newly formed swamps where once the sea had been. In turn, the mosquitoes began to transmit malaria.[7]

Perhaps the best known example of the misuse of water resources in the U.S.S.R. has been what happened to Lake Baikal. This magnificent lake is estimated to be over 20 million years old. There are over 1200 species of living organisms in the lake, including freshwater seals and 700 other organisms that are found in few or no other places in the world. It is one of the largest and deepest freshwater lakes on earth, over 1½ kilometers deep in some areas.[8] It is five times as deep as Lake Superior and contains twice the volume of water. In fact, Lake Baikal holds almost one-fortieth of all the world's fresh water. The water is low in salt content and is highly transparent; one can see as far as 36 meters under water.[9]

In 1966, first one and then another paper and pulp mill appeared on Lake Baikal's shores. Immediately limnologists and conservationists protested this assault on an international treasure. Nonetheless, new homes were built in the vicinity of the paper and pulp mills, and the plant at the nearby town of Baikalsk began to dump 60 million cubic meters of effluent a year into the lake. A specially designed treatment plant had been erected in the hope that it would maintain the purity of the lake. Given the unique quality of the water, however, it soon became apparent that almost no treatment plant would be good enough. Even though the processed water is drinkable, it still has a yellowish tinge and a barely perceptible odor. As might be expected, a few months after this effluent had been discharged into the lake, the Limnological Institute reported that animal and plant life had decreased by one-third to one-half in the zone where the sewage was being discharged.

Several limnologists have argued that the only effective way to prevent the mill's effluent from damaging the lake is to keep it out of the lake entirely. They suggest that this can be done if a 67-kilometer sewage conduit is built over the mountains to the Irkut River, which does

7. *Turkm. Iskra* 1969, 3 (16 Sept. 1969).
8. O. Volkov, *Soviet Life* 1966, 6 (Aug. 1966).
9. L. Rossolimo, *Baikal* (Nauka, Moscow, 1966), p. 91.

not flow into the lake. So far the Ministry of Paper and Pulp Industries has strongly opposed this, since it would cost close to $40 million to build such a bypass. They argue that they have already spent a large sum on preventing pollution. Part of their lack of enthusiasm for any further change may also be explained by the fact that they have only had to pay fines of $55 for each violation. It has been cheaper to pay the fines than to worry about a substantial cleanup operation.

Amid continuing complaints, the second paper and pulp mill, at Kamensk, was told that it must build and test its treatment plant before production of paper and pulp would be allowed. Moreover, the lake and its entire drainage basin have been declared a "protected zone," which means that in the future all timber cutting and plant operations are to be strictly regulated. Many critics, however, doubt the effectiveness of such orders. As far back as 1960, similar regulations were issued for Lake Baikal and its timber, without much result. In addition, the Ministry of Pulp and Paper Industries has plans for constructing yet more paper and pulp mills along the shores of Lake Baikal and is lobbying for funds to build them.

Many ecologists fear that, even if no more paper mills are built, the damage may already have been done. The construction of the mills and towns necessitated the cutting of trees near the shoreline, which inevitably increased the flow of silt into the lake and its feeder streams. Furthermore, instead of being shipped by rail, as was originally promised, the logs are rafted on the water to the mill for processing. Unfortunately about 10 percent of these logs sink to the lake bottom in transit. Not only does this cut off the feeding and breeding grounds on the bottom of the lake but the logs consume the lake's oxygen, which again reduces its purity.

There are those who see even more dire consequences from the exploitation of the timber around the lake. The Gobi Desert is just over the border in Mongolia. The cutting of the trees and the intrusion of machinery into the wooded areas has destroyed an important soil stabilizer. Many scientists report that the dunes have already started to move, and some fear that the Gobi Desert will sweep into Siberia and destroy the taiga and the lake.

## Air

The misuse of air resources in the U.S.S.R. is not very different from the misuse of water. Despite the fact that the Russians at present produce less than one-tenth the number of cars each year that we produce in the United States, most Soviet cities have air pollution. It can be quite serious, especially when the city is situated in a valley or a hilly region. In the hilly cities of Armenia, the established health norms for carbon monoxide are often exceeded. Similarly Magnitogorsk, Alma Ata, and Chelyabinsk, with their metallurgical industries, frequently have a dark blue cap over them. Like Los Angeles, Tbilisi, the capital of the Republic of Georgia, has smog almost 6 months of the year. Nor is air pollution limited to hilly regions. Leningrad has 40 percent fewer clear daylight hours than the nearby town of Pavlovsk.[10]

Of all the factories that emit harmful wastes through their stacks, only 14 percent were reported in 1968 to have fully equipped air-cleaning devices. Another 26 percent had some treatment equipment. Even so, there are frequent complaints that such equipment is either operating improperly or of no use. There have been several reported instances of factories' spewing lead into the air.[11] In other cases, especially in Sverdlovsk and Magnitogorsk, public health officials ordered the closing of factories and boilers. Nevertheless, there are periodic complaints that some public health officials have yielded to the pleadings and pressures of factory directors and have agreed to keep the plants open "on a temporary basis."

One particularly poignant instance of air pollution is occurring outside the historic city of Tula. Not far away is the site of Leo Tolstoy's former summer estate, Yasnaya Polyana, now an internationally known tourist attraction with lovely grounds and a museum. Due to some inexcusable oversight a small coal-gasification plant was built within view of Yasnaya Polyana in 1955. In 1960 the plant was expanded as it began to produce fertilizer and other chemicals. Now known as the Shchkino Chemical Complex, the plant has over 6000 employees and produces a whole range of chemicals, including formaldehyde and synthetic fibers. Unfortunately the prevailing winds from this extensive complex blow across the street onto the magnificent forests at Yasnaya Polyana. As a result, a prime oak forest is reported near extinction and a pine forest is similarly affected.

## Land

As in other nations of the world, environmental disruption in the U.S.S.R. is not limited to air and water. For example, the Black Sea coast in the Soviet Republic of Georgia is disappearing. Since this is a particularly desirable resort area, a good deal of concern has been expressed over what is happening. At some places the sea has moved as much as 40 meters inland. Near the resort area of Adler, hospitals, re-

10. I. Petrov, *Kommunist* 1969, No. 11, 74 (1969).
11. *Rabochaia Gaz.* 1969, 4 (27 June 1969); *Ekon. Gaz.* 1968, No. 4, 40 (1968); *Lit. Gaz.* 1967, No. 32, 10 (1967).

sort hotels, and (of all things) the beach sanitarium of the Ministry of Defense collapsed as the shoreline gave way. Particular fears that the mainline railway will also be washed away shortly have been expressed.

New Yorkers who vacation on Fire Island have had comparable difficulties, but the cause of the erosion in the U.S.S.R. is unique. Excessive construction has loosened the soil (as at Fire Island) and accelerated the process of erosion. But, in addition, much of the Black Sea area has been simply hauled away by contractors. One contractor realized that the pebbles and sand on the riviera-type beach were a cheap source of gravel. Soon many contractors were taking advantage of nature's blessings. As a result, as much as 120,000 cubic meters a year of beach material has been hauled away. Unfortunately the natural process by which those pebbles are replaced was disrupted when the state came along and built a network of dams and reservoirs on the stream feeding into the sea. This provided a source of power and water but it stopped the natural flow of pebbles and sand to the seacoast. Without the pebbles, there is little to cushion the enormous power of the waves as they crash against the coast and erode the shoreline.

In an effort to curb the erosion, orders have been issued to prevent the construction of any more buildings within 3 kilometers of the shore. Concrete piers have also been constructed to absorb the impact of the waves, and efforts are being made to haul gravel material from the inland mountains to replace that which has been taken from the seacoast. Still the contractors are disregarding the orders—they continue to haul away the pebbles and sand, and the seacoast continues to disappear.

Nor is the Black Sea coast the only instance of such disregard for the forces of nature. High in the Caucasus is the popular health resort and spa of Kislovodsk. Surrounded on three sides by a protective semicircle of mountains which keep out the cold winds of winter, the resort has long been noted for its unique climate and fresh mountain air. Whereas Kislovodsk used to have 311 days of sun a year, Piatagorsk on the other side of the mountain had only 122.[12] Then, shortly after World War II, an official of the Ministry of Railroads sought to increase the volume of railroad freight in the area. He arranged for the construction of a lime kiln in the nearby village of Podkumok. With time, pressure mounted to increase the processing of lime, so that now there are eight kilns in operation. As the manager of the lime kiln operation and railroad officials continued to "fulfill their ever-increasing plan" in the name of "socialist competition," the mountain barrier protecting Kislovodsk from the northern winds and smoke of the lime kilns has been gradually chopped away. Consequently, Kislovodsk has almost been transformed into an ordinary industrial city. The dust in the air now exceeds by 50 percent the norm for a *nonresort* city.

12. *Izv*. 1966, 5 (3 July 1966).

Much as some of our ecologists have been warning that we are on the verge of some fundamental disruptions of nature, so the Russians have their prophets of catastrophe. Several geographers and scientists have become especially concerned about the network of hydroelectric stations and irrigation reservoirs and canals that have been built with great fanfare across the country. They are now beginning to find that such projects have had several unanticipated side effects. For example, because the irrigation canals have not been lined, there has been considerable seepage of water. The seepage from the canals and an overenthusiastic use of water for irrigation has caused a rise in the water table in many areas. This has facilitated salination of the soil, especially in dry areas. Similarly, the damming of water bodies apparently has disrupted the addition of water to underground water reserves. There is concern that age-old sources of drinking water may gradually disappear. Finally, it is feared that the reduction of old water surfaces and the formation of new ones has radically altered and increased the amount of water evaporation in the area in question. There is evidence that this has brought about a restructuring of old climate and moisture patterns.[13] This may mean the formation of new deserts in the area. More worrisome is the possibility of an extension of the ice cap. If enough of Russia's northward-flowing rivers are diverted for irrigation purposes to the arid south, this will deprive the Arctic Ocean of the warmer waters it receives from these rivers. Some scientist critics also warn that reversing the flow of some of the world's rivers in this way will have disruptive effects on the rotation of the earth.

## Reasons for Pollution

Because the relative impact of environmental disruption is a difficult thing to measure, it is somewhat meaningless to say that the Russians are more affected than we are, or vice versa. But what should be of interest is an attempt to ascertain why it is that pollution exists in a state-owned, centrally planned economy like that of the Soviet Union. Despite the fact that our economies differ, many if not all of the usual economic explanations for pollution in the non-Communist world also hold for the Soviet Union. The Russians, too, have been unable to adjust their accounting system so that each enterprise pays not only its direct costs of production for labor, raw materials, and equipment but also its social costs of production arising from such byproducts as dirty air and water. If the factory were charged for these social costs and had to take them into account when trying to make a profit on its opera-

13. *Soviet News* 1969, 105 (11 March 1969).

tions, presumably factories would throw off less waste and would reuse or recycle their air and water. However, the precise social cost of such waste is difficult to measure and allocate under the best of circumstances, be it in the United States or the U.S.S.R. (In the Ruhr Valley in Germany, industries and municipalities are charged for the water they consume and discharge, but their system has shortcomings.)

In addition, almost everyone in the world regards air and water as free goods. Thus, even if it were always technologically feasible, it would still be awkward ideologically to charge for something that "belongs to everyone," particularly in a Communist society. For a variety of reasons, therefore, air and water in the U.S.S.R. are treated as free or undervalued goods. When anything is free, there is a tendency to consume it without regard for future consequences. But with water and air, as with free love, there is a limit to the amount available to be consumed, and after a time there is the risk of exhaustion. We saw an illustration of this principle in the use of water for irrigation. Since water was treated virtually as a free good, the Russians did not care how much water they lost through unlined canals or how much water they used to irrigate the soil.

Similarly, the Russians have not been able to create clear lines of authority and responsibility for enforcing pollution-control regulations. As in the United States, various Russian agencies, from the Ministry of Agriculture to the Ministry of Public Health, have some but not ultimate say in coping with the problem. Frequently when an agency does attempt to enforce a law, the polluter will deliberately choose to break the law. As we saw at Lake Baikal, this is especially tempting when the penalty for breaking the law is only $55 a time, while the cost of eliminating the effluent may be in the millions of dollars.

The Russians also have to contend with an increase in population growth and the concentration of much of this increase in urban areas. In addition, this larger population has been the beneficiary of an increase in the quantity and complexity of production that accompanies industrialization. As a result, not only is each individual in the Soviet Union, as in the United States, provided with more goods to consume, but the resulting products, such as plastics and detergents, are more exotic and less easily disposed of than goods of an earlier, less complicated age.

Like their fellow inhabitants of the world, the Russians have to contend with something even more ominous than the Malthusian Principle. Malthus observed that the population increased at a geometric rate but that food production grew at only an arithmetic rate. If he really wants to be dismal, the economist of today has more to worry about. It is true that the population seems to be increasing at accelerated rates, but, whereas food production at least continues to increase, our air, water, and soil supplies are relatively constant. They can be renewed, just as crops can be replanted, but, for the most part, they cannot be expanded. In the long run, this "Doomsday Principle" may prove to be of more consequence than the Malthusian doctrine. With time and pollution we may simply run out of fresh air and water. Then, if the damage is not irreversible, a portion of the population will be eliminated and those who remain will exist until there is a shortage once again or until the air, water, and soil are irretrievably poisoned.

## Incentives to Pollute under Socialism

In addition to the factors which confront all the people of the earth, regardless of their social or economic system, there are some reasons for polluting which seem to be peculiar to a socialist country such as the Soviet Union in its present state of economic development. First of all, state officials in the Soviet Union are judged almost entirely by how much they are able to increase their region's economic growth. Thus, government officials are not likely to be promoted if they decide to act as impartial referees between contending factions on questions of pollution. State officials identify with the polluters, not the conservationists, because the polluters will increase economic growth and the prosperity of the region while the antipolluters want to divert resources away from increased production. There is almost a political as well as an economic imperative to devour idle resources. The limnologists at Lake Baikal fear no one so much as the voracious Gosplan (State Planning) officials and their allies in the regional government offices. These officials do not have to face a voting constituency which might reflect the conservation point of view, such as the League of Women Voters or the Sierra Club in this country. It is true that there are outspoken conservationists in the U.S.S.R. who are often supported by the Soviet press, but for the most part they do not have a vote. Thus the lime smelters continued to smoke away behind the resort area of Kislovodsk even though critics in *Izvestiia, Literaturnaya Gazeta, Sovetskaia Rossiia, Trud,* and *Krokodil* protested long and loud.

At one time state governments in our country often reflected similar onesidedness. Maine, for example, was often cited as an area where industry did what it wanted to do to nature. Now, as the conservationist voting bloc has grown in size, the Maine state government finds itself acting as referee. Accordingly it has passed a far-reaching law which regulates the location and operation of all new industry. Failure to have voted for such legislation may have meant defeat at the polls for many politicians. No such device for transmitting voting pressure exists at present in the U.S.S.R.

Second, industrialization has come relatively recently to the U.S.S.R.

and so the Russians continue to emphasize the increase in production. Pollution control generally appears to be nonproductive, and there is usually resistance to the diversion of resources from productive to nonproductive purposes. This is even reflected in the words used to describe the various choices. "Conserve" generally seems to stand in opposition to "produce."

Third, until July 1967, all raw materials in the ground were treated by the Russians as free goods. As a result, whenever the mine operator or oil driller had exploited the most accessible oil and ore, he moved on to a new site where the average variable costs were lower. This has resulted in very low recovery rates and the discarding of large quantities of salvageable materials, which increase the amount of waste to be disposed of.

Fourth, as we have seen, it is as hard for the Russians as it is for us to include social costs in factory-pricing calculations. However, not only do they have to worry about social cost accounting, they also are unable to reflect all the private cost considerations. Because there is no private ownership of land, there are no private property owners to protest the abuse of various resources. Occasionally it does happen that a private property owner in the United States calculates that his private benefits from selling his land for use in some new disruptive use is *not* greater than the private cost he would bear as a result of not being able to use the land any more. So he retains the land as it is. The lack of such private property holders or resort owners and of such a calculation seems to be the major reason why erosion is destroying the Black Sea coast. There is no one who can lay claim to the pebbles on the shore front, and so they are free to anyone who wants to cart them away. Of course private landowners do often decide to sell their land, especially if the new use is to be for oil exploitation rather than pebble exploitation. Then the private benefits to the former owner are high and the social costs are ignored, as always. The Russians, however, under their existing system, now only have to worry about accounting for social costs, they lack the first line of protection that would come from balancing private costs and private benefits.

Fifth, economic growth in the U.S.S.R. has been even more unbalanced, and in some cases more onesided, than in the United States. Thus, occasionally change takes place so rapidly and on such a massive scale in a state-run economy that there is no time to reflect on all the consequences. In the early 1960's, Khrushchev decided that the Soviet Union needed a large chemical industry. All at once chemical plants began to spring up or expand all over the country. In their anxiety to fulfill their targets for new plant construction, few if any of the planners were able to devote much attention to the disruptive effects on the environment that such plants might have. We saw one result at Yasnaya Polyana. In fact, the power of the state to make fundamental changes may be so great that irreversible changes may frequently be inflicted on the environment without anyone's realizing what is happening until it is too late. This seems to be the best explanation of the meteorological disruption that is taking place in Siberia. It is easier for an all-powerful organism like the state than for a group of private entrepreneurs to build the reservoirs and reverse the rivers. Private enterprises can cause their own havoc, as our own dust bowl experience or our use of certain pesticides or sedatives indicates, but in the absence of private business or property interests the state's powers can be much more far-reaching in scope. In an age of rampant technology, where the consequences of one's actions are not always fully anticiapted, even well-intentioned programs can have disastrous effects on the environmental status quo.

## Advantages of a Socialist System

Amidst all these problems, there are some things the Russians do very well. For example, the Russians have the power to prevent the production of various products. Thus, the Soviet Union is the only country in the world that does not put ethyl lead in most of the gasoline it produces. This may be due to technical lag as much as to considerations of health, but the result is considerably more lead-free gasoline. Similarly, the Russians have not permitted as much emphasis on consumer-goods production as we have in the West. Consequently there is less waste to discard. Russian consumers may be somewhat less enthusiastic about this than the ecologists and conservationists, but in the U.S.S.R. there are no disposable bottles or disposable diapers to worry about. It also happens that, because labor costs are low relative to the price of goods, more emphasis is placed on prolonging the life of various products. In other words it is worthwhile to use labor to pick up bottles and collect junk. No one would intentionally abandon his car on a Moscow street, as 50,000 people did in New York City in 1969. Even if a Russian car is 20 years old, it is still valuable. Because of the price relationships that exist in the U.S.S.R., the junkman can still make a profit. This facilitates the recycling process, which ecologists tell us is the ultimate solution to environmental disruption.

It should also be remembered that, while not all Russian laws are observed, the Russians do have an effective law enforcement system which they have periodically brought to bear in the past. Similarly, they have the power to set aside land for use as natural preserves. The lack of private land ownership makes this a much easier process to implement than in the United States. As of 1969, the Soviet Government had set aside 80 such preserves, encompassing nearly 65,000 square kilometers.

Again because they own all the utilities as well as most of the buildings, the Russians have stressed the installation of centrally supplied steam. Thus, heating and hot water are provided by central stations, and this makes possible more efficient combustion and better smoke control than would be achieved if each building were to provide heat and hot water for itself. Although some American cities have similar systems, this approach is something we should know more about.

In sum, if the study of environmental disruption in the Soviet Union demonstrates anything, it shows that not private enterprise but industrialization is the primary cause of environmental disruption. This suggests that state ownership of all the productive resources is not a cure-all. The replacement of private greed by public greed is not much of an improvement. Currently the proposals for the solution of environmental disruption seem to be no more advanced in the U.S.S.R. than they are in the United States. One thing does seem clear, however, and that is that, unless the Russians change their ways, there seems little reason to believe that a strong centralized and planned economy has any notable advantages over other economic systems in solving environmental disruption.

# 19

J. H. DALES

# *The Property Interface*

*J. H. Dales is Professor of Economics at the University of Toronto.*

"Interface" is current academic slang for "boundary." It is along physical interfaces, when air meets water, or water meets land, or prairie meets woodland, that many of the most mysterious and exciting phenomena of the physical and biological world occur. Similarly in social science. The dividing line between work and leisure forms an interface between economics and sociology; behavioural studies at this margin throw considerable light on social attitudes on the one hand and economic performance on the other. And every observer of politics knows how important the "floating vote" is, how behaviour along the margins between voters and non-voters, and between party A supporters and party B supporters, affects the outcomes of elections.

"Property rights" form interfaces between law and several social sciences, especially economics, political science, and sociology. It is with property rights as the dividing line between law and economics that we shall be chiefly concerned, and our first task is to survey this boundary from both sides of the fence.

In everyday conversation we usually speak of "property" rather than

*"The Property Interface," by J. H. Dales. Reprinted from* Pollution, Property and Prices *by J. H. Dales, by permission of the University of Toronto Press. © 1968. University of Toronto Press. Excerpt from "Fish and Wildlife Values in Pollution" by C. H. D. Clarke, from the papers of the Ontario Pollution Control Conference of December 1967, reprinted by permission of The Queen's Printer and Publisher.*

"property rights," but the contraction is misleading if it tends to make us think of property as *things* rather than as *rights,* or of ownership as outright rather than circumscribed. The concepts of property and ownership are created by, defined by, and therefore limited by, a society's system of law. When you own a car, you own a set of legally defined rights to use the vehicle in certain ways and not in others; you may not use it as a personal weapon, for example, nor may you leave it unattended beside a fire hydrant. Among the most important rights you do have are the right to prevent others from using the vehicle, except with your permission and on your terms, and the right to divest yourself of your ownership rights in the vehicle by selling them to someone else. We may say, then, that ownership always consists of (1) a set of rights to use property in certain ways (and a set of negative rights or prohibitions, that prevent its use in other ways); (2) a right to prevent others from exercising those rights, or to set the terms on which others may exercise them; and (3) a right to sell your property rights.

What economics deals with is the buying and selling, or leasing, or using, of property rights. It could hardly be otherwise. You can only buy, sell, lease, rent, lend, or borrow things that are owned; and the only things that are owned are property rights. The prices of the things you buy and sell are prices for property rights to those things.

We can see immediately, then, the interaction that is constantly going on at the interface between law and economics. Consider further the example of automobile ownership. If property rights change so that automobiles cannot be driven unless they are equipped with exhaust-control devices the price of automobiles is likely to rise; if property rights are changed so that automobiles may not be used in cities, the price of cars is likely to fall. On the other hand, if the price of cars rises or falls, so that people own fewer or more of them, there will be social and political pressures on the legal system to change the prescription of property rights in automobiles—to change the law about where they may be driven, how fast they may be driven, where and when they may be parked, and so on. There is always, then, an interface where the law of automobile ownership and the economics of automobile ownership meet, but the boundary may be shifted by a change in either the legal or the economic aspect of the situation. So long as such shifts are small and gradual the interface remains relatively peaceful. But if a major change occurs in either the law or the economics of automobile ownership, the shift in the boundary may be large and abrupt; the legal and economic aspects of the situation tend to become disjointed, and the interface then becomes seriously disturbed. This may be as good a way as any of describing the emergence of a new social problem or set of social problems. But before we attempt to analyse such problems, we must pursue the concept of property rights somewhat further.

## More Property Rights

We have discussed property rights to physical objects. The concept can be easily extended to such things as money, stocks, and bonds. These documents have no particular uses in themselves, but they give their owners certain rights; these rights are exclusive—others can be prevented from using them—and they are transferable. In our society, too, individuals may be said to have property rights to their own labour: a person can use his time for any legal purpose; he has exclusive rights to the fruits, monetary or otherwise, of his efforts; and he can sell his services (and often use them as security for borrowing). In the "extended family" system that is found in several African societies people do *not* have full property rights to their labour since one's distant relatives have a customary right to share in one's earnings; the lack of exclusive property rights to income naturally results in some unusual features of the labour market in extended family societies.

It is when we come to various forms of public property that the concept of property rights becomes rather more tenuous—and yet, paradoxically, even more enlightening. A publicly owned building is very much like a privately owned building; the government can use it for whatever purpose it wishes, can prevent others from using it, and can sell or rent it. A public road system is a rather different matter; since it is built for public use (not, like a government building, solely for the use of government employees) there is no question of exclusive use, and in practice there is only very limited transferability of the asset. Thus while a government "owns" a road system, and can set general rules about its use, its ownership is clearly of a very special kind, reflecting the special public nature of the asset.

We also say that a government "owns" the air and water systems within its jurisdiction. Air and water create special problems partly because they are "natural" assets—unlike roads, they are not man-made and the quantity of them cannot be altered—and partly because they are mobile, "flowing" resources that move around from one area to another. About the only things in this world that are not owned in any sense are the high seas and their animal inhabitants. We shall call these special types of property—roads, water, air, public parks and so on—*common property*. The term covers all property that is both owned in common (or unowned as in the case of oceans) and used in common; and it is to a study of common property that I now turn.

Before I do so, however, I want to try to avoid a possible misconception. I have been talking about property rights and not "private property." That sadly overworked and ill-defined phrase, "private property," has become an ideological concept that I want nothing to do with. I think that in some cases property rights should be vested in individuals; in other cases in groups of individuals, such as firms; and in some cases in governments. Different types of situations, it is reasonable to suppose, call for different forms of ownership. But in any event: no ideology! No "private property"! Just "property rights," by whomever exercised.

## Common Property, Restricted and Unrestricted

The first question to ask is why some property is owned in common. If we think of the history of this continent we remember that at one time virtually all the land was owned by some government. The land, however, was sold off to private owners, except for some public domain that the government wished to keep for its own use, and for the land in far northern areas in Canada that no private party wished to buy. On the other hand, many road systems were at one time privately owned toll roads, and these have all been brought under common ownership. What makes it easy to arrange for full property rights to land, and thus for private ownership of land, is that land is easily divisible (you can buy a few square yards or a few square miles) and not mobile (by means of fences, exclusivity of use can be enforced at a reasonable cost). Roads are land, of course, and the existence of common property in road systems is therefore a matter of choice; people have made a collective decision that it is more convenient to build and operate roads in common than to have private owners run the "road industry."

Air and water (except for a few non-navigable streams), however, seem to be owned in common because there is no alternative. There is no feasible way of separating a cubic yard, or an acre, of water from other cubic yards or acres; there is therefore no way of ensuring exclusive use; fences simply don't work. (Major drainage basins provide for a physical separation of waters up to the point where they enter oceans. These large units might be bought privately by one owner; but even if a private buyer came forward, society would undoubtedly decide that it would be undesirable to put all the water in a major drainage basin under the control of a private monopolist.) Thus it seems to be the physical characteristics of air and water, the fact that they are fluids and are naturally mobile over the face of the earth, that make it inevitable that they be owned in common or "vested in the right of" some government as the constitutional lawyers put it.

The nominal owner of a common property asset (i.e., some government) has, of course, an undoubted right to lay down rules for the use of the property. Rules for the use of man-made common property, such as parks or roads, are usually promulgated by the owner; as is nearly always true of stationary property—the French, with reason, refer to real estate as *immeubles*—enforcement of rules about use is practicable at reasonable cost. Where specific rules about the use of common property are laid down, we can call it "restricted" common property. Until recently, however, most governments have *not* made rules about the use of air or water. Implicitly, the government policy or rule has been that anyone could use air and water for whatever purpose he wished, without charge, permission, or hindrance. Such property we shall refer to as unrestricted common property.

In the past most people, at least in Canada, would no doubt have agreed that air and water resources were so vast that no rule about their use was needed. That day has now passed. But even if it be agreed that some regulation of use would be beneficial, we must still ask if the desired rules would be enforceable at reasonable cost. A "no-policy" policy makes sense if the cost of enforcing a positive policy is greater than its benefits. The costs of enforcing a policy fall, however, with improvement in administrative techniques, including such administrative hardware as computers and automatic monitoring devices. Such improvements therefore involve the possibility that rules about the use of common property resources that were impracticable in the past may be practicable now. Moreover, as larger and larger populations press against our fixed resources of air and water, the benefits to be gained by rules regulating their use increase, while enforcement costs are likely to become more easily bearable as they are spread over larger populations. These considerations suggest why our political scientists and our lawyers ought to be on a continuous look-out both for new legislative methods and for old ones that become newly practicable, in order to help control the use of our two most important common property resources, air and water.

The economic effect of making common property available for use on a no-rule basis, so that it may be freely used by anyone for any purpose at any time, is crystal clear. Common property will be over-used relative to both private property and to public property that *is* subject to charges for its use or to rules about its use; and if the unrestricted common property resource is depletable, over-use will in time lead to its depletion and therefore to the destruction of the property.

There is an old saying that "everyone's property is no one's property," the inference being that no one looks after it, that everyone over-uses it, and that the property therefore deteriorates. History bears out the truth of this saying in many sad ways. Property that is freely available to all is unowned except in a purely formal, constitutional, sense, and lack of effective ownership is almost always the source of much mischief. The inefficiency of medieval farming resulted in large part from the fact that ownership rights were usually poorly defined; in particular, the commons were unowned (i.e., owned by everybody—and nobody), and common pastures were so overstocked that their productivity fell to the vanishing point. Not until ownership concepts had evolved to a point where something like a modern view of property rights in land became accepted was it possible to use the land efficiently and increase agricultural output.

Another example: The sad list of animal species that have been extinguished by man's predation results purely from the fact that property rights in these animals did not exist, perhaps because they could not have been enforced if they had been established, but in any event because they did not exist. If animals are sought after they are valuable, and if they are owned those who seek them will have to pay their owners for the right to kill or capture them. Owners will charge a high enough price for the right to kill their animals that some stock of animals will always remain; you don't have to be an economist to know that it doesn't pay to kill the goose that lays the golden egg. No domestic animal has ever been threatened with extinction simply because domestic animals are owned. Nobody owned the buffalo or the passenger pigeon; and in recent years whales and kangaroos have been sadly victimized by the absence of ownership. If in the past Canadian governments had said of trees, as they said of buffaloes and passenger pigeons, that they belonged to everybody and everybody could cut them down free of charge, we may be sure that there would be no lumber industry or pulp and paper industry in Canada today.

With the rise of the automobile, the treating of road systems as unrestricted common property has accentuated congestion problems and public toll-roads may be the best way of relieving them. At any rate cities are beginning to learn that freeways seldom make for free-flowing traffic, and that the building of freeways soon increases traffic to the point where more freeways have to be built. Medieval men who witnessed the overstocking of unrestricted common pastures would understand automobile congestion on unrestricted common roads. (Knowing that common pasturing led to the deterioration of the livestock as well as of the pasture, they might also observe with interest the deterioration of the automobile stock resulting from, say, a hundred-car pile-up on a California freeway.)

Air and water in this country, and in most other countries, have been treated as unrestricted common property; so long as they are so treated air and water pollution will increase and the physical condition of our air and water assets will continue to deteriorate. Moreover, as has already been pointed out, we can manufacture more roads, but we cannot manufacture more air or water; all we can do is to use existing supplies as wisely as possible. It is time, I believe, that we took air and water out of the category of unrestricted common property, and began to establish some specific rules about their use or, to put it another way, to establish something more sophisticated in the way of property rights to their use than the rule that "anything goes." That rule may have been quite sensible in the past, when the demands made by human populations on the services of air and water were very small compared to the volumes of these assets; the benefits of controlling use would probably have been small and the costs of enforcing restrictions would no doubt have been large. All I am arguing is that growth in population, production, and urbanization inexorably changes the balance of the benefit-cost analysis against the policy of doing nothing and in favour of some positive policy. But to say that is not to say or in any way to imply that it is an easy matter to choose a *wise* positive policy, or to establish wise new property rights to the use of air and water.

## Social Problems

Like benefit-cost analysis, an analysis based on property rights provides a way of looking at pollution problems (and other social problems); but unlike the economic analysis, which is confined to the study of solutions that have been proposed, the legal analysis often generates proposals for solutions to social problems. In this section, I propose to look at, and comment on, a few examples of the relationships between property rights and social problems. Two points should be kept in mind throughout. First, there is no perfect legal solution to social problems, any more than there is a perfect economic solution. Second, a given legal definition of property rights in an asset has not only economic consequences (as we have seen in the previous section of this chapter) but also social and political consequences; there are interfaces between law and sociology and law and political science, as well as between law and economics.

Consider, first, a frivolous example. You are driving in a city after a heavy rain, and inadvertently drive through a large puddle of water so fast that you thoroughly drench some unfortunate pedestrian who happened to be in the wrong place at the wrong time. When you notice through your rear-view mirror that the victim is taking down your licence number, you stop and, after a brief conversation, pay him $10, shake hands, and go on your way.

This problem, then, was settled expeditiously and with a minimum of social friction. The reason is that the legal situation was clear and known to both parties. Ownership of an automobile did not confer the right to dirty other peoples' clothing, and ownership of clothing did confer the right not to have it dirtied by inattentive motorists; and neither party was interested in having a judge tell them what they both knew. The law might, of course, have said the reverse—that pedestrians had to look out for splashing motorists, rather than saying that motorists had an obligation to avoid splashing pedestrians. Had this been the case, there would have been no more social friction, but the economic outcome would have been different; the cost of the incident would have been borne by the pedestrian rather than the motorist. If the law had not been clear, there would have been bad feelings, there might have been a court case, and the economic outcome would have been unpredictable. There is much to be said for definiteness especially where the law is concerned.

Notice, too, that the existing law about splashing problems imposes the cost of injury on the active party, the motorist, rather than the passive pedestrian. The probable rationale for this policy is not the *social* view that the pedestrian is more important than the motorist, but the *technological* consideration that motorists can more easily avoid splashing pedestrians than pedestrians can avoid being splashed by motorists, and the *economic* consideration that it is cheaper to persuade motorists that it does not pay to splash pedestrians than to protect pedestrians by building a six-foot wall along the interface between sidewalks and roads. As a pedestrian, a motorist, and a taxpayer, I think the present law is very sensible.

Imagine, now, that you own a factory in Toronto and that you have been dumping your untreated factory wastes into Lake Ontario for forty years. Until recently not a single person complained of your practices and you are breaking no law by continuing to do what you have always done. Yet in the last couple of years it seems that you have become a villainous polluter, a heartless despoiler of nature, and a sneak thief robbing the children of Toronto of their natural right to swim in Lake Ontario; the press is after your scalp and trying to put the government on you; even your best friends seem to think that "something ought to be done about pollution." You object to such rough treatment, and reply that your lawyers advise that you have as much right to dump your garbage in Lake Ontario as any kid has to swim in it. You are probably right, legally, but you are in for a lot of trouble with your public relations.

Two comments suffice. Unrestricted common property rights are bound to lead to all sorts of social, political, and economic friction, especially as population pressure increases, because, in the nature of the case, individuals have no legal rights with respect to the property when its government owner follows a policy of "anything goes." Notice, too, that such a policy, though apparently neutral as between conflicting interests, in fact always favours one party against the other. Technologically, swimmers cannot harm the polluters, but the polluters can harm the swimmers; when property rights are undefined those who wish to use the property in ways that deteriorate it will inevitably triumph every time over those who wish to use it in ways that do not deteriorate it.

Economically and socially the question is always which set of interests *should* prevail, or rather what sort of accommodation should be made among the various interests concerned. The question is always, and inescapably, the great question of social justice.

Questions of social justice can be answered in many ways. Consider carefully the following example of an actual solution to water pollution problems in Britain. For this example, which I find utterly fascinating, I am indebted to Douglas Clarke who has recently described it in the following words.

> The island of Great Britain is moist and verdant, and blessed with innumerable cool streams that once were all haunts of trout and salmon. Most of them still are, even though they now flow through an industrialized countryside. The total poundage of fine game fish taken would put any accessible part of Canada to shame. We are so used to the idea that the waters of any industrial area are a write-off, so far as quality angling is concerned, that one cannot help but be curious as to how all that fishing is maintained.
>
> It is not because they do not have to watch out for pollution. There is an organization called the Anglers' Cooperative Association which has been in existence for nineteen years, which has taken over the watch dog functions formerly left to individuals. It is an interesting organization. It has a fluctuating and rather small list of members and subscribers, barely enough to keep an office open, but it is able to call on some powerful help, especially legal. It has investigated nearly 700 pollution cases since it started and very rarely does it fail to get abatement or damages, as the case requires. These anglers have behind them a simple fact. Every fishery in Britain, except for those in public reservoirs, belongs to some private owner. Many of them have changed hands at high prices and action is always entered on behalf of somebody who has suffered real damage. It has been that way from ancient times. Over here the fishing belongs to everybody—and thus to nobody. The A.C.A. exists merely to take action where individuals may not act themselves.
>
> Two cases from some time back well explain why the Derwent, which flows through the industrial city of Derby, still has its trout. Action was entered against the city because its effluent was harmful to trout, and the city, through its legal representatives, claimed in the highest court in the land that it was completely unreasonable to expect them to maintain the standards of a trout stream. The A.C.A., incidentally, acted on behalf of the "Pride of Derby Angling Club," which leased the fishery from the titled gentleman who owned it. The law lords said that the city had no more right to put its muck in the river than the citizens had to put theirs on the property of their neighbours. About the same time, and for the same city and river, an injunction was obtained against British Electric, a public corporation. All they had been doing was to run warm water directly into the river. Trout like it cool. The A.C.A. also deals with such—to us—trivia as mud running into a stream from a new road grade, or a ditch. It doesn't have to and the anglers are willing to go to court. This is

actually a good example of a common form of pollution which we accept but which is quite unnecessary and not hard to avoid.

What it amounts to is that you can have good fishing, which means good water, in a river in a populated British countryside if you make it your business to have it. It is not only Britain. We get an anglers' magazine from Germany and there are lovely illustrations showing good fishing on the Ruhr river, of which you may have heard, and on the Binnen, or inner, Alster, . . . in the industrial part of Hamburg. . . .

I will be the first to admit that there are geological and climatic differences between Ontario and western Europe which have influenced the impact of European settlement on our area, so that some of our streams have, inevitably, a less constant flow and a warmer temperature than they used to have. Within these limitations, however, we ask ourselves why we have to sacrifice water quality still further by deliberate pollution.

Some time ago the A.C.A. analyzed their comparatively few failures. In some cases the polluter could not be identified. In some other cases the polluter was insolvent, hence no damages. They call this failure. However, and this underlines the comparison between them and us, the most important single cause of failure was when the anglers who suffered from the pollution had no concrete evidence of interest, such as a valid lease, and had only tacit consent or a gentleman's agreement with the owner, who refused to become involved in the action. That sounds familiar. We, as individuals, fish the waters that we all own, collectively. As individuals we have sustained no damages at law. Collectively—as owners—well, forget it. In Britain, when a truck involved in an accident spills chemicals into a stream, the public liability insurance pays for the fish, for all the costs of clean-up and restocking, and for the loss of use and enjoyment during the period between kill and restoration because property damage has been done. Who looks after us?

Officially we have tried to do by statute what the British have done by the Common Law, but never, apparently, have we really meant what we said. Our first legislation, in 1865, had its teeth pulled in 1868. It is interesting that one simply cannot conceive of a judgment or an injunction obtained through legal action by the A.C.A. being set aside. Part of the explanation may be social. The A.C.A. has the Duke of Edinburgh for Patron. Apparently it is quite all right for him to be honorary keeper of a watch dog that has sunk its teeth into government corporations such as British Electric and the Coal Board, municipalities big and small, industries and private individuals, without fear or favour. I notice that His Grace the Duke of Devonshire is President, and there are two more dukes among the vice-presidents, (that is over ten per cent of the total number of non-royal dukes), as well as two additional peers, [and] a couple of knights.

There are many worthwhile comments to be made about this passage, but let me mention only a few of them. Note, first, that the solution results from a particular set of property rights (based in this case on Common Law) that are enforced by the courts, and that the property rights seem to be in the fish, or the fishing, not the water; there is no administrative agency that is concerned either with the fish or the water. Second, the solution may seem simple but in fact it isn't; Mr. Clarke is careful to suggest that the workability of the system may depend in important ways on such apparently irrelevant factors as the English climate, English history, and the particular social status enjoyed by the English nobility.

Third, there is no way of knowing whether the solution is a "good" one or not. At one level of analysis it can be argued that the solution favours the fishermen over industrialists and municipalities who have to bear the costs either of disposing of their wastes in such a way as to avoid polluting the rivers or of buying up the fishing rights to a river and then using it for waste disposal purposes; and there seems no obvious reason why the shoe should not be on the other foot—why polluters should not own property rights in the waste disposal capability of the river, in which case the fishermen, if they wanted to fish, would have to buy out the polluters' rights. Note, however, that if the government "owner" of the river follows a policy of "anything goes" neither party can buy out the other because neither has anything to sell! Under a system of unrestricted common property, groups that have opposing interests in the use of the property cannot negotiate because they have nothing to negotiate with; all they can do is yell interminably at each other.

In my opinion, however, it is often misleading to think of pollution problems in terms of groups rather than in terms of the society as a whole. In the present example fishermen no doubt live in cities and buy manufactured products, and industrialists and residents of cities no doubt sometimes go fishing. The groups are, in fact, all mixed up together. And it is not true that fishermen pay nothing for their good fishing; they pay higher prices for manufactured goods and higher municipal taxes than they would pay if the law favoured polluters and if fishing were not so good. Similarly, polluters get better fishing for their higher expenditures on waste disposal. From an over-all, social point of view the whole British population in effect buys good quality fishing (and other water-based recreation) by paying higher taxes and higher prices for goods; in Ontario, we have in effect accepted water pollution in return for cheaper goods and lower taxes. Which is the better policy? A silly question deserves a silly answer: whichever policy is preferred is better.

Mr. Clarke makes it quite clear that he prefers the British solution, or something like it. So do I. It would help if we could let our provincial member of parliament know roughly where each of us stands on the question of better quality air and water versus higher taxes and higher costs of goods. But it wouldn't help very much. The important question is *how much* "better quality environment" we would be willing to buy at different "prices" in terms of higher taxes and higher costs of goods, and most of us are not sure about this. The only way to answer the question may be to have the politicians start charging us for better quality air and water and then keep "upping the ante" until we say "Enough! No more!"

The trouble is that when we call a halt about half of us will think we are already spending too much to improve the environment, and about half of us will want to spend more; therefore very few of us will be very happy with the outcome. In some cases there is nothing more that can be done. In many other cases, however, there *is* a better solution; we have in fact often adopted it, but it is only recently that Professor Mishan, an economist, has generalized the argument that underlies it. The point is that it is often possible to avoid the sort of fifty-fifty compromise that we have been discussing. Take the question of smoking on a train, for example. If all passengers, half of whom are smokers, are required to come to a single decision, they may decide to allow smoking or not to allow smoking (in which case half of them are going to be unhappy all of the time) or they may decide to allow smoking during half the journey (in which case all of them will be unhappy half of the time). The sensible solution in this case is to provide what Mishan calls "separate facilities," i.e., to provide both smoking cars and no-smoking cars; everyone should then be happy all of the time. This solution, of course, is not applicable in a single-cabin vehicle such as a bus; "separate facilities" may not always be practicable. Zoning laws in cities are another common example of the "separate facilities" type of solution.

Is this solution applicable to pollution problems? Not perfectly, certainly, but to some considerable extent. Although air and water move around they do not mix thoroughly; over a large area such as Ontario it is certainly practicable to provide for different air and water qualities in different regions. When pollution matters come under municipal control different municipalities, or groups of municipalities, are likely to provide different quality environments. Under provincial control, the same variety is possible if the provincial authorities choose to follow some variation of the "separate facilities" or "zoning" principle. This principle will not always be applicable, and may not always be desirable. It is, however, always worth considering, if only because it offers some possibility of meeting a variety of demands and opinions with a variety of solutions. Instead of giving property rights in water use to polluters *or* fishermen, it may be thought desirable to assign the rights to fishermen in one area and to polluters in another.

## Soil Pollution and Inappropriate Property Rights

As new products, new practices, and new situations emerge, property rights need continual redefinition. Torontonians with trees on their lots used commonly to burn their leaves every fall. As general air pollution increased, it was considered unwise to add to the burden of the air by the widespread burning of leaves. Accordingly Torontonians agreed to reduce their property rights in their leaves; they can now do almost anything they like with them, except burn them; normally they pay the city, through their taxes, to haul them away.

New technology, especially in the form of chemical fertilizers and pesticides, has posed certain serious problems of soil pollution. Too much fertilizer can render soil infertile, and unless it is carefully applied large quantities of it run off the land and add to the "excess nutrient" type of water pollution. The new pesticides, based on chlorinated hydrocarbons, raise new problems; because they are chemically stable, they accumulate through time. It is feared, though so far as I know not firmly established, that in sufficient concentrations they might kill enough micro-organisms to disrupt normal bacteriological processes and seriously reduce soil fertility.

It is certainly not the absence of property rights that is responsible for the threat of soil pollution, since full property rights exist both in the land and in the chemicals. Nevertheless property rights are not static things, and we may at least describe the soil pollution problem in terms of *inappropriate* property rights in the new chemical substances. When first introduced in the late 1940s, these toxic materials were considered to raise no particular question of property rights. They *might,* of course, have been considered potentially dangerous products—like guns, narcotics, some medicines, and radioactive isotopes, for example —and it would then have been appropriate to define explicitly the property rights attached to them, i.e., what an owner could and could not do with them. (Remember that ideological overtones of such concepts as ownership or property are completely irrelevant to the present argument; ideological commitment to private ownership, though fairly strong in North American society, has not constituted much of a barrier to a fairly prompt, and restrictive, definition of property rights in hallucinatory drugs.)

It was almost certainly as a result of ignorance that when chlorinated hydrocarbons first appeared on the market their owners were given full property rights in them. Evidence of their unwanted effects on the environment built up, however, and after the publication of Rachel Carson's

*Silent Spring,* which gave the public a few lessons in elementary ecology, the tide turned, and a general movement toward a narrower definition of property rights in these materials has been apparent. In the meantime, an impressive array of vested interests had quickly built up around both the manufacture and use of these products. It is hardly too much to say that the whole structure of agriculture had changed to take advantage of the great increases in productivity that their use made possible, and it seemed unthinkable to do anything that might jeopardize such recently won progress. The difficulty of *reducing* property rights in the interests of combatting soil pollution seemed as great as the difficulty of *establishing* property rights in the interests of combatting air and water pollution.

Nevertheless, in Ontario and elsewhere, we are now witnessing early attempts to tackle both problems. Not much has been done in Ontario about reducing property rights in pesticides; but publicity about their harmful effects and increasing awareness of their dangers on the part of government officials have already resulted in greater caution in their use. The pesticides case is indeed an interesting one from the standpoint of public administration. Except perhaps for radioactivity, which was recognized to be deadly from the start, mankind had never before confronted himself with man-made wastes that were both toxic and persistent. Chlorinated hydrocarbons were something new in history, and governments had never before been faced with the types of problems they raised. It is not surprising that we made mistakes. I think it likely, though, that we have learned something by those mistakes. At least we can hope that as new man-made chemical wonders become available we and our governments will give long thought to the question of what property rights their owners should be granted. In the other direction, that of *establishing* property rights in air and water, much more has been done. For some years the province of Ontario has been pre-eminent in tackling water pollution problems; and in January 1968 provincial authorities took over jurisdiction from the municipalities in matters of air pollution, no doubt as a prelude to a more active policy in this field.

## Summary

This chapter has tried to suggest that legal definitions of property rights lie at the heart of social decision-making and problem-solving. Property rights are clearly antecedent to economics, since it is property rights that define the economist's "goods and services," and we have seen, particularly in the discussion of unrestricted common property, how property rights affect individual and social behaviour. A study of property rights gives us no magic key for the solution of social problems, but it does lead to suggestions for solutions that are refreshingly different from those offered by economists and other social scientists.

The main substantive conclusion I wish to draw from this chapter is that to treat air and water as unrestricted common property is socially indefensible. A policy of "anything goes" is defensible if the cost of enforcing a positive policy exceeds the benefits to be gained from it. This may be true now of polar ice, and it may have been true in the past of air and water, although I doubt it; English courts have apparently long enforced property rights in fishing, and in the process have enforced one solution to the problem of water pollution. In any event, it is perfectly clear on both theoretical and historical grounds that, as population grows, unrestricted common property will be over-used and deteriorate physically to the point of uselessness. On the assumption that we don't want that to happen to our air and water, it is high time that we began to devise some new forms of property rights, not to air and water, but to the *use* of air and water. In Ontario, during the last dozen years, we have begun to move in that direction, at least to the extent of changing the status of air and water from unrestricted to restricted common property. But the field for new ideas and social experimentation is still wide open.

# 20

## E. J. MISHAN

# *Growthmania*

*E. J. Mishan is a Reader in Economics at The London School of Economics.*

## I

Revolutions from below break out not when material circumstances are oppressive but, according to a popular historical generalization, when they are improving and hope of a better life is in the air. So long as toil and hardship was the rule for the mass of people over countless centuries, so long as economic activity was viewed as a daily struggle against the niggardliness of nature, men were resigned to eke out a living by the sweat of their brows untroubled by visions of ease and plenty. And although economic growth was not unheard of before this century—certainly the eighteenth century economists had a lively awareness of the opportunities for economic expansion, through innovation, through trade and through the division of labour—it was not until the recent post-war recovery turned into a period of sustained economic advance for the West, and the latest products of technological innovation were everywhere visible, and audible, that countries rich and poor became aware of a new phenomenon in the calendar of events, since watched everywhere with intentness and anxiety, the growth index.[1] While his



1. Like a national flag and a national airline, a national plan for economic growth is deemed an essential item in the paraphernalia of every new nation state.

father thought himself fortunate to be decently employed, the European worker today expresses resentment if his attention is drawn to any lag of his earnings behind those of other occupations. If, before the war, the nation was thankful for a prosperous year, today it is urged to chafe and fret on discovering that other nations have done perhaps better yet.

Indeed with the establishment of the National Economic Development Council in 1962 economic growth has become an official feature of the Establishment. To be *with* growth is manifestly to be 'with it' and, like speed itself, the faster the better. And if NEDC, or 'Neddy' as it is affectionately called, is to be superseded, it will be only to make way for larger and more forceful neddies. In the meantime every businessman, politician, city editor or writer, impatient to acquire a reputation for economic sagacity and no-nonsense realism is busy shouting giddy-up in several of two-score different ways. If the country was ever uncertain of the ends it should pursue, that day has passed. There may be doubts among philosophers and heart-searchings among poets, but to the multitude the kingdom of God is to be realized here, and now, on this earth; and it is to be realized via technological innovation, and at an exponential rate. Its universal appeal exceeds that of the brotherhood of man, indeed it comprehends it. For as we become richer, surely we shall remedy all social evils; heal the sick, comfort the aged and exhilarate the young. One has only to think with sublime credulity of the opportunities to be opened to us by the harvest of increasing wealth: universal adult education, free art and entertainment, frequent visits to the moon, a domesticated robot in every home and, therefore, woman forever freed from drudgery; for the common man, a lifetime of leisure to pursue culture and pleasure (or, rather, to absorb them from the TV screen); for the scientists, ample funds to devise increasingly powerful and ingenious computers so that we may have yet more time for culture and pleasure and scientific discovery.

Here, then, is the panacea to be held with a fervour, indeed with a piety, that silences thought. What conceivable alternative could there be to economic growth? Explicit references to it are hardly necessary. When the Prime Minister talks with exaltation of a 'sense of national purpose' it goes without saying that he is inspired by a vision, a cornucopia of burgeoning indices.

But to be tediously logical about it, there is an alternative to the postwar growth-rush as an overriding objective of economic policy: the simple alternative, that is, of not rushing for growth. The alternative is intended to be taken seriously. One may concede the importance of economic growth in an indigent society, in a country with an outsize population wherein the mass of people struggle for bare subsistence. But despite ministerial twaddle about the efforts we must make to 'survive in a competitive world', Britain is just not that sort of country. Irrespective of its 'disappointing' rate of growth, or the present position of the gold reserves, it may be reasonably regarded, in view of its productive capacity and skills, as one of the more affluent societies of the West, a country with a wide margin of choice in its policy objectives. And it is palpably absurd to continue talking, and acting, as if our survival—or our 'economic health'—depended upon that extra one or two per cent growth. At the risk of offending financial journalists and other fastidious scrutinizers of economic statistics, whose spirits have been trained to soar or sink on detecting a half per cent swing in any index, I must voice the view that the near-exclusive concern with industrial growth is, in the present condition of Britain, unimaginative and unworthy.

The reader, however, may be more inclined to concede this point and to ponder on a more discriminating criterion of economic policy if he is reminded of some of the less laudable consequences of economic growth over the last twenty years.

Undergraduate economists learn in their first year that the private enterprise system is a marvellous mechanism. By their third year, it is to be hoped, they have come to learn also that there is a great deal it cannot do, and much that it does very badly. For today's generation in particular, it is a fact of experience that within the span of a few years the unlimited marketing of new technological products can result in a cumulative reduction of the pleasure once freely enjoyed by the citizen. If there is one clear policy alternative to pressing on regardless, it is the policy of seeking immediate remedies against the rapid spread of disamenities that now beset the daily lives of ordinary people. More positively, there is the alternative policy of transferring resources from industrial production to the more urgent task of transforming the physical environment in which we live into something less fit for machines, perhaps, but more fit for human beings.

Since I shall illustrate particular abuses of unchecked commercialism in later chapters and criticize them on grounds familiar to economists, I refrain from elaboration at this point. However, it is impossible not to dwell for a moment on the most notorious by-product of industrialization the world has ever known: the appalling traffic congestion in our towns, cities and suburbs. It is at this phenomenon that our political leaders should look for a really outstanding example of post-war growth. One consequence is that the pleasures of strolling along the streets of a city are more of a memory than a current pastime. Lorries, motor-cycles and taxis belching fumes, filth and stench, snarling engines and unabating visual disturbance have compounded to make movement through the city an ordeal for the pedestrian at the same time as the mutual strangulation of the traffic makes it a purgatory for motorists. The formula of mend-and-make-do followed by successive transport ministers is culminating in a maze of one-way streets, peppered with parking meters, with massive signs, detours, and weirdly shaped junctions and circuses across which traffic pours from several directions, while penned-in pedestrians jostle each other along narrow pavements. Think, for instance, of Piccadilly Circus, the hub of a capital city, imprisoned in its traffic.

Towns and cities have been rapidly transmogrified into roaring workshops, the authorities watching anxiously as the traffic builds up with no policy other than that of spreading the rash of parking meters to discourage the traffic on the one hand, and, on the other, to accommodate it by road-widening, tunnelling, bridging and patching up here and there; perverting every principle of amenity a city can offer in the attempt to force through it the growing traffic. This 'policy'—apparently justified by reckoning as social benefits any increase in the volume of traffic and any increase in its average speed—would, if it were pursued more ruthlessly, result inevitably in a Los Angeles-type solution in which the greater part of the metropolis is converted to road space; in effect a city buried under roads and freeways. The once-mooted alternative, a Buchanan-type plan—'traffic architecture' based on the principle of multi-level separating of motorized traffic and pedestrians—may be an improvement compared with the present drift into chaos, but it would take decades to implement, would cost the earth, and would apparently remove us from contact with it. The more radical solution of prohibiting private traffic from town and city centres, resorts, and places of recreation, can be confidently expected to meet with the organized hostility of the motoring interests and 'friends of freedom'. Yet, short of dismembering our towns and cities, there is no feasible alternative to increasing constraints on the freedom of private vehicles.

## II

Other disagreeable features may be mentioned in passing, many of them the result either of wide-eyed enterprise or of myopic municipalities, such as the post-war 'development' blight, the erosion of the countryside, the 'uglification' of coastal towns, the pollution of the air [2] and of rivers with chemical wastes, the accumulation of thick oils on our coastal waters, the sewage poisoning our beaches, the destruction of wild life by the indiscriminate use of pesticides, the change-over from animal farming to animal factories and, visible to all who have eyes to

2. According to Professor L. J. Battan, of Arizona, *The Unclean Sky: A Meteorologist looks at Air Pollution,* the air above is treated as a vast sewer. Gases have been poured into the atmosphere in the mistaken belief that the wind, like a river, would not only carry the wastes away but somehow purify them in the process. As a result, some ten million tons of solid pollutants are now floating around in the sky. There is, however, a limit to what the finite atmosphere can safely disperse: what goes up must eventually come down.

see, a rich heritage of natural beauty being wantonly destroyed—a heritage that cannot be restored in our lifetime.

For it is the unprecedented speed and scale of developments that have caught us off our guard. Such is the expanding power of modern technology, such the opportunism of man's enterprise, that the disposal of the waste products of industry, which for thousands of years were absorbed into the cycle of nature, has suddenly, it seems, broken all ecological bounds. And the general public, its attention continually distracted by technological wonders, has simply no notion of the extent or gravity of the situation. It does not, therefore, stop to reflect on the generations it would take to undo the damage wrought over the last half century; to regrow forests on the hundreds of thousands of acres stripped of timber, to remove some of the thousands of square miles of concrete laid over the earth, to purify lakes and rivers reeking of sewage, to revive the wild life of Africa, to restore to its original magnificence several thousand miles of Mediterranean coastline ruined in the post-war tourist spree, to cleanse the atmosphere of millions of tons of floating pollutants and of the uncertain but growing amount of radioactive matter.

If anything, one detects an 'eat, drink, and be merry spirit' abroad, inadvertently perhaps fostered by the historian whose cultivated detachment forbids him to be ruffled by current events. Rather than project uncouth magnitudes into the future, it is simpler to respond to present dangers by observing patiently that similar alarms were sounded in ages gone by. Did not Malthus warn the world of the mounting pressure of population at the close of the eighteenth century—when population was a small fraction of what it is now? Science, like love, will surely find a way! And the disappearance of the countryside? Go back a century, and we shall also find men lamenting the vanishing countryside! Go back to the eighteenth century, or farther back to the seventeenth, and always we shall find men who, despising the new towns, bemoan the passing of the rustic virtues! *Ergo,* nothing has really changed; all alarm is groundless. This time-hallowed anti-Jeremiad ploy was particularly in evidence after the explosion of the two atomic bombs over Japan in 1945. Some readers may recall a particular cartoon representing a conversation between two cave-men watching a third walking away with a rough-hewn stone axe. The caption read: 'With this dreadful new weapon that Smith has invented, the survival of mankind itself is in the greatest danger.' Find a precedent, invent a precedent if needs be, and, apparently, the danger is effectively exorcized!

## III

Our political leaders, all of them, have visited the United States, and all of them seem to have learned the wrong things. They have been impressed by the efficient organization of industry, the high productivity, the extent of automation, and the new one-plane, two-yacht, three-car, four-television-set family. The spreading suburban wilderness, the near traffic paralysis, the mixture of pandemonium and desolation in the cities, a sense of spiritual despair scarcely concealed by the frantic pace of life—such phenomena, not being readily quantifiable, and having no discernible impact on the gold reserves, are obviously not regarded as agenda.

Indeed, the jockeying among party leaders for recognition as the agents of modernization, of the new, the bigger and better, is one of the sadder facts of the post-war world, in particular as their claim to the title rests almost wholly on a propensity to keep their eyes glued to the speedometer without regard to the direction taken. Our environment is sinking faster into a welter of disamenities, yet the most vocal part of the community cannot raise their eyes from the trade figures to remark the painful event. Too many of us try not to notice it, or if occasionally we feel sick or exasperated we tend to shrug in resignation. We hear a lot about 'the costs of progress', and since the productivity figures over the years tend to rise we assume that on balance, and, in some sense, we must be better off.

In the endeavour to arrest this mass flight from reality into statistics I hope to persuade the reader that the chief sources of social welfare are to be found not in economic growth *per se* but in a more selective form of development, one that includes a radical reshaping of our physical environment with the needs of civilized living—and not the needs of motorized traffic or industry—foremost in mind. Indeed the social accommodation to technological advance is, in any case, almost sure to reduce our experience of welfare.

Something must be said about the widespread misconception which persuades us that, as a nation, we have no real choice: that living in the twentieth century we are compelled to do all sorts of things we might otherwise not wish to do. Since childhood, I imagine, all too many of us have lived in awe of the balance of payments. And now that economic growth is all the rage we have come unthinkingly to link faster economic growth with the prospect of improved trade figures, a proposition for which there is no economic warrant. There is a great deal more choice in our domestic affairs and in the conduct of our foreign trade than is usually conveyed in the financial columns of the press; at least enough to free us from the imagined compulsion to expand rapidly and from popular bogies of 'not surviving' or of 'being left behind in the race', or of 'stagnating in an amiable backwater'.

# 21

ROGER W. WEISS

# *Mishan on Progress*

*Roger W. Weiss is an associate professor at the University of Chicago.*

In the avalanche of criticism of the modern market economy, including that of Galbraith, Heilbroner, the Triple Revolutionists, Marxians, Veblenians, and ecologists, a solution is always held out. Government activity, control, and reorganization are usually sufficient to keep us in progress and to avoid the catastrophies of public parsimony, unemployment, exploitation, and the silent spring. Those who warn us against the death of our cities and the pollution of streams and the atmosphere usually have a solution. Occasionally, an economist has held out a "middle way" as a superior method of organizing mass industrial societies (Röpke, 1950).[1] Some (but curiously, not Mishan) have suggested a control of the number of births. It must be very rare that an economist turns against progress itself, but E. J. Mishan, in his recent book, *The Costs of Economic Growth* (1967),[2] announces his almost complete disaffection with the twentieth century and its characteristic feature, progress. Its resulting technology is destroying the amenities of civilized life through the destruction and pollution of the landscape. The social pat-



1. Röpke, Wilhelm. *The Social Crisis of our Time*. Chicago: Univ. of Chicago Press, 1950.
2. Mishan, Ezra J. *The Costs of Economic Growth*. New York: Frederick A. Praeger, 1967. [See Selection 20.]

terns of pride of workmanship, religious devotion, and family cohesiveness are rapidly deteriorating.

Progress is a statistical fallacy, in Mishan's view. Its costs are not counted in the usual national income accounts. They are omitted either because compensation is not paid to the losers or because the neighborhood costs of pollution, noise, loss of satisfaction in work, and loss of meaningful social relationships are not included in the statistical accounts of income.

If any economic change in which someone is made worse off is to be defined as economically an inferior position, then indeed "less is more" and we are better off with the status quo. But if this logic were relevant today, it would have been relevant fifty years ago, and buggy makers, blacksmiths, railway equipment manufacturers, small grocers, and cotton pickers would either still be following their occupations or would have been compensated. We might still use horses and railways. Mishan leaves us no alternatives to holding the status quo, even though he realizes that his economic argument doesn't justify the status quo more than some alternative. The book's greatest deficiency is in not developing criteria and a mechanism with which society can make decisions about change.

Mishan's second economic ground for opposing progress is that it generates (as it has been working) very destructive neighborhood effects. The formal conditions of private and social cost are obviously not met in many areas. But because the distribution of traffic on two "free" roads will not be optimal, we cannot infer that the roads should be closed. This seems to be Mishan's solution—particularly when he would prohibit the automobile altogether in cities and nationalize road transport as a solution to rational allocation of resources to this industry. The traffic horrors of London, Paris, and Rome result from cities built up before the automobile. The horrors of Los Angeles are not traffic horrors—they result from a group of suburbs being organized by the automobile. But for most of the places touched by the automobile, one might believe that the public agency that builds roads would build them up to the point where the economic rent of the more convenient roads was equalized, where the marginal social product of additional dollars spent on roads equaled their marginal cost, and where user taxes on gasoline or toll charges on the use of roads approximated the social costs of building the roads, perhaps including the loss of amenity of open spaces. There would remain a special problem in heavily built-up areas, where the ratio of existing population to highways is very high, where the costs of adding to the number of roads is much higher than the average for the country, and so on. Here special measures will be needed to assure that the very high marginal social costs of driving are borne by the motorists. Special licensing of London autos, heavy tolls on the arterial roads bringing rush-hour traffic into London, and other measures, including an improvement of the competing transportation systems, are called for.

The use of coal in fireplaces in London blackened the skies in the nineteenth century. One could plausibly have argued that coal should have been prohibited from the use in the city, owing to the discrepancy between its private and social cost, in the absence of excise taxes to make them equal. Perhaps the population would have grown more slowly, and perhaps the pollution of the atmosphere would have been less. The railroad brought such undesirable neighborhood effects that great landlords often determined that stations would not be located near their houses. Electricity has brought its undesirable neighborhood effects: The ugly sight of long-distance wires, the noise level created by window air-conditioning units, and, when compensation is taken into account, the losses to employees in kerosene-lamp manufacturing must all be considered. But we would not prohibit electricity, and the neighborhood effects of window air-conditioners have been greatly reduced as these appliances have become ubiquitous. And electricity has reduced many of the neighborhood effects of coal.

Mishan's economic indictment of progress is the protest of the uncompensated minority whose budget includes large proportions of contemplation of the countryside (while living in the inner city), leisure-time conversation, and a taste for old ways. The commuter, the week-end working-class vacationer, and the municipalities who build the blocks of tall apartment buildings are his natural enemies. But he seems not to have discovered that, when the many live in the apartment complexes, it is possible to find a quiet apartment in St. John's Wood. The mass market requires standardization and large output. Mishan's minority has done quite well in the abandoned country farmhouses and charming city mews. Of course, prosperity drives up their prices. But Mishan's generation has gotten the capital gains (*pace* Malthus, whose neighborhood effects Mishan ignores).

The minority outlook found in this book is not a new one. It echoes the reservations of John Stuart Mill expressed in his essay on "Civilization" and even Mill's recommendation of nationalization of railways. Mill's argument that two railways connecting the same places were a social waste was urged partly by the "violent irritation . . . at the continual spoiling of [his] favourite rambles by the random tracks of the juggernaut." In his essay on "Civilization" he saw advance bringing power to the masses, and with this change, of "Quackery and especially of puffing . . . the inevitable fruits of immense competition; of a state of society, where any voice, not pitched in an exaggerated key, is lost in the hubbub. . . . Literature has suffered more than any other human production by the common disease." Intellectual distinction would be lost in an egalitarian world. Although enjoying a comfortable mediocrity, the society of the future would not be happy. For Tocqueville, from whom Mill took many of his opinions on the future, saw that "a cloud habitually hung upon [the Americans'] brow, and I thought them serious and almost sad, even in their pleasures." Progress would not bring happiness. But Mishan writes, deliberately ignoring what others have said. In overlooking Mill and Tocqueville, he has not properly diagnosed the blight he finds. It is not the result of progress or neighborhood effects; rather, it is the result of the consensus for automobiles and portable radios. Mishan is in the uncompensated minority. Tocqueville called them the aristocracy. Mill called them the educated middle classes. At least Mill and Tocqueville would not have had the illusion that they were in the majority. Mishan seems to think so, otherwise his economics would have not assumed the place in his argument that they have.

# 22

## JOHN KENNETH GALBRAITH

# *The Further Dimensions*

*John Kenneth Galbraith is Paul M. Warburg Professor of Economics at Harvard University.*

> ". . . for the first time since his creation man will be faced with his . . . permanent problem—how to use his freedom from pressing economic cares, how to occupy the leisure, which science and compound interest will have won for him, to live wisely and agreeably and well."
>
> John Maynard Keynes
> *Essays in Persuasion*

> "We should not fall prey to the beautification extremists who have no sense of economic reality."
>
> Fred L. Hartley
> President of the Union Oil Company

1

The industrial system identifies itself with the goals of society. And it adapts these to its needs. The adaptation would not be so successful were those who comprise society aware of it—did they know, in effect, how they are guided. It is the genius of the industrial system that it makes the goals that reflect its needs—efficient production of goods, a steady expansion in their output, a steady expansion in their consumption, a powerful preference for goods over leisure, an unqualified commitment to technological change, autonomy for the technostructure, an adequate supply of trained and educated manpower—coordinate with social virtue and human enlightenment. These goals are not thought to be derived from our environment. They are assumed to be original with human personality. To believe this is to hold a sensibly material view of mankind. To question it is to risk a reputation for eccentricity or asceticism.

Or so it has been. Few things are so appealing as reflection on the novelty or originality of one's own position. In recent times there has, in fact, been a persistent questioning of conventional economic and social goals. Economic values have been especially questioned. Alienation has been evident among youth. It has manifested itself in rejection of conventional attitudes on work, career, clothing, and foreign policy. But this unease is not confined to youth. It has been widespread in the educational and scientific estate. And it has invaded, even, the great philanthropic foundations where it has led to grants to groups duly constituted to re-examine the purposes of the society. Such re-examination has invariably led to a strong affirmation of the goals that serve the needs of the industrial system.

* * *

## 2

The industrial system generally ignores or holds unimportant those services of the state which are not closely related to the system's needs. National defense, support to research and technological development, such collateral needs of industrial growth as highways and air traffic management are not neglected. Nor is education. With the passage of time, support for education reflects not alone the needs of the industrial system but also the increasing political power of the educational and scientific estate. Educators, in pursuit of political interest, differ from others only in the impression of exceptional purity of motive which they are able to convey.

Services of the state that are not directly related to the needs of the industrial system are much less favored. Two factors operate here. Services that are unneeded by the industrial system and which, unavoidably, the state must render, suffer from a negative discrimination. Soap and dentifrices are accorded importance by the industrial system by the advertising by which it manages demand. Public clinics, which may do more for health, are the beneficiaries of no similar promotion. They suffer accordingly. Still other activities of the state are inimical to the industrial system, or to the goals it avows and the priorities it enjoys. They encounter the active opposition of the industrial system. Both cases require brief examination.

Such services of the state as the care of the ill, aged and physically or mentally infirm, the provision of health services in general, the provision of parks and recreation areas, the removal of rubbish, the provision of agreeable public structures, assistance to the impoverished and many other services are not of particular importance to the industrial system. And they are in competition for funds with the wants that result from the aggressive management of the consumer by the industrial system. In consequence, hospitals do badly in competition for resources with automobiles. Expenditures for parks for outdoor play do poorly in competition with those for chromatic television. And so forth.

And belief is extensively, although imperfectly, accommodated to this discrimination. Private virtue consists in producing more for more money. Public virtue still lies, on the whole, not with the politician who proposes to accomplish more for the same expenditure, but with the one who proposes to do more for less. And the voice of the man who wishes government to do less for less is still heard. By especially accommodating philosophers it is still held that the state should minimize its services. Otherwise, it abridges the right of the individual to decide his purchases for himself.

Doctrine thus supports organic tendency to create a natural imbalance between the goods produced and the services supplied by the industrial system and those which are supplied by the state and which do not serve the needs of the industrial system. But these are matters on which I have written elsewhere and at length.[1] With no views is it so pleasant to agree in detail as with one's own. But the temptation must be put down.

I turn now from the negative discrimination against public services to the much stronger opposition that arises from the association of the state with goals that are alien or inimical to the industrial system.

## 3

Beyond the area of goods and services, however supplied, and the demand for them, however contrived, is the further world of aesthetic experience. This is served not by factories or engineers but, in one manifestation or other, by artists. Enjoyment of the experience owes something to preparation; no more than the response to a lighter, smoother, blended whiskey is it original in the soul of man.

The aesthetic experience was once a very large part of life—unimaginably large, given the values of the industrial system. The traveler from the United States or the industrial cities of Europe or Japan goes each summer to visit the remnants of preindustrial civilizations.

1. In *The Affluent Society* [Boston: Houghton Mifflin, 1958; 2*nd* edition, 1969].

That is because Athens, Florence, Venice, Seville, Agra, Kyoto and Samarkand, though they were infinitely poor by the standards of modern Nagoya, Düsseldorf, Dagenham, Flint or Magnitogorsk, included, as part of life, a much wider aesthetic perspective. No city of the post-industrial era is, in consequence, of remotely comparable artistic interest. Indeed, no traveler of predominantly artistic interest ever visits an industrial city and he visits very few of any kind which owe their distinction to architecture and urban design postdating the publication of Adam Smith's *Wealth of Nations* in 1776.

One of the terms of disapprobation in the industrial system is aesthete. This is because aesthetic achievement is beyond the reach of the industrial system and, in substantial measure, in conflict with it. There would be little need to stress the conflict were it not part of the litany of the industrial system that none exists.

The conflict derives partly from a conflict in goals and partly because aesthetic goals are beyond the reach of the technostructure, which is to say that it cannot identify itself with them. So, if they are strongly asserted, they will be viewed as a constraint.

Thus, in an obvious case, if aesthetic goals are strongly asserted, this will affect the location of industrial plants. These will be placed not where they are most efficient but where they are least offensive. Their mode of operation, including the odors they dispense into the atmosphere and the wastes they deposit into streams, lakes and subsoil, will also be controlled. This means higher cost, smaller output or both. Questions would be asked about products—about the shape, number and construction of automobiles that is consistent with a pleasant urban aspect or an agreeably neutral air.

Such constraints would be inconvenient. Social thought in the industrial system does not allow of inquiry as to whether increased or more efficient production of a particular product is a good thing. It is, *per se,* a good thing.

Aesthetic goals contest the claims of power lines over landscape, of power development over natural streams or national parks, of highways over urban open spaces, strip mining over virgin mountainsides, shopping centers over antique squares and high-speed air travel over tranquillity below. Many of these claims to industrial priority have, in fact, been contested. The contests are seen, however, as episodic and accidental rather than organic, and the burden of proof remains heavily upon those who assert the aesthetic priority. If economic advantage—the effect on output, income and cost—is clear, it will usually be decisive. Aesthetic considerations must usually prove that they yield economic advantage in the long run.

Were aesthetic goals accorded priority, their assertion would be normal and there would be a presumption that industrial efficiency would be subordinate to them. This too would be inconvenient.

To assert aesthetic goals is also to interfere seriously with the management of the consumer. This, in many of its manifestations, requires dissonance—a jarring of the aesthetic sensibilities. An advertising billboard that blends gracefully into the landscape is of little value; it must be in sharp contrast with surroundings. This jarring effect then becomes competitive. The same principles of planned dissonance are even more spectacularly in evidence in the radio and television commercial. They also characterize the design or packaging of numerous industrial products. And an effort is made to bring this dissonance within the ambit of social goals. It is defended interestingly by the contention that it "gives the consumer what he wants." If he did not approve, he would not respond. A man who comes to a full stop because he is hit over the head with an ax proves similarly by his response that it was what he was yearning for.

## 4

The industrial system has a yet further and more fundamental conflict with the aesthetic dimension. The industrial system, as we have sufficiently seen, depends urgently on organization. Fragments of information, each associated with a person, are combined to produce a result which is far beyond the capacity of any one of the constituent individuals. But while this is a procedure which lends itself admirably to technological development and to the less inspired levels of scientific research, it does not lend itself to art. Artists do not come in teams. The greatest industrial achievement, myth to the contrary, emerges from committees. But not the greatest painting, sculpture or music. The artist may be more of a social being than the legend holds. It is noticeable that he regularly eschews, in practice, the cruel isolation which, as a deeply creative being, he is supposed to suffer in principle. His flocking and nesting tendencies are no less convivial than those of accountants, engineers and high executives. But he does, in his work, enfold the whole of his task within himself. He cannot work on or with a team. We have here a principal explanation of why the high technical and productive achievements of the industrial system are so regularly combined with banal or even offensive design.

The aesthetic dimension being beyond the ready reach of the industrial system, members of that system are led naturally to assert its unimportance. Juveniles who do not like Latin, economists who do not like mathematics and men who do not like women manifest precisely the same tendency.

But this is not all. Cultivation of the aesthetic dimension accords a new and important role to the state and one to which, by virtue of its handicaps, the industrial system is unrelated. Part of this role has already been implied. Where there is a conflict between industrial and aesthetic priorities, it is the state which must assert aesthetic priority against the industrial need. Only the state can defend the landscape against power lines, advertisers, lumbermen, coal miners, and, on frequent occasions, its own highwaymen. Only it can rule that some patterns of consumption—the automobile in the downtown areas of the modern city is a prominent possibility—are inconsistent with aesthetic goals. The state alone can protect radio and television from contrived dissonance—or provide alternatives that are exempt. And were aesthetic priority asserted, the state would be required to come to its defense not, as now, episodically and in response to some exceptional outrage of aesthetic sensibilities. It would have to do so normally and naturally as the defender of goals in which aesthetic considerations were consistently important. Such goals, it must be added, will not occasionally but usually be achieved at some cost to industrial expansion—to economic growth. That one must pause to affirm that beauty is worth the sacrifice of some increase in the Gross National Product shows how effectively our beliefs have been accommodated to the needs of the industrial system.

## 5

But the role of the state on the aesthetic dimension is not merely protective; it is also affirmative. While art is an expression of individual personality, important branches of the arts can only flourish within a framework of order. This must be provided by the state. Specifically, painting, sculpture and music, although not within the ambit of the industrial system, do reasonably well on the patronage that it provides. There is need for instruction in appreciation and enjoyment. (In keeping with the ethos of the system this is considered a much inferior employment of educational funds to their devotion to science, mathematics and engineering.) But, while there is much that the state can do by way of encouragement here, its role is not decisive.

In the case of architecture and urban and environmental design, its role is decisive. Art is one manifestation of order. And it is the first casualty of disorder. Florence, Seville, Bloomsbury and Georgetown are beautiful because each part is in orderly relation to the whole. The modern commercial highway, the sprawling fringe of any city, the route into town from any airport is hideous because no part is related to a larger design. This order is rarely if ever achieved permissively; it must always be imposed by the state or by social pressure.

Good architecture is also mostly meaningless unless it is within a consistent framework. The Taj Mahal would lose much of its queenly elegance if surrounded by modern service stations. This has been the fate of quite a few distinguished modern buildings. Nineteenth-century Paris owes its excellence not to the brilliance of the individual buildings but to the consistency of the overall design.

Further, there is much architecture of which the state must always be the patron. It is the natural source of distinguished buildings, interesting monuments, agreeable gardens and fountains, long vistas, imposing squares, soaring towers and rich facades. Only as nations have become very rich and the industrial system has made economic growth identical with life, have we ceased to suppose that such patronage is a fit function of government. Quite commonly it is said that it cannot be afforded.

## 6

It would be foolish to insist that government in the United States—whether of cities, states, or the Federal government—is a good custodian of aesthetic goals. Politicians may well have a special penchant for banality. Those who do not urge it out of personal preference will think it necessary as a concession to the popular taste. Although the world owes a greater debt to public architecture than to private, it owes more to the taste of talented despots—Shah Jahan, Cosimo and Lorenzo, Peter the Great, Louis XIV—than to democrats. It is part of the case against public sponsorship of aesthetic goals by modern democratic governments that they will be strongly biased in favor of what is bad.

This cannot be denied. It is only that for asserting aesthetic priorities and providing the essential framework for artistic effort there is no alternative to the state. Those who say that in consequence of its shortcomings the state must forgo all concern for art thereby reject aesthetic priority. They become advocates of environmental disorder.

For even when the state exercises artistically imperfect control over environment, the result will be better than when there is none at all. In the late twenties and early thirties the planners and architects of Washington, D.C. swept clear an area between Pennsylvania and Constitution Avenues to build a vast block of buildings called the Federal Triangle. The Triangle is unimaginative, derivative and pretentious. Artists rightly condemned it. But it is far better than the cabbage patch of buildings it replaced. In its general cohesiveness it has come to be admired in comparison with those parts of the city where no similar effort was ever made.

And the state can be expected to do better in support of the aesthetic dimension in the future than in the modern past. For this will be recognized as a high public responsibility. What is done as an afterthought is rarely done well. Something better can be expected when a task is seen to be central, not marginal, to life. It is worth hoping that the educational and scientific estate, as it grows in power, will encourage and enforce more exacting aesthetic standards. Nothing would more justify its intrusion on public life.

## 7

For many years, politicians completing a term of office and seeking another have taken, as the measure of their deservedness, whether their constituency is more prosperous than when they began. If it is, and larceny has not been palpable, they consider themselves to have a good claim for re-election. It is a test which it has been hard for even the most negligible statesman to fail. All, the intelligent and stupid, diligent and idle, have been swept along on a current of increased output that, in the usual case, owed nothing whatever to their efforts.

The aesthetic dimension introduces a new and much stronger test. It means that mayors completing a term at city hall, governors at the state capital, Presidents in the White House, Prime Ministers at 10 Downing Street will be asked whether they have left their city, state or country more beautiful than before. This test will not be so easy. Few if any in this century would have passed. The fact of universal failure is another reason for insisting on the unimportance of the aesthetic dimension. No one likes an examination which he surely flunks. But far more than the test of production, which is far too easy, the test of aesthetic achievement is the one that, one day, the progressive community will apply.

# V

# *MEASURING COSTS AND BENEFITS*

The theme that runs most insistently through this volume is the need to oblige users of the environment to "internalize" the social costs of their activities. The government is admonished to "put a price on pollution"; but what price? We come in this final section to what may be the most perplexing problem that environmental economists have to contend with: how to measure the social costs of environmental damage or, alternatively, how to value the social benefits of environmental improvement. Assessing the costs of abating environmental damage is an equally important matter, but one that can be dealt with more straightforwardly.

Whatever device a government chooses for regulating the extent of damages to the environment—whether direct controls, effluent charges, or undertaking abatement measures of its own—its decision will have to rest on some form of quantification of the benefits its constituents may be expected to derive and the costs they may be expected to incur in connection with each of the range of relevant alternatives. Some such accounting is essential to the choice of an optimal level of abatement, one which will yield the greatest surplus of social benefits over social costs. In the introductory piece in this section, Paul Davidson discusses some of the theoretical and procedural problems that are likely to be encountered in carrying out the task of evaluating costs and benefits, particularly where goods such as environmental resources, which are enjoyed by large numbers of users simultaneously, are involved.

The next two papers are presented here to give the reader a taste of the kind of workmanship required in undertaking to quantify even a small segment of a social-welfare function. Lester B. Lave and Eugene P. Seskin attempt to measure the costs of damages to human health associated with varying degrees of air pollution. Jack L. Knetsch and Robert K. Davis are concerned with valuing the recreational benefits of environmental improvement. The most serious problem to contend with is the absence of a pricing system for most of the kinds of benefits which improvements in the condition of the environment yield, since the bulk of them represent enhancements of health and personal amenities for which there is no market. Thus, although Lave and Seskin achieve a good measure of the association between human morbidity and mortality and varying degrees of air pollution, the difficulties in trying to value pollution costs in dollars and cents are manifest. Knetsch and Davis's main concern is finding a proxy for the price people would be willing to pay for increases in recreation benefits.

In the last article in this collection, Wassily Leontief shows how the input-output technique can be used to add up the total pollution costs of final outputs by following the outputs through the stages of production and cumulating the amounts of pollution generated at all stages.

# 23

PAUL DAVIDSON

# *The Valuation of Public Goods*

*Paul Davidson is Professor of Economics at Rutgers, the State University of New Jersey.*

Cost-benefit methodology has been widely adopted by many governmental agencies for deciding on the desirability of specific projects. In general, this approach suggests that, in the absence of budgetary constraints, a project should be undertaken whenever the present *value* of the associated stream of net benefits from a project (discounted at the appropriate social rate of discount) is greater than the present cost of facilities.

Cost-benefit analysis is a method of collecting information, processing it, and providing a systematic approach to choosing between alternatives. It requires two things: clearly established alternatives, and a net-benefits or welfare function, which allows one to quantify the benefits from the various available choices. This second requisite is the most difficult and most arbitrary for it requires the specification of the order of priority of goals for society, i.e., it requires us to define quantitatively social welfare. Given the choices and the welfare functions, it is possible to array the alternatives in an order of priority. When a large number of choices and / or elements in the welfare function must be

*"The Valuation of Public Goods," by Paul Davidson, from* Social Sciences and the Environment, *Morris E. Garnsey and James R. Hibbs, eds., University of Colorado Press (Boulder: 1967).*

considered, it is essential that a sophisticated systematic approach such as cost-benefit analysis be utilized if rational choices are to be made.

The success or failure of such a methodical approach to public policy decision making depends largely on the ability of the welfare function to correctly specify the value of the choices to our society and on our ability to correctly enumerate and evaluate, with the help of the welfare function, the associated benefits and costs. The costs are normally taken to be the opportunity costs or supply prices at full employment of inputs. The money outlays on capital constructions are assumed to be equal to these supply prices.

The enumeration and evaluation of benefits are much more complex. The existence of externalities means that not all beneficiaries may be enumerated by observing similar market transactions. Benefits, as measured by market prices, will tend to be incorrectly estimated since not all beneficiaries are included in the observed market demand price and buyers will falsify their true demand price if a public good externality is involved. Moreover, the relevant measure of benefits should be total utility provided by the project or value in use and not value in exchange, and, as Adam Smith pointed out, these two value magnitudes are rarely equal. Consequently, even if market prices do correctly measure value in exchange, they are still likely to be an underestimate of total benefits.

Professor Eckstein, one of the leading exponents of the cost-benefit approach, has argued that "the resultant absolute benefits of a commodity is equal to the price which the consumer pays," [1] i.e., market price is the measure of benefits. If a particular project yields a block of units of output (that is, a production indivisibility exists) rather than a single incremental unit, Eckstein suggests that benefits be measured by total revenue derived, i.e., the market price multiplied by the associated output. Thus, Eckstein indicates that, if a change in a design of a dam yields 10,000 units of output worth $3 each, the benefit of the change will be $30,000 and the change should be undertaken if the benefit equals or exceeds the costs of the change.[2,3]

Eckstein's use of the market price of a commodity as a measure of benefits means utilizing Smith's value in exchange or Marshall's demand

1. O. Eckstein, *Water Resource Development, The Economics of Project Evaluation* (Cambridge: Harvard University Press, 1958), pp. 24–25.

2. The cost-benefit analysis is often likened to the decision-making process undertaken by a firm in judging which investment projects should be developed. The firm's calculations of discounted future profits are based on some "objective" facts and hence the demand price for investments is a readily calculable figure, while the supply price is merely the cost of production. In reality, of course, there is little objective evidence for either the expected future income from a private investment or for the expected future stream of benefits from a public good. Nevertheless, the facade of accounting for these benefits is comparable.

3. Eckstein, *op. cit.*, p. 25.

price and ignoring Smith's value in use or Marshall's venerable consumer surplus. The schedule of market demand prices for alternative volumes of output is obtained by a lateral summation of individual demand curves such as $d_1$ and $d_2$ into the market demand curve D (Figure 1). Thus, $P_1$ would be the market demand price associated with output level $Q_1$. The total revenue individuals would pay for $Q_1$ output would, in a competitive market, be equal to the rectangle $OP_1AQ_1$, but the

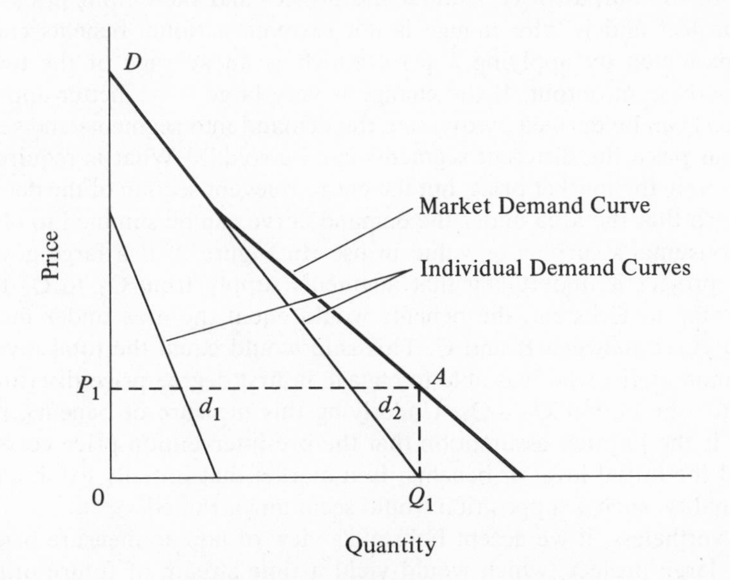


FIG. 1. Demand for a Private Good

total utility or what Smith called "value in use" is equal to the area under the demand curve between points D and A, that is, the polygon $ODAQ_1$. This difference between total revenue and the value in use is equal to the polygon $P_1DA$, which is Marshall's consumers' surplus. Thus the market price measure of benefits, which Eckstein advocates as a measure of benefits, would seriously underestimate total benefit unless the consumer surplus was zero—unless the demand curve for the output was horizontal. A horizontal demand curve, however, would be relevant only if the project would provide an insignificant change in the total output in a particular market, a situation unlikely to be true, almost by definition, for any government project. If government directly intervenes by augmenting supply (often from a zero or near zero level of output) in a market that initially exhibited an externality, it should be obvious that neither the initial market price nor the post-intervention price is, by itself, going to properly reflect benefits. Nor are the two prices likely to be equal since the *raison d'etre* for government intervention is to alter the price-quantity relationship existing in the private market. It is absurd to assume that government entry has the same impact on price as the entry of a perfectly competitive firm.

Eckstein, of course, does recognize the possibility of omitting the consumer surplus if market prices are used as a measure of benefits when a "large" project is undertaken, since the project itself will increase supply sufficiently to drive down the market price. Accordingly, Eckstein argues that the actual benefit will be somewhere between the limit of the output price without the project and the output price with the project and if "the change is not extreme national benefits can be approximated by applying a price which is an average of the two to the increase in output. If the change is very large . . . better approximations can be derived by dividing the demand into segments and seeing at what price the different segments can be sold." [4] What is required is not merely the market price, but the entire relevant section of the demand curve so that the area under the demand curve can be summed to obtain the consumer's surplus or value in use. In Figure 2, if a large government project is undertaken that augments supply from $Q_1$ to $Q_2$ then, according to Eckstein, the benefits would equal the area under the demand curve between B and C. This sum would equal the total revenue of a monopolist who was able to engage in first degree price discrimination for the output $Q_2$—$Q_1$. Underlying this measure of benefits, however, is the implicit assumption that the pre-intervention price correctly stated the initial level of benefits. In a market that initially exhibited an externality, such a supposition would seem unwarranted.

Nevertheless, if we accept Eckstein's view of how to measure benefits for a large project (which would yield a time stream of future utility), then the total increment in the commonweal would be estimated by adding up the discounted relevant areas under separate and different demand curves for each period over the life of the project.[5] In reality, however, such integrations are rarely performed since we rarely have sufficient information and the typical case is that the discounted expected market price is used as a valuation of the future stream of benefits. This generally leads to a downward bias in our estimate of benefits.

Cost-benefit analysis tends to follow the utility theorist approach of defining the goal of maximization of national welfare as synonymous with the maximization of national income. Eckstein, for example, notes "if it were the desire of all participants in the political process to maximize total national real income, they would simply abide by the results of the benefits-cost criterion." [6] This maximizing national income view, however, results in a downward bias in measuring benefits since na-

4. *Ibid.*, p. 37.

5. The demand curves would differ in each period as consumer tastes, income levels, income distribution, the underlying population, and the prices of complements and substitutes changed over time.

6. Eckstein, *op. cit.*, p. 39.

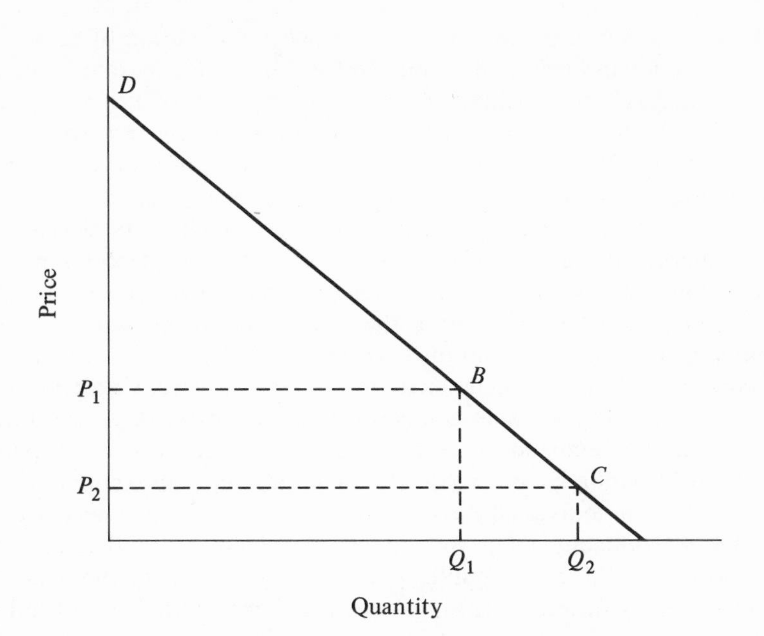


FIG. 2. Total Benefits for Large Project

tional income is measured at market prices or the "value in exchange" multiplied by the quantity of goods produced. This can lead to a value that is less than the integrated area under the demand curve or the value in use of the total national product. Cleaning up slums, improving public education, and preserving wild country are not normally profitable productive activities. In the national accounts, as in most cost-benefit studies, goods are entered in the computations at their exchange values rather than their value in use.

This bias of equating money payments for commodities and services with the benefits derived from these goods can, in the extreme, lead to absurd positions. No self-respecting businessman would invest in the development of the Garden of Eden, unless, of course, he could have a monopoly on its output. Nevertheless, from a historical view, the Garden of Eden was the first public good, since the output consumed by Adam in no way limited the availability of commodities to Eve. Such a real estate project may, to the uninitiated, look like a highly desirable social investment project. The Garden of Eden output, however, would have zero market prices, and, unless one knew what the entire areas under the relevant demand curves for each good were, the benefits might well be understated by taking some average market price between zero and pre-Garden of Eden prices.[7]

7. The benefits from a world of abundance may indeed be very small. For a frightening description see J. M. Keynes, *Essays in Persuasion* (New York: Norton, 1963), pp. 368 ff.

If we limit ourselves only to market prices in valuation of commodities and national wealth, then, as Mrs. Robinson has pointed out, the issue is not likely to be what is useful to society and what is not; rather, it is often between what creates a field for profitable enterprise and what does not. Implicit in this market valuation approach is the belief that buying goods is a pleasure and paying taxes is a burden.[8]

So far we have implicitly assumed that the valuation of benefits involved marketable goods. The existence of a market price for private goods allows observable quantifications to be made even if the market is not a competitive one and even if the price is an underestimate of total benefits, since it ignores consumer surpluses. The problem of valuation of benefits for public goods is likely to be less objective, since, although there may be an observable competitive supply price, there can be no efficient market demand price because of the very nature of public goods. Since public goods are those "which all enjoy in common in the sense that each individual's consumption of such a good leads to no subtraction from any other individual's consumption of that good." [9] There is no need to ration public goods among consumers and hence no efficient market demand price can exist, and the market would fail by existence.

A graphical interpretation of this public good condition involves a vertical rather than a horizontal addition of individuals demand curve such as $d^1$ and $d^2$ to get the total demand curve D in Figure 3. It is important to note that what are being vertically added in the public goods case are areas under individual demand curves. This involves the summation of consumer surpluses or the value in use for each consumer.

The hypothetical demand price of public goods is equal to the total sum of all that individuals would pay for the good rather than do without it. This collective sum is akin to the desired monetary measurement of total benefits, particularly if we are discussing the production of a project producing single or large indivisible units of output of a particular good such as national defense, recreation, etc. It should therefore be obvious that a method of quantifying benefits without appeal to market demand prices must be adopted. Either some observable index beside market demand prices must be substituted for benefits or independent judgment about the demand price must be exercised.

For example, in the cost-benefit analysis of a weapons system, the extimated number of enemy killed by the weapon may act as an index of benefits. Thus the cost-benefit analyst merely has to search for that weapon which will kill the greatest number of the enemy per dollar of expenditures, or to obtain the most "bang for the buck." On the other

8. J. Robinson, *Economic Philosophy* (Chicago: Aldine, 1962), p. 111.

9. P. A. Samuelson, "The Pure Theory of Public Expenditure," *Review of Economics and Statistics,* 38 (November 1958), p. 387.

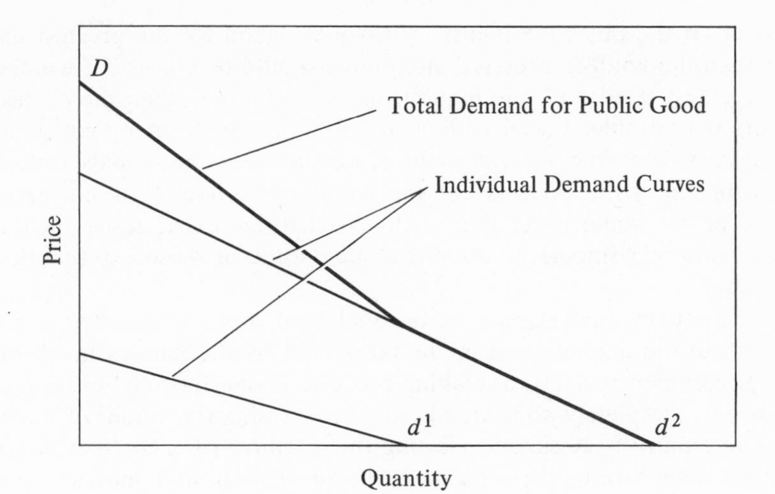


FIG. 3. Demand for a Public Good

hand, if it is felt that all dead enemies are not equally valuable—it is more valuable to kill military than civilian personnel, or more valuable to kill Chinese than Russian Communists—then in the absence of some exogenous valuating system for quantifying the value of dead civilians or dead soldiers, or dead Chinese or dead Russians, some discretionary judgment of values must be made. Only if one accepts the dictum "that the only good commie is a dead commie" can subjective valuation by the policy maker be completely avoided.

Furthermore, under such a weapons-benefit function, it is not clear how one would be able to uniquely and objectively evaluate the benefits of a system that increased the probability of converting enemies into friends *vis-à-vis* a weapon that increased the probability of killing enemies. How shall we evaluate converted enemies per dollar of expenditures versus killed enemies per dollar? Surely there is an externality in eliminating an enemy by making him a friend rather than merely eliminating him.

In cases other than weapons, it is even more obvious that valuation of public goods is going to involve judgment on the parts of policy makers. The social objective, "the greatest good for the greatest number," involves one too many greatest to provide an objective standard of valuation. If a million people would be willing to pay one dollar each for the use of a land as a wildlife preserve, while ten wealthy individuals were willing to spend a hundred thousand dollars and one cent each to form a company to produce an exclusive high price vacation resort in this area, then which use of this scarce land resource will provide the greatest good for the greatest number? It is my contention that neither cost-benefit analysis nor a market test really provides a uniquely satisfying answer. If the objective is just to provide a good for the greatest number then the wildlife preserve alternative should be chosen; if we desire the greatest good, i.e., the highest market valuation, then the exclusive resort is desirable. Faced with such market evidence many cost-benefit analysts will search for externalities that prevent individuals from expressing the "true benefits" of the wildlife preserve. It is not enough however to demonstrate that such externalities exist; some measurement, however imperfect, about the magnitude of these externalities is essential.

Cost-benefit analysis can be an analytical tool for a systematic approach to the enumeration of the large field of alternatives involved in the governmental decision-making process. It does not allow the policy maker to completely abdicate his role in assessing the values of the various alternatives. A skeptic viewing the attempts of a cost-benefit analyst to demonstrate the existence of externalities and measure social benefits of things such as wildlife preserves to frustrate commercial interests may state (perhaps with some justification) that "if one scratches a planner deep enough one finds an economic dictator." In the absence of a market price, one requires a judgment about value—a judgment which someone must make. Just as the 19th century economists used the differential calculus to justify *laissez-faire,* there may be a tendency of the present-day social scientists to use the matrix algebra to justify government intervention.

In certain cost-benefit analysis, where the absence of market demand price has been well recognized, the index used for the demand price has been the damage costs involved if the project was not undertaken. Thus one could estimate the probable annual damages from floods in a flood plain and use the savings in these costs as the measure of benefits that would accrue from a flood control program. In a cost-benefit analysis of highway beautification which we at Rutgers are now engaged in, we have tried to use a measure of the differential accident rate between scenically enhanced and non-enhanced highways as an index of benefits. The reduction in certain diseases can be used as an index of benefits for a program of air pollution control. Nevertheless, the monetary equivalent of these measures alone will tend to systematically underestimate total benefits, the benefit of stepping outside to get a breath of fresh air, the benefit of an aesthetically pleasing road, the benefit of not having to fear a flood.

It is the cost-benefit analyst who sets the benefits and costs in juxtaposition, since often we individually do not fully realize the totality of costs involved. For example, if society, 60 years ago, was presented with the following cost-benefit analysis, what would our choice be? We could have a scientific breakthrough that would allow people to travel from place to place at speeds in excess of 60 miles per hour. This transportation instrument will kill 50,000 and maim 3.5 billions annually.

Would society have thought the automobile was worth it—that is, is the demand price significantly greater than the value of so many lives?

The need to make a subjective valuation about the demand price or benefits from a public goods project can be illustrated from the findings of our Delaware Estuary Study where we used survey data and population projections to estimate the increase in fishing, boating, and swimming activity days, which would be revealed by consumer actions (through the year 1990) if the Delaware River estuary was to have an improvement in water quality sufficient to allow each of these activities to occur. The total cost of improving the water in the estuary was provided by a study done by the Public Health Service under the direction of Dr. Robert Thomann.[10]

Our analysis showed that there is no evidence that swimming activity would increase at all if the Delaware was cleaned up. If this result is correct, then it is clear that there is a zero demand price for improved water for swimming purposes and hence zero benefit to swimmers. Consequently, on the basis of swimming, there would be no justification for undertaking an improvement in water quality. Nevertheless, our study did show an increase in boating activity days (discounted at 5 percent) of almost 9 million days through 1990 and also almost 900,000 additional fishing activity days for the same period. Given the cost estimates, the supply price for raising water quality up to a level sufficient to permit recreational boating activity, we found that if each boating activity day was valued at approximately $2.55 then the value of such a project would equal the costs. This does not mean that each boat user should have to pay $2.55 for each day of boating; it implies that the total sum all boaters would pay had a present value of an amount equal to the supply price of improving water quality sufficiently to provide boating as estimated by readily available engineering criteria. The demand price can, in this case, only be resolved by judgment of the policy-making authorities.

## Summary and Conclusions

It is exceedingly difficult to provide some objective normative guidelines for public investment. Yet decisions on environmental utilization will be made either by the market or the political process. It is just as much a national policy decision, for example, to allow land utilization to go to the highest bidder as nationalizing the land for the preserving of a wildlife preserve for yet unborn generations of citizens to enjoy. A

10. R. Thomann, *Delaware Estuary Comprehensive Study* (Philadelphia: Federal Water Pollution Control Administration, 1966).

decision must be, and will be, made on the basis of the available evidence and the existing political institutional structures. Those economists not encumbered with the belief that existing markets (with their inequitable distribution of income and market power) [11] are ideal allocators of resources have used existing economic theory to show under what circumstances even idealized markets will fail to efficiently allocate resources. Characteristics of particular markets producing externalities, which will result in a failure to allocate resources in line even with consumer idealized market valuations, can be identified. In certain cases these externalities can be offset by government imposing either cost incentives or simple rules that either internalize these externalities into the private decision-making process or prevent these externalities from being effective. In the case of clean air or water environment, where dumping the pollutants may be costless to the polluters but weigh heavily in cost to others, we can either internalize the cost by imposing a variable tax on the polluters or we may simply impose a rule which makes pollution illegal.[12]

In the case of public goods externalities, no simple rule or individual cost incentive program is by itself likely to lead to an optimal system. Thus for public goods in general, and especially for environmental quality problems involving public goods, cost-benefit analysis can be used to systematically enumerate beneficiaries and types of benefits and place these in juxtaposition to the supply prices of various projects. Except in the simple and unrealistic case where all benefits from all projects are equal, however, there can be no substitutes for informed judgment in the evaluation of benefits. Given the nebulous objectives of maximizing social welfare in a democratic society, some quantitative estimating procedure will, in a world of uncertainty, lead reasonable men into subjectively weighing alternatives and into making thoughtful policy decisions.

If the net benefit function cannot be defined in a rigorous, readily measurable and unambiguous manner, and in many cases it cannot, then the results of a cost-benefit analysis cannot be completely rigorous and unambiguous. It is easy to tell the individual businessman that in choosing between alternative investment projects he should choose that project that maximizes the net money income stream (given the rate of dis-

11. Even costs are misleading when factor owners have market power (see Eckstein, *op. cit.,* p. 27).

12. Although economists normally suggest that a system of monetary incentives is more efficient than simple rules (see A. V. Kneese, "What Are We Learning from Economic Studies of Environmental Quality?" [mimeo]), and I tend to agree with this generalization, there may be times when rules are more effective. For example, I think the simple rule of requiring everyone to drive on the right side of the road results in a fairly optimal system, and offhand I cannot suggest a cost incentive system that would be as effective and simple. Hence we should not neglect the possibility of simple rules in our search for offsetting some types of externalities.

count). Of course, in a world of uncertainty he may make the wrong choice, one which from hindsight does not maximize profits. Nevertheless, at the end of the life of the project, even if it can be proved that he did choose the net revenue maximizing project, it is still unlikely that it can be demonstrated, except under very certain stringent accepted conventions, that the project has maximized the benefits of society.

It is the maximization of social benefits that is normally the desirable end product of most cost-benefit analysis. To compete in the political market place the cost-benefit analyst has adopted the guise of simulating the market demand price—a shadow price measuring how much consumers would be willing to pay if the good was marketable [13]—even for public goods, where the investigator knows that no efficient market price can exist.[14]

There is no acceptable substitute for informed value judgments in the evaluation of benefits of public goods. In fact, the only substitute is uninformed value judgments. Moreover, we should not deceive policy makers into believing that cost-benefit analysis relieves them of the necessity of making some judgments. The present prosperity of the American economy is due, in no small measure, to the benefits derived by investing ten percent of our resources in digging holes and filling them with missiles which we all hope need never be utilized. The benefits of these silos exceed the costs, according to the Pentagon. Yet the opportunity cost of national defense is staggering. In 1966, for example, we could have more than doubled our annual net investment by removing these resources from hole digging and placing them in capital goods production.

The Declaration of Independence declared that it was self-evident that "all men are created equal," but the notion of "the greatest good for the greatest number" firmly asserts that all benefits and all beneficiaries are not equal. The present state of the economic arts cannot provide a very sturdy bridge for linking these two democratic dicta but it can provide a reasonable guide to approaching both these objectives.

13. Eckstein, *op. cit.*, p. 39.
14. *Ibid.*, pp. 40–41.

# 24

LESTER B. LAVE AND
EUGENE P. SESKIN

# *Air Pollution and Human Health*

*Lester B. Lave is Professor of Economics and Head of the Department of Economics and Eugene Seskin is a Research Associate in the Graduate School of Industrial Administration, both at Carnegie-Mellon University.*

Air pollution is a problem of growing importance; public interest seems to have risen faster than the level of pollution in recent years. Presidential messages and news stories have reflected the opinion of scientists and civic leaders that pollution must be abated. This concern has manifested itself in tightened local ordinances (and, more importantly, in increased enforcement of existing ordinances), in federal legislation, and in extensive research to find ways of controlling the emission of pollutants from automobiles and smokestacks. Pollutants are natural constituents of the air. Even without man and his technology, plants, animals, and natural activity would cause some pollution. For example, animals vent carbon dioxide, volcanic action produces sulfur oxides, and wind movement insures that there will be suspended particulates; there is no possibility

*"Air Pollution and Human Health," by Lester B. Lave and Eugene P. Seskin, from* Science, *Vol. 169, August 21, 1970, pp. 723–733.* 

of removing all pollution from the air. Instead, the problem is one of balancing the need of polluters to vent residuals against the damage suffered by society as a result of the increased pollution.[1] To find an optimum level, we must know the marginal costs and marginal benefits associated with abatement. This article is focused on measuring one aspect of the benefit of pollution abatement.

Polluted air affects the health of human beings and of all animals and plants.[2] It soils and deteriorates property, impairs various production processes (for example, the widespread use of "clean rooms" is an attempt to reduce contamination from the air), raises the rate of automobile and airline accidents,[3] and generally makes living things less comfortable and less happy. Some of these effects are quite definite and measurable, but most are ill-defined and difficult to measure, even conceptually. Thus, scientists still disagree on the quantitative effect of pollution on animals, plants, and materials. Some estimates of the cost of the soiling and deterioration of property have been made, but the estimates are only a step beyond guesses.[4] We conjecture that the major benefit of pollution abatement will be found in a general increase in human happiness or improvement in the "quality of life," rather than in one of the specific, more easily measurable categories. Nonetheless, the "hard" costs are real and at least theoretically measurable.

In this article we report an investigation of the effect of air pollution on human health; we characterize the problem of isolating health effects; we derive quantitative estimates of the effect of air pollution on

1. For a general discussion of inherent problems in handling residuals, see R. U. Ayres and A. V. Kneese, *Amer. Econ. Rev.* 59, 282 (1969).
2. For summaries of studies relating air pollution to health, see J. R. Goldsmith, in *Air Pollution,* vol. 1; *Air Pollution and Its Effects,* A. Stern, Ed. (Academic Press, New York, 1968), p. 547; E. C. Hammond, paper presented at the 60th annual meeting of the Air Pollution Control Association, 1967; H. Heimann, *Arch. Environ. Health* 14, 488 (1967). For more general reviews of the literature, see A. G. Cooper, "Sulfur Oxides and Other Sulfur Compounds," *U.S. Public Health Serv. Publ. No. 1093* (1965); ———, "Carbon Monoxide," *U.S. Public Health Serv. Publ. No. 1503* (1966); "The Oxides of Nitrogen in Air Pollution," *Calif. Dep. Public Health Publ.* (1966); "Air Quality Criteria for Sulfur Oxides," *U.S. Public Health Serv. Publ. No. 1619* (1967); *Effects of Chronic Exposure to Low Levels of Carbon Monoxide on Human Health, Behavior, and Performance* (National Academy of Sciences and National Academy of Engineering, Washington, D.C., 1969).
3. *Public Health (Johannesburg)* 63, 30 (1963); D. M. Johnson, *Good Housekeeping* 1961, 49 (June 1961).
4. See A. V. Kneese, in *Social Sciences and the Environment; Conference on the Present and Potential Contribution of the Social Sciences to Research and Policy Formulation in the Quality of the Physical Environment,* M. E. Garnsey and J. R. Hibbs, Eds. (Univ. of Colorado Press, Boulder, 1967), p. 165; R. G. Ridker, *Economic Costs of Air Pollution* (Praeger, New York, 1967); "Air Quality Criteria for Sulfur Oxides," *U.S. Public Health Serv. Publ. No. 1619* (1967), pp. 54–57.

various diseases and point out reasons for viewing some earlier estimates with caution; we discuss the economic costs of ill health; and we estimate the costs of effects attributed to air pollution.

## The Effect of Air Pollution on Human Health

In no area of the world is the mean annual level of air pollution high enough to cause continuous acute health problems. Emitted pollutants are diluted in the atmosphere and swept away by winds, except during an inversion; then, for a period that varies from a few hours to a week or more, pollutants are trapped and the dilution process is impeded. When an inversion persists for a week or more, pollution increases substantially, and there is an accompanying increase in the death rate.

Much time has been spent in investigating short-term episodes of air pollution.[5] We are more concerned with the long-term effects of growing up in, and living in, a polluted atmosphere. Few scientists would be surprised to find that air pollution is associated with respiratory diseases of many sorts, including lung cancer and emphysema. A number of studies have established a qualitative link between air pollution and ill health.

A qualitative link, however, is of little use. To estimate the benefit of pollution abatement, we must know how the incidence of a disease varies with the level of pollution. The number of studies that allow one to infer a quantitative association is much smaller.

### *Quantifying the Relation*

Our objective is to determine the amount of morbidity and mortality for specific diseases that can be ascribed to air pollution. The state of one's health depends on factors (both present and past) such as inherited characteristics (that cause a predisposition to certain diseases), personal habits such as smoking and exercise, general physical condition, diet (including the amount of pollutants ingested with food), living conditions, urban and occupational air pollution, and water pollution.[6,7] Health is a complex matter, and it is exceedingly difficult to sort out the contributions of the various factors. In trying to determine the contribu-

5. L. Greenburg, M. B. Jacobs, B. M. Drolette, F. Field, M. M. Braverman, *Public Health Rep.* 77, 7 (1962); M. McCarroll and W. Bradley, *Amer. J. Public Health Nat. Health* 56, 1933 (1966); J. Firket, *Trans. Faraday Soc.* 32, 1192 (1936); H. H. Schrenk, H. Heimann, G. D. Clayton, W. M. Gafafer, H. Wexler, "Air Pollution in Donora, Pennsylvania," *Public Health Bull. No. 306* (1949).
6. See J. R. Goldsmith, *Med. Thoracalis* 22, 1 (1965).
7. B. G. Ferris, Jr., and J. L. Whittenberger, *N. Engl. J. Med.* 275, 1413 (1966).

tion of any single factor one must be careful neither to include spurious effects nor to conclude on the basis of a single insignificant correlation that there is no association. Laboratory experimentation is of little help in the sorting process.[8]

The model implicit in the studies we have examined is a simple linear equation wherein the mortality or morbidity rate is a linear function of the measured level of pollution and, possibly, of an additional socioeconomic variable. In only a few cases do the investigators go beyond calculating a simple or partial correlation.

A number of criticisms can be leveled at this simple model. No account is taken of possibly important factors such as occupational exposure to air pollution and personal habits. These and other factors influencing health must be uncorrelated with the level of pollution, if the estimated effect of pollution is to be an unbiased estimate. In addition, the linear form of the function is not very plausible, except insofar as one considers it a linear approximation over a small range.

Both because of the rather crude nature of the studies and because of the statistical estimation, there is a range of uncertainty concerning the quantitative effect of pollution on human health. This range is reflected in the estimate of the benefit of pollution abatement, discussed below.

### *Epidemiological Studies*

Epidemiological data are the kind of health data best adapted to the estimation of air pollution effects. These data are in the form of mortality (or morbidity) rates for a particular group, generally defined geographically.[9] For example, an analyst may try to account for variations in the mortality rate among the various census tracts in a city. While these vital statistics are tabulated by the government and so are easily available, there are problems with the accuracy of the classification of the cause of death (since few diagnoses are verified by autopsy and not all physicians take equal care in finding the cause of death). Other problems stem from unmeasured variables such as smoking habits, occupations, occupational exposure to air pollution, and genetic health factors. Whenever a variable is unmeasured, the analyst is implicitly assuming either that it is constant across groups or else that it varies randomly with respect to the level of air pollution. Since there are many unmea-

8. For a summary of laboratory experiments see "Air Quality Criteria for Sulfur Oxides," *U.S. Public Health Serv. Publ. No. 1619* (1967), pp. 79–93.

9. Chronic effects, where the incidence of the disease is small, can be studied only for large samples (millions of man-years of exposure); see J. R. Goldsmith, *Arch. Environ. Health* 18, 516 (1969); J. Rumford, *Amer. J. Public Health* 51, 165 (1961). Morbidity data would be more useful than mortality data, since death may result from a cause having no direct relationship to the original pollution-induced disease.

sured variables, one should not be surprised to discover that some studies fail to find a significant relationship or that others find a spurious one. For the same reason, one should not expect the quantitative effect to be identical across various groups, even when the relationship in each group is statistically significant.

Sample surveys are a means of gathering a more complete set of data. For example, a retrospective analysis might begin with a sample of people who died from a particular disease. Through questionnaires and interviews, the smoking habits and residence patterns of the deceased can be established. The analysis would then consist of an attempt to find the factors implicated in the death of these individuals. Two types of problems arising from such a study are the proper measurement of variables such as exposure to air pollution (there are many pollutants and many patterns of lifetime exposure) and the possible contributions of variables which still are unobserved, such as occupational hazards, socioeconomic characteristics, and personal habits.

Whatever the source of data, the investigators must rest their cases by concluding that the associations which they find are so strong that it is extremely unlikely that omitted variables could have given rise to the observed correlations; they cannot account for all possible variables.

## *Episodic Relationships*

Another method of investigating the effects of pollution involves an attempt to relate daily or weekly mortality (or morbidity) rates to indices of air pollution during the interval in question.[10] The conclusions of

10. For example, M. McCarroll and W. Bradley [*Amer. J. Public Health Nat. Health* 56, 1933 (1966)] correlate the daily mortality rate in New York City with daily pollution indices. See also J. R. McCarroll, E. J. Cassell, W. A. B. Ingram, D. Wolter, "Distribution of families in the Cornell air pollution study" and "Health profiles vs. environmental pollutants," papers presented at the 92nd annual meeting of the American Public Health Association, New York, 1964; ———, *Arch. Environ. Health* 10, 357 (1965); W. Ingram, J. R. McCarroll, E. J. Cassell, D. Wolter, *ibid.*, p. 364; E. J. Cassell, J. R. McCarroll, W. Ingram, D. Wolter, *ibid.*, p. 367. Other workers have attempted to explain daily variations in hospital admissions [see L. Greenburg, F. Field, J. I. Reed, C. L. Erhardt, *J. Amer. Med. Ass.* 182, 161 (1962); W. W. Holland, C. C. Spicer, J. M. G. Wilson, *Lancet* 1961-II, 338 (1961); G. F. Abercrombie, *ibid.* 1953-I, 234 (1953); A. E. Martin, *Mon. Bull. Min. Publ. Health Lab. Serv. Directed Med. Res. Counc.* 20, 42 (1961); R. Lewis, M. M. Gilkeson, Jr., R. O. McCaldin, *Public Health Rep.* 77, 947 (1962); T. D. Sterling, S. V. Pollack, D. A. Schumsky, I. Degroot, *Arch. Environ. Health* 13, 158 (1966); T. D. Sterling, S. V. Pollack, J. Weinkam, *ibid.* 18, 462 (1969)]; absence rates (see J. Ipsen, F. E. Ingenito, M. Deane, *ibid.*, p. 462); symptoms in school children [see B. Paccagnella, R. Pavanello, R. Pesarin, *ibid.*, p. 495; T. Toyama, *ibid.* 8, 153 (1964)]; the incidence of asthma attacks [see L. D. Zeidberg, R. A. Prindle, E. Landau, *Amer. Rev. Resp. Dis.* 84, 489

these studies are of limited interest, for two reasons. First, someone who is killed by an increase in air pollution is likely to be gravely ill. Air pollution is a rather subtle irritant, and it is unlikely that a healthy 25-year-old will succumb to a rise in pollution levels. Our interest should be focused on the initial cause of illness rather than on the factor that is the immediate determinant of death. Thus, morbidity data are more useful than mortality data. Second (and more important for the morbidity studies), there are many factors that affect the daily morbidity rate or daily rate of employee absences. Absence rates tend to be high on Mondays and Fridays for reasons that have nothing to do with air pollution or illness. One would expect little change in these absence rates if air pollution were reduced. Other factors, such as absence around holidays, give rise to spurious variation; this can be handled by ignoring the periods in question or by gathering enough data so that this spurious variation is averaged away. Some of these factors (such as high absence rates on Fridays and seasonal absence rates) may be correlated with variations in air pollution and no amount of data or of averaging will separate the effects. We have chosen to disregard the results of these episodic studies, with a few exceptions, cited below.

It is difficult to isolate the pollutants that have the most important effects on health on the basis of the studies we survey here. Measurement techniques have been crude, and there has been a tendency to base concentration figures on a single measurement for a large area. A more important problem is the fact that in most of these studies only a single pollutant was reported. Discovering which pollutants are most harmful is an important area, where further exploration is necessary. We have tried, nevertheless, to differentiate among pollutants in the survey that follows.[11] The problem is complicated, since pollution has increased over time, and since lifetime exposure might bear little relation to currently measured levels. These problems are discussed elsewhere.[12]

(1961); C. E. Schoettlin and E. Landau, *Public Health Rep.* 76, 545 (1961); R. Lewis, *J. La. State Med. Soc.* 115, 300 (1963)]; and other morbidity [see J. T. Boyd, *Brit. J. Prev. Soc. Med.* 14, 123 (1960); R. G. Loudon and J. F. Kilpatrick,. *Arch. Environ. Health* 18, 641 (1969)].

11. The most complete investigation of various pollutants was that of the Nashville studies. See L. D. Zeidberg, R. A. Prindle, E. Landau, *Amer. Rev. Resp. Dis.* 84, 489 (1961); L. D. Zeidberg and R. A. Prindle, *Amer. J. Public Health* 53, 185 (1963); L. D. Zeidberg, R. A. Prindle, E. Landau, *ibid.* 54, 85 (1964); L. D. Zeidberg, R. J. M. Horton, E. Landau, *Arch. Environ. Health* 15, 214 (1967); ———, *ibid.*, p. 225; R. M. Hagstrom, H. A. Sprague, E. Landau, *ibid.*, p. 237; H. A. Sprague and R. Hagstrom, *ibid.* 18, 503 (1969). It is conceptually possible to differentiate among pollutants, since, for example, the correlation between mean level of suspended particulates and mean level of sulfates for 114 U.S. Standard Metropolitan Statistical Areas is only .20.

12. L. B. Lave, "Air pollution damage" in *Research on Environmental Quality*, A. Kneese, Ed. (Johns Hopkins Press, Baltimore, in press).

## A Review of the Literature

We will proceed with a detailed review of studies made in an attempt to find an association between mortality or morbidity and air pollution indices.

### *Air Pollution and Bronchitis*

Studies link morbidity and mortality from bronchitis to air pollution in England,[13] the United States,[14,15] Japan,[16] and other countries.[17] Mortality rates by county boroughs in England and Wales have been correlated with pollution (as measured by the sulfation rate, total concentration of solids in the air, a deposit index, and the density of suspended particulates) and with socioeconomic variables (such as population density and social class). The smoking habits of the individuals studied have also been investigated. The conclusion of these studies is that air pollution accounts for a doubling of the bronchitis mortality rate for urban, as compared to rural, areas.

We took data reported by Stocks [18,19] and by Ashley [20] and performed a multiple regression analysis, as shown in Table 1. We fit the following equation to the data

$$MR_i = a_0 + a_1P_i + a_2S_i + e_i \tag{1}$$

where $MR_i$ is the mortality rate for a particular disease in country borough $i$, $P_i$ is a measure of air pollution in that borough, $S_i$ is a mea-

13. D. J. B. Ashley, *Brit. J. Cancer* 21, 243 (1967); C. Daly, *Brit. J. Prev. Soc. Med.* 13, 14 (1959); J. Pemberton and C. Goldberg, *Brit. Med. J.* 2, 567 (1954); P. Stocks, *ibid.* 1, 74 (1959); R. E. Waller and P. J. Lawther, *ibid.* 2, 1356 (1955); ———, *ibid.* 4, 1473 (1957); P. J. Lawther, *Proc. Roy. Soc. Med.* 51, 262 (1958); ———, *Nat. Acad. Sci. Nat. Res. Counc. Publ. No. 652* (1959), pp. 88–96; ———, *Instrum. Pract.* 11, 611 (1957); J. Pemberton, *J. Hyg. Epidemiol. Microbiol. Immunol. (Prague)* 5, 189 (1961); J. L. Burn and J. Pemberton, *Int. J. Air Water Pollut.* 7, 5 (1963); E. Gorham, *Lancet* 1958-I, 691 (1958); P. Stocks, *Brit. J. Cancer* 14, 397 (1960). These studies are updated and summarized in S. F. Buck and D. A. Brown, *Tobacco Res. Counc. Paper No. 7* (1964).
14. W. Winkelstein, Jr., S. Kantor, E. W. Davis, C. S. Maneri, W. E. Mosher, *Arch. Environ. Health* 14, 162 (1967).
15. International Joint Commission U.S. and Canada, "Report on the pollution of the atmosphere in the Detroit River Area" (Washington and Ottawa, 1960).
16. T. Toyama, *Arch. Environ. Health* 8, 153 (1964).
17. F. L. Petrilli, G. Agnese, S. Kanitz, *ibid.* 12, 733 (1966); A. Bell, in *Air Pollution by Metallurgical Industries,* A. Bell and J. L. Sullivan, Eds. (Department of Public Health, Sydney, Australia, 1962), pp. 2:1–2:144.
18. P. Stocks, *Brit. Med. J.* 1, 74 (1959).
19. ———, *Brit. J. Cancer* 14, 397 (1960).
20. D. J. B. Ashley, *ibid.* 21, 243 (1967).

sure of socioeconomic status in borough *i*, and $e_i$ is an error term with a mean of zero. (We also fit other functional forms, as discussed below.) Under general assumptions, the estimated coefficients ($a_0$, $a_1$, $a_2$) will be best linear, unbiased estimates.[21] Only if we want to perform significance tests must we make an assumption about the distribution of the error term (for example, the assumption that it is distributed normally).

The first regression in Table 1 relates the bronchitis mortality rate for men to a deposit index (see Table 1, footnote †), and the population density in each of 53 country boroughs. Thirty-nine percent of the variation in the mortality rate (across boroughs) is "explained" by the regression. It is estimated that a unit increase in the deposit index (1 gram per 100 square meters per month) leads to an increase of 0.18 percent in the bronchitis mortality rate (with population per acre held constant). An increase of 0.1 person per acre in the population density is estimated to lead to an increase of 0.02 percent in the mortality rate (with air pollution held constant). As indicated in Table 1 by the *t* statistics (the values in parentheses below the estimated coefficients), the air pollution variable is extremely important, whereas the socioeconomic variable contributes nothing to the explanatory power of the regression.

The first ten regressions in Table 1 are an attempt to explain the bronchitis death rate. Four different data sets are used, along with three measures of pollution and two socioeconomic variables. The coefficient of determination, $R^2$ (the proportion of the variation in the mortality rate explained by the regression), ranges from .3 to .8. Air pollution is a significant explanatory variable in all cases. In only three cases is the socioeconomic variable significant.

The implication of the first regression is that a 10 percent decrease in the deposit rate (38 g 100 $m^{-2}$ month $^{-1}$) would lead to a 7 percent decrease in the bronchitis death rate. Another way of illustrating the effect of air pollution on health is to note that, if all the boroughs were to improve the quality of their air to that enjoyed by the borough having

21. That the least-squares method provides the best linear unbiased estimates is the conclusion of the Gauss-Markov theorem, for which $E(ee') = \sigma^2 I$ and $E(e) = 0$ are the basic assumptions. These assumptions are that the basic model must be linear and that the distribution of the errors must have an expected value of zero, have finite variance, have a constant distribution over the various observations, and be independent. In addition, no explanatory variables may be omitted which are correlated with included variables. It is also convenient to assume that the explanatory variables are measured without error, although the framework can easily be adjusted to handle errors. In order to perform significance tests, one must make an assumption about the distribution of the error term. For all the relations we estimated, we plotted the residuals and discovered that all distributions were unimodel, symmetric, and basically consistent with the normal distribution. Thus, in the discussion that follows, we have assumed that the error term is distributed normally.

TABLE 1. Multiple Regressions Based on Data from England. Numbers in Parentheses Are the *t* statistic.

| Category | $R^2$ * | Index † Air pollution | Index † Socio-economic |
|---|---|---|---|
| *Bronchitis mortality rate* | | | |
| 1. Males, 53 county boroughs ‡ | .386 | .182 | .016 |
| (deposit index, persons / acre) | | (4.80) | (.22) |
| 2. Females | .332 | .182 | −.031 |
| | | (4.55) | (−.42) |
| 3. Males, 28 county boroughs § | .433 | 1.891 | .180 |
| (smoke, persons / acre) | | (3.79) | (1.86) |
| 4. Females | .412 | 1.756 | .252 |
| | | (3.23) | (2.40) |
| 5. Males, 26 areas ‖ | .766 | .310 | .062 |
| (smoke, persons / acre) | | (3.77) | (.53) |
| 6. Females | .559 | .303 | −.038 |
| | | (2.85) | (−.25) |
| 7. Males, 26 areas | .783 | .301 | .176 |
| (smoke, social class) | | (5.86) | (1.44) |
| 8. Females | .601 | .213 | .248 |
| | | (3.31) | (1.59) |
| 9. Both sexes, 53 urban areas ¶ | .377 | .199 | .159 |
| (smoke, persons / acre) | | (4.07) | (3.02) |
| 10. Both sexes, 53 urban areas | .300 | .161 | .151 |
| ($SO_2$ persons / acre) | | (3.05) | (2.64) |
| *Lung cancer mortality rate* | | | |
| 11. 53 County boroughs | .445 | .041 | .154 |
| (deposit index, persons / acre) | | (2.09) | (4.23) |
| 12. 28 County boroughs | .576 | .864 | .161 |
| (smoke, persons / acre) | | (4.08) | (3.89) |
| 13. Male, 26 areas | .781 | .137 | .115 |
| (smoke, persons / acre) | | (2.86) | (1.70) |
| 14. Male, 26 areas | .805 | .161 | .172 |
| (smoke, social class) | | (5.62) | (2.47) |
| 15. 53 Urban areas | .344 | −.086 | .184 |
| (smoke, persons / acre) | | (−2.42) | (4.83) |
| 16. 53 Urban areas | .378 | −.105 | .197 |
| ($SO_2$, persons / acre) | | −3.00 | (5.23) |
| *Other cancers* | | | |
| 17. Stomach, male, 53 county boroughs | .167 | .070 | .005 |
| (deposit index, persons / acre) | | (3.08) | (.12) |
| 18. Stomach, female | .175 | .070 | −.023 |
| | | (3.08) | (−.56) |

| Category | $R^2$ * | Index † Air pollution | Index † Socioeconomic |
|---|---|---|---|
| 19. Stomach, male, 28 county boroughs | .257 | .714 | .065 |
| (smoke, persons / acre) | | (2.57) | (1.21) |
| 20. Stomach, female | .454 | .883 | .066 |
| | | (4.13) | (1.60) |
| 21. Intestinal, 53 county boroughs | .041 | .018 | −.012 |
| (deposit index, persons / acre) | | (1.45) | (−.52) |
| 22. Intestinal, 28 county boroughs | .129 | .174 | .036 |
| (smoke, persons / acre) | | (1.26) | (1.35) |
| 23. Other cancer, male, 26 areas | .454 | .019 | .073 |
| (smoke, persons / acre) | | (.59) | (1.60) |
| 24. Other cancer, female, 26 areas | .044 | .039 | −.062 |
| (smoke, persons / acre) | | (.93) | (−1.03) |
| 25. Other cancer, male, 26 areas | .396 | .060 | .017 |
| (smoke, social class) | | (2.75) | (.33) |
| 26. Other cancer, female, 26 areas | .002 | .005 | −.013 |
| (smoke, social class) | | (.17) | (−.19) |
| *Pneumonia mortality rate* | | | |
| 27. Male, 26 areas | .477 | .118 | .121 |
| (smoke, persons / acre) | | (1.34) | (.97) |
| 28. Female | .253 | .068 | .137 |
| | | (.58) | (.83) |
| 29. Male, 26 areas | .475 | .158 | .126 |
| (smoke, social class) | | (2.82) | (.93) |
| 30. Female | .242 | .124 | .106 |
| | | (1.65) | (.58) |

* The coefficient of determination: a value of .386 indicates a multiple correlation coefficient of .62, and indicates that 39 percent of the variation in the death rate is "explained" by the regression. † The *t* statistic: for a one-tailed *t*-test with 23 degrees of freedom, a value of 1.71 indicates significance at the .05 level; for 25 or 50 degrees of freedom, the critical values are 1.71 and 1.68. ‡ Data for 53 county boroughs in England and Wales as reported by Stocks (*18*). Air pollution is measured by a deposit index (in grams per 100 square meters per month) whose observed range is 96 to 731, with a mean of 375. The socioeconomic index is expressed in numbers of persons per acre (multiplied by 10); the range is 69 to 364, and the mean is 163. Death rates are measured as index numbers, with the mean for all boroughs in England and Wales equal to 100. Ranges within this sample are as follows: bronchitis (males), 73 to 259; bronchitis (females), 72 to 268; lung cancer, 70 to 159; stomach cancer (males), 67 to 168; stomach cancer (females), 84 to 161; intestinal cancer, 87 to 123. § Data for 28 county boroughs in England and Wales as reported by Stocks (*18*). Air pollution is measured by a smoke index (suspended matter, in

the best air of all those in the sample (a standard deposit rate for all boroughs of 96 g 100 $m^{-2}$ month $^{-1}$), the average mortality rate (for this sample) would fall from 129 to 77. Thus, cleaning the air to the level of cleanliness enjoyed by the area with the best air would mean a 40 percent drop in the bronchitis death rate among males. In the fifth regression the pollution index is a smoke index (Table 1, footnote §), and a different set of areas is considered. This is a more successful regression in terms of the percentage of variation explained. As before, the air pollution coefficient is extremely significant, and the implication is that cleaning the air to the level of cleanliness currently enjoyed by the area with the best air (15 mg / 100 $m^3$) would lower the average bronchitis mortality rate from 106 to 30, a drop of 70 percent. Results of the other regression analyses based on bronchitis mortality data have similar implications. Note that the effect is almost the same for males and females. This indicates reliability and suggests that the effect is independent of occupational exposure.

Winkelstein *et al.* collected data on 21 areas in and around Buffalo, New York. A cross tabulation of census tracts by income level and pollution level shows that the mortality rate for asthma, bronchitis, and emphysema (in white males 50 to 69 years old) increases by more than 100 percent as pollution rises from level 1 to level 4.[22]

---

*Table note continued*

milligrams per 100 cubic meters); the range is 6 to 49. Again, the socioeconomic index is expressed in numbers of persons per acre (× 10); the range is 83 to 342. || Data for 26 areas in northern England and Wales as reported in Stocks (*19*). Air pollution is measured by a smoke index, as for category 3; the range is 15 to 562 mg/1000 $m^3$ and the mean is 260. One socioeconomic variable is the number of persons per acre (× 10); the range is 1 to 342 and the mean is 102. The other socioeconomic variable is social class; the range is 61 to 295. Death rates are measured as for category 1; within this sample, the range for lung cancer is 23 to 165; for other cancer, 6 to 122 (males) and 88 to 154 (females); for bronchitis, 18 to 259 (males) and 12 to 240 (females); for pneumonia, 61 to 227 (males) and 40 to 245 (females). ¶ Data for 53 areas as reported by Ashley (*20*). Air pollution is measured (i) by a smoke index (as for category 3), with a range of 23 to 261 $\mu g/m^3$ and a mean of 124, or (ii) by an $SO_2$ index (apparently in the same units), with a range of 33 to 277 and a mean of 124. Death rates are measured as for category 1; within this sample, the range for lung cancer is 70 to 146, and for bronchitis, 64 to 186.

22. For example, for economic level 1 (defined below), the death rates (per 100,000) for pollution levels 2 to 4 (defined below) are 126, 271, and 392. For economic level 2, the death rates for air pollution levels 1 to 4 are 136, 154, 172, and 199. For economic level 4, the death rates for pollution levels 1 to 3 are 70, 80, and 177. The five economic levels, based on median family income in a census tract, are as follows: \$3005–\$5007; \$5175–\$6004; \$6013–\$6614; \$6618–\$7347; and \$7431–\$11,792. The four air pollution levels (in micrograms) of suspended particulates (per cubic meter per 24 hours) are as follows: less than 80, 80–100, 100–135, and more than 135.

These studies indicate a strong relationship between bronchitis mortality and a number of indices of air pollution. We conclude that bronchitis mortality could be reduced by from 25 to 50 percent depending on the particular location and deposit index, by reducing pollution to the lowest level currently prevailing in these regions. For example, if the air in all of Buffalo were made as clean as the air in those parts of the area that have the best air, a reduction of approximately 50 percent in bronchitis mortality would probably result.

## *Air Pollution and Lung Cancer*

The rate of death from lung cancer has been correlated with several indices of pollution and socioeconomic variables in studies that provided controls for smoking habits and other factors. For English nonsmokers, Stocks and Campbell [23,24] found a tenfold difference between the death rates for rural and urban areas. Daly,[25] in comparing death rates in urban and rural areas of England and Wales, found the urban rate twice as high. Evidence for other parts of Europe also shows an association between lung cancer and air pollution.[26]

Regressions 11 through 16 (Table 1) show our reworking of the data for lung cancer mortality for England and Wales (there is no control for smoking). Regressions 11 through 14 imply that, if the quality of air of all boroughs were improved to that of the borough with the best air, the rate of death from lung cancer would fall by between 11 and 44 percent. Regressions 15 and 16 show a relationship between air pollution and lung cancer which is either insignificant or inverse. The only contrary results come from Ashley's data. In the absence of more complete evidence, we must remain curious about these results. Use of such small

23. P. Stocks and J. M. Campbell, *Brit. Med. J.* 2, 923 (1955).
24. In most of the early studies, pollution measures were not available, and so urban mortality rates were contrasted with rural rates. In these studies a substantial "urban factor" was found, which, unfortunately, was a compound of air pollution and many other factors. In the later studies the portion ascribable to air pollution is separated out.
25. C. Daly, *Brit. J. Prev. Soc. Med.* 13, 14 (1959).
26. Buck and Brown [*Tobacco Res. Counc. Res. Paper No. 7* (1964)], in examining data from England, control for population per acre, for social class, and for smoking habits. They find no relationship between smoking and lung cancer, and a relationship between $SO_2$ and lung cancer that is not consistent. Stocks uses three sets of data to isolate the effect of air pollution on lung cancer. Contrasting data for eight northern European cities, he finds a correlation between lung cancer and air pollution of .60, and correlations between lung cancer and smoking that range between .27 and .36. Contrasting data for 19 countries, he finds that an index of solid fuel consumption is a much stronger variable than cigarette consumption per capita. Finally, with data from northern England, he finds confirmation of an association between lung cancer and air pollution. See P. Stocks, *Brit. J. Prev. Soc. Med.* 21, 181 (1966).

samples and inadequate controls is certain to lead to some contrary results, but they are disconcerting when they appear.

In a study of 187,783 white American males (50 to 69 years old), Hammond and Horn [27] reported that the age-standardized rate of death due to lung cancer was 34 (per 100,000) in rural areas as compared to 56 in cities of population over 50,000. When standardized with respect both to smoking habits and to age, the rate was 39 in rural areas and 52 in cities of over 50,000.

Haenszel *et al.*[28] analyzed 2191 lung cancer deaths among white American males, that had occurred in 46 states, and data for a control group consisting of males who died from other causes. They found the crude rate of death from lung cancer to have been 1.56 times as high in the urban areas of their study as in the rural areas in 1958 and 1.82 times as high in the period 1948–49 (in subjects 35 years and older, with adjustments made for age). When adjustments are made for both age and smoking history, the ratio is 1.43. Also the ratio increased with duration of residence in the urban or rural area, from 1.08 for residence of less than 1 year to 2.00 for lifetime residence. Haenszel and Taeuber [29] report similar results for white American females. In a number of additional studies the association between air pollution and lung cancer is examined.[30]

27. E. C. Hammond and D. Horn, *J. Amer. Med. Ass.* 166, 1294 (1958).
28. W. Haenszel, D. B. Loveland, M. G. Sirken, *J. Nat. Cancer Inst.* 28, 947 (1962).
29. Haenszel and Taeuber analyzed data for 683 white American females who died of lung cancer, and for a control group. They found the crude rate of death from lung cancer to be 1.32 times as high in urban areas as in rural areas for 1958–1959 and 1.29 times as high for 1948–1949 (in subjects 35 years and older, with adjustments made for age). When adjustments were made for both age and smoking history, the ratio was 1.27. This ratio increased with the duration of residence in the urban or rural area, from 0.80 for residence of less than 1 year to 1.76 for lifetime residence. See W. Haenszel and K. E. Taeuber, *J. Nat. Cancer Inst.* 32, 803 (1964).
30. L. D. Zeidberg, R. J. Horton, and E. Landau [*Arch. Environ. Health* 15, 214 (1967)] are not able to isolate an air pollution effect on mortality from lung cancer from data for Nashville for the years 1949 through 1960; C. A. Mills [*Amer. J. Med. Sci.* 239, 316 (1960)] investigated rates of death from lung cancer in Ohio. Stratifying according to the amount of driving done by the deceased, he found that the death rate varied with driving and urban exposure; L. Greenburg, F. Field, J. I. Reed, M. Glasser [*Arch. Environ. Health* 15, 356 (1967)] investigated 1190 cancer deaths that occurred on Staten Island between 1959 and 1961 and found a relationship between lung cancer and air pollution; M. L. Levin, W. Haenszel, B. E. Carroll, P. R. Gerhardt, V. H. Handy, S. C. Ingraham II [*J. Nat. Cancer Inst.* 24, 1243 (1960)] found significant differences between urban and rural mortality rates (for periods around 1950) in New York State, Connecticut, and Iowa. For males, the death rates were 41 percent higher in urban areas in New York, 57 percent higher in Connecticut, and 184 percent higher in Iowa. For females, the differences were 7 percent, 24 percent, and 47 percent, respectively; P. Buell, J. E. Dunn, L. Breslow [*Cancer* 20, 2139 (1967)] utilized 69,868

Buell and Dunn [31] review the evidence on lung cancer and air pollution; a summary of their findings is given in Table 2. For smokers, death rates (adjusted for age and smoking) ranged from 25 to 123 percent higher in urban areas than in rural areas. For non-smokers, all differences exceeded 120 percent. "The etiological roles for lung cancer of urban living and cigarette smoking seem each to be complete," they say,

TABLE 2. A Summary of Lung Cancer Mortality Studies. Number of Deaths From Lung Cancer per 100,000 Population [from Buell and Dunn (*31*)].

| Standardized for age and smoking | | | Nonsmokers | | | Studv |
|---|---|---|---|---|---|---|
| Urban | Rural | Urban/ rural | Urban | Rural | Urban/ rural | |
| 101 | 80 | 1.26 | 36 | 11 | 3.27 | Buell, Dunn, and Breslow * (*67*) |
| 52 | 39 | 1.33 | 15 | 0 | ∞ | Hammond and Horn (*68*) † |
| 189 | 85 | 2.23 | 50 | 22 | 2.27 | Stocks (*69*) ‡ |
| | | | 38 | 10 | 3.80 | Dean (*70*) § |
| 149 | 69 | 2.15 | 23 | 29 | .79 | Golledge and Wicken (*71*) ‖ |
| 100 | 50 | 2.00 | 16 | 5 | 3.20 | Haenszel *et al.* (*72*) ¶ |

* California men; death rates by counties. † American men. ‡ England and Wales. § Northern Ireland. ‖ England; no adjustment for smoking. ¶ American men.

"in that the urban factor is evident when viewing nonsmokers exclusively, and the smoking factor is evident when viewing rural dwellers exclusively." They argue that differences in the quality of diagnosis could not account for the observed differences for urban and rural areas.

## *Nonrespiratory-Tract Cancers and Air Pollution*

Our reworking of data from England on rates of death from nonrespiratory-tract cancer is presented in Table 1 (regressions 17 through 26). In the regressions, stomach cancer is significantly related to a deposit index

questionnaires covering 336,571 man-years, in their study of lung cancer in California veterans. They found rates of death from lung cancer (adjusted for differences in age and smoking habits) to be 25 percent higher in the major metropolitan areas than in the less urbanized areas. Among nonsmokers, the rates of death from lung cancer were 2.8 to 4.4 times as high for major metropolitan areas as for more rural areas.

31. P. Buell and J. E. Dunn, Jr., *Arch. Environ. Health* 15, 291 (1967).

and a smoke index. The effects are nearly identical for males and females. Intestinal cancer appears to be only marginally related to indices of either deposit or smoke. For 26 areas in northern England and Wales, there appears to be little relationship between nonrespiratory-tract cancers and a smoke index. The single exception in the four regressions occurs for males when the socioeconomic variable is social class; here the smoke index explains a significant amount of the variation in the cancer mortality rate. (Apparently population density and smoke index are so highly related in these 26 areas that neither has significant power to explain such variation.)

Winkelstein and Kantor [32] investigated rates of mortality from stomach cancer in Buffalo, New York, and the immediate environs. Their measure of pollution is an index of suspended particulates averaged over a 2-year period. They found the rate of mortality due to stomach cancer to be more than twice as great in areas of high pollution as in areas of low pollution.[33]

Hagstrom *et al.*[34] tabulated rates of death from cancer among middle class residents of Nashville, Tennessee, between 1949 and 1960, using four measures of air pollution. They found the cancer mortality rate to be 25 percent higher in polluted areas than in areas of relatively clean air.[35] They also found significant mortality-rate increases associated with individual categories of cancer, such as stomach cancer, cancer of the esophagus, and cancer of the bladder. The individual mortality rates are more closely related to air pollution after the data are broken down by sex and race.

Levin *et al.*[36] report, for all types of cancer, these relationships: The

32. W. Winkelstein, Jr., and S. Kantor, *ibid.* 18, 544 (1969).
33. For economic level 2 (see 22), the mortality rate per 100,000 for gastric cancer in white males 50 to 69 years old changed from 45 to 41, 48, and 84 as the pollution level (see 22) rose. For economic level 4, the rates were 15, 38, and 63 for the first three pollution levels. For white women 50 to 69 years old, the death rates for economic level 2 were 8, 18, 25, and 40 per 100,000. For economic level 4, the death rates were 5 and 21 for the first two pollution levels.
34. R. M. Hagstrom, H. A. Sprague, E. Landau, *Arch. Environ. Health* 15, 237 (1967).
35. The four measures of pollution are suspended particulates (soiling), dustfall, $SO_3$ and $SO_2$. For all cancer deaths, the number per 100,000 for middle class residents (defined to include about 75 percent of all residents) fell from 153 for high-pollution areas, to 130 for moderate-pollution areas, to 124 for low-pollution areas when a soiling index (concentration of haze and smoke per 1000 linear feet) was used to classify air pollution. When $SO_3$ (milligrams per 100 square centimeters per day) was used as a basis for classification, the corresponding death rates were 150, 129, and 145, respectively. With dustfall as a measure, the figures were 145, 130, and 131, and with 24-hour $SO_2$, in parts per million, they were 141, 129, and 138.
36. M. L. Levin, W. Haenszel, B. E. Carroll, P. R. Gerhardt, V. H. Handy, S. C. Ingraham II, *J. Nat. Cancer Inst.* 24, 1243 (1960).

age-adjusted cancer-incidence rates for urban males was 24 percent higher than that for rural males in New York State (exclusive of New York City) (1949–51), 36 percent higher in Connecticut (1947–51), and 40 percent higher in Iowa (1950); the incidence rate for urban females was 14 percent higher than that for rural females in New York State, 28 percent higher in Connecticut, and 34 percent higher in Iowa. For both males and females, the incidence rate for each of 16 categories of cancer was higher in urban than in rural areas.

## *Cardiovascular Disease and Air Pollution*

Enterline *et al.*[37] found that mortality from heart disease is higher in central-city counties than in suburban counties, and, in turn, higher in suburban counties than in nonmetropolitan counties. Zeidberg *et al.*[38] found that both morbidity and mortality rates for heart disease are associated with air pollution levels in Nashville. The morbidity rate was about twice as high in areas of polluted air as in areas of clean air. The mortality rate was less closely associated; it was 10 to 20 percent higher in areas of polluted air than in areas of clean air.[39]

Friedman[40] correlated the rate of mortality from coronary heart disease in white males aged 45 to 64 with the proportion of this group living in urban areas. The simple correlation for 33 states is .79. When cigarette consumption is held constant, the partial correlation is .67.

On the basis of these studies we conclude that a substantial abatement of air pollution would lead to 10 to 15 percent reduction in the mortality and morbidity rates for heart disease. We caution the reader that the evidence relating cardiovascular disease to air pollution is less comprehensive than that linking bronchitis and lung cancer to air pollution.

37. P. E. Enterline, A. E. Rikli, H. I. Sauer, M. Hyman, *Public Health Rep.* 75, 759 (1960).
38. L. D. Zeidberg, R. J. M. Horton, E. Landau, *Arch. Environ. Health* 15, 225 (1967).
39. When air-pollution level was measured on the basis of sulfation ($SO_3$, in milligrams per 100 square centimeters per day), the morbidity rates (for white, middle-class males aged 55 and older) were 64.0 man-years per 1000 man-years for high-pollution areas, 34.1 for moderate-pollution areas, and 36.8 for low-pollution areas. Measurement of air pollution on the basis of 24-hour concentrations of $SO_2$ gave morbidity rates of 47.2, 36.8, and 22.2, respectively. For these same white, middle-class males, in areas of high atmospheric concentrations of $SO_3$, the mortality rate was 425.6 per 100,000 population; in moderate-concentration areas, 327.41; and in low-concentration areas, 361.97. With $SO_2$ concentrations as a measure, the corresponding figures were 424.87, 319.19, and 364.93. When soiling (smoke or suspended particles) was used as the pollution index, the figures were 376.65, 339.13, and 399.88, respectively.
40. G. Friedman, *J. Chronic Dis.* 20, 769 (1967).

### *Total Respiratory Disease* [41]

Daly found significant correlations between air pollution and death rates for all respiratory diseases (and for nonrespiratory diseases as well) in England. Douglas and Waller [42] found significant relationships between air pollution and respiratory disease in 3866 British school children. Fairbairn and Reid [43] found significant correlations between air pollution and morbidity rates (for bronchitis, pneumonia, pulmonary tuberculosis, and lung cancer) in England. Regressions 27 through 30 in Table 1 show the pneumonia mortality to be related only marginally to a smoke index.

41. The effect of air pollution on pneumonia, tuberculosis, and asthma has also been investigated. C. Daly (see 25) reports simple correlations of .60 for pneumonia mortality and pollution from domestic fuel and .52 for pneumonia mortality and pollution from industrial fuel. For tuberculosis mortality the correlations are .59 and .22, respectively. The death rates for pneumonia rise from 30 to 52 per 100,000, and those for tuberculosis rise from 47 to 89, as one goes from rural settings to conurbations. Stocks (*19*) reports data on pneumonia mortality, by sex, for 26 areas of northern England and Wales. As shown by regressions 27 through 30 in Table 1, there appears to be a strong relationship between a smoke index and pneumonia mortality. The relationship is much stronger for men than for women. C. A. Mills [*Amer. J. Hyg.* 37, 131 (1943)], in a classic study of wards in Pittsburgh and Cincinnati for 1929–30, reports substantial correlation between pneumonia death rates and local pollution indices. He found the correlation between dustfall and rates for pneumonia mortality in white males to be .47 for Pittsburgh and .79 for Cincinnati. The actual variation in these death rates is 41 to 165 per 100,000 population for Cincinnati and 0 to 7852 for Pittsburgh. Mills argues that omitted socioeconomic variables could not account for these correlations, but he made no attempt to control for such variables in his studies. He also found that death rates fell significantly as the altitude of an individual's residence increased; there was a drop of approximately 10 percent in death rate for every 100 feet (30 meters) of elevation [see also C. A. Mills, *Amer. J. Med. Sci.* 224, 403 (1952); E. Gorham, *Lancet* 1959-II, 287 (1959)]. Zeidberg. Prindle, and Landau [*Amer. Rev. Resp. Dis.* 84, 489 (1961)] studied 49 adult and 35 child asthma patients for a year. They found that the attack rate (attacks per person per day) for adults rose from .070 during days when atmospheric concentrations of sulfates were low to .216 when concentrations were high. In children, the effect of increased concentrations of sulfates was insignificant. Schoettlin and Landau [*Public Health Rep.* 76, 545 (1961)] investigated 137 asthma patients in Los Angeles during the fall months. They found that 14 percent of the variance in daily attacks ($n = 3435$) could be explained by the maximum atmospheric concentrations of oxidants for that day. These two studies document a strong relationship between asthma and air pollution; Lewis, Gilkeson, and McCaldin [*Public Health Rep.* 77, 947 (1962)] found no association between the daily frequency of visits to charity hospitals for treatment of asthma attacks and measures of air pollution.

42. J. W. B. Douglas and R. E. Waller, *Brit. J. Prev. Soc. Med.* 20, 1 (1966).

43. A. S. Fairbairn and D. D. Reid, *ibid.* 12, 94 (1958).

Zeidberg *et al.*[44] questioned 9313 Nashville residents about recent illnesses. Among males aged 55 and older from white middle-class families, the numbers of illnesses per respondent during the past year were 1.92, 1.15, and 1.26 for areas of high, moderate, and low pollution, respectively. There are a number of other comparisons, based on other measures of air pollution and on data for females and nonwhites (some of these are given below).[45] However, we should add a word of caution: although the sample size in this study was large and controls for many socioeconomic variables were included, many important factors were ignored—for example, smoking habits and length of residence. Nonetheless, the finding is extremely strong and seems unlikely to be an artifact of unmeasured variables.

Hammond[46] studied over 50,000 men to find the relationships between emphysema, age, occupational exposure to pollution, urban exposure, and smoking. His results indicated that the effect of air pollution is significant and that heavy smokers have a much higher morbidity rate in cities than in rural areas; the effect becomes more marked as age increases.

Ishikawa *et al.*[47] estimated the incidence of emphysema in Winnipeg (Canada) and St. Louis. They examined the lungs of 300 corpses in each city (the samples were comparable). Findings for each age group (over 25 years old) indicated that the incidence and severity of emphysema is higher in St. Louis, the city with the more polluted air. (In the 45-year-old group 5 percent of those in Winnipeg and 46 percent of those in St. Louis showed evidence of emphysema.)

A number of studies have been made in England on homogeneous occupational groups, such as postmen. The results are relatively pure in that all members of the sample have comparable incomes, working con-

44. L. D. Zeidberg, R. A. Prindle, E. Landau, *Amer. J. Public Health* 54, 85 (1964).
45. Morbidity rates associated with a soiling index were 140, 122, and 96, respectively, for high, moderate, and low pollution; corresponding rates associated with an $SO_2$ index were 177, 117, and 81. For white females, morbidity rates associated with an $SO_3$ index were 169, 134, and 160; with a soiling index, 158, 139, and 127; and with an $SO_2$ index, 172, 136, and 116. For nonwhite males, the morbidity rates associated with an $SO_3$ index were 86 for high concentrations and 84 for moderate and low concentrations; corresponding rates associated with a soiling index were 94 and 67, and with an $SO_2$ index, 84 and 88. For nonwhite females, morbidity rates of 136 and 140 were associated with high and with moderate and low $SO_3$ concentrations, respectively; corresponding rates associated with soiling were 140 and 129, and with $SO_2$ concentrations, 145 and 126. The effects for working women and for housewives, between the ages of 14 and 65, were similar.
46. E. C. Hammond, paper presented at the 60th annual meeting of the Air Pollution Control Association, 1967.
47. S. Ishikawa, D. H. Bowen, V. Fisher, J. P. Wyatt, *Arch. Environ. Health* 18, 660 (1969).

ditions, and social status. Holland and Reid [48] found that the rates of occurrence of severe respiratory symptoms were 25 to 50 percent higher for London postmen than for small-town postmen (sample size, 770). Reid [49] found that, in the postmen of his study, absences due to bronchitis rose from an index number of 100 for the area of least air pollution, to 120 for an area of moderate pollution, to 250 and 283 for the areas of highest pollution. Corresponding figures for absences due to other respiratory illness were 100, 100, 150, and 151, respectively, and for absences due to infectious and parasitic diseases, 100, 115, 130, and 140. Cornwall and Raffle [50] made a similar study of bus drivers in London. They found that 20 to 35 percent of absences due to sickness of any kind could be ascribed to air pollution (they used a fog index as a measure of pollution). Fairbairn and Reid tabulated absences due to sickness for postmen, for males working indoors, and for females working indoors. They found that the age-standardized morbidity rate for bronchitis and pneumonia in the postmen of their study rose from 40 man-years, per 1000 man-years, for the area of lowest air pollution (of the four areas studied) to 122 for the area of highest air pollution. Corresponding figures for morbidity from colds were 75 and 171 man-years, and for morbidity from influenza, 131 and 184 man-years. For males working indoors, the low and high morbidity rates were as follows: bronchitis and pneumonia, 32 and 39; colds, 53 and 64; influenza, 88 and 102.

Dohan [51] studied absences (of more than 7 days) of female employees in eight Radio Corporation of America plants. He found a correlation of .96 between atmospheric concentrations of $SO_3$ and absences due to respiratory disease in the five cities for which complete data were available. During Asian flu epidemics there was a 200 percent increase in illness in cities with polluted air and only a 20 percent increase in those with relatively unpolluted air.

## *Infant Mortality and Total Mortality Rates*

Sprague and Hagstrom [52] compared air-pollution data for Nashville with fetal and infant mortality rates for Nashville as given in census tracts (for 1955 through 1960). Controls for socioeconomic factors were not included. For infant death rates (ages 28 days to 11 months), the highest correlation was with atmospheric concentrations of $SO_3$ (in milligrams per 100 square centimeters per day) and was .70. For the neo-

48. W. W. Holland and D. D. Reid, *Lancet* 1965-I, 445 (1965).
49. D. D. Reid, *ibid*. 1958-I, 1289 (1958).
50. C. J. Cornwall and P. A. B. Raffle, *Brit. J. Ind. Med.* 18, 24 (1961).
51. F. C. Dohan, *Arch. Environ. Health* 3, 387 (1961); ——— and E. W. Taylor, *Amer. J. Med. Sci.* 240, 337 (1960).
52. H. A. Sprague and R. Hagstrom, *Arch. Environ. Health* 18, 503 (1969).

natal death rates (ages 1 day to 27 days), the highest correlation was with dustfall and was .49. For infants dying during their first day whose death certificate includes mention of immaturity, the highest correlation was with dustfall and was .45. The correlation of the fetal death rate with dustfall was .58.

In a study just being completed,[53] we have collected data for 114 Standard Metropolitan Statistical Areas in the United States and have attempted to relate total death rates and infant mortality rates to air pollution and other factors. Socioeconomic data, death rates, and air-pollution data were taken from U.S. government publications.[54] Regression 1 (Table 3) shows how the total death rate in 1960 varies with air pollution levels and with socioeconomic factors. As the (biweekly) minimum level of suspended particulates increases, the death rate rises significantly. Moreover, the death rate increases with (i) the density of population of the area, (ii) the proportion of nonwhites, (iii) the proportion of people over age 65, and (iv) the proportion of poor families. Eighty percent of the variation in the death rate across these 114 statistical areas is explained by the regression.

Regression 3 shows how the 1960 infant death rate (age, less than 1 year) varies. A smaller proportion (55 percent) of the variation in the death rate is explained by the regression, although the minimum air-pollution level, the percentage of nonwhites, and the proportion of poor families continue to be significant explanatory variables. Regression 5 is an attempt to explain variation in the neonatal death rate. The results are quite similar to those of regression 3. The fetal death rate is examined in regression 7. Here the minimum air-pollution level, population density, the percentage of nonwhites, and the percentage of poor families are all significant explanatory variables.

Regressions 2, 4, 6, and 8 are an attempt to relate these death rates to the atmospheric concentrations of sulfates for the 114 statistical areas of the study. Regression 2 shows that the total death rate is significantly related to the minimum level of sulfate pollution, to population density, and to the percentage of people over age 65; 81 percent of the variation is explained. Regressions 4, 6, and 8 show that the minimum atmospheric concentration of sulfates is a significant explanatory variable in three categories of infant death rates.

One might put these results in perspective by noting estimates on how small decreases in the air-pollution level affect the various death rates. A 10 percent decrease in the minimum concentration of measured

53. L. B. Lave and E. P. Seskin, in preparation.

54. "County and City Data Book," *U.S. Dep. Commerce Publ.* (1962); "Analysis of Suspended Particulates, 1957–61," *U.S. Public Health Serv. Publ. No. 978* (1962); "Vital Statistics of the United States (1960)," *U.S. Dep. Health Educ. Welf. Publ.* (1963); "Vital Statistics of the United States (1961)," *U.S. Dep. Health Educ. Welf. Publ.* (1963).

TABLE 3. Regressions Relating Infant and Total Mortality Rates for 114 Standard Metropolitan Statistical Areas in the United States to Air Pollution and Other Factors. Values in Parentheses Are the *t* statistic.* For Means and Standard Deviations (S.D.) of the Variables, see †

| Category | $R^2$ ‡ | Air pollution (minimum concentrations) | Socioeconomic | | | |
|---|---|---|---|---|---|---|
| | | | P/m² § | Non-white (%) | Over 65 (%) | Poor (%) |
| | | *Total death rate* | | | | |
| 1. Particulates | .804 | 0.102 | 0.001 | 0.032 | 0.682 | 0.013 |
| | | (2.83) | (2.58) | (3.41) | (18.37) | (0.93) |
| 2. Sulfates | .813 | 0.085 | 0.001 | 0.033 | 0.652 | 0.006 |
| | | (3.73) | (1.86) | (3.56) | (17.60) | (0.49) |
| | | *Death rate for infants of less than 1 year* | | | | |
| 3. Particulates | .545 | 0.393 | | 0.190 | | 0.150 |
| | | (3.07) | | (6.63) | | (3.28) |
| 4. Sulfates | .522 | 0.150 | | 0.200 | | 0.123 |
| | | (1.91) | | (6.83) | | (2.70) |
| | | *Death rate for infants less than 28 days old* | | | | |
| 5. Particulates | .260 | 0.273 | | 0.089 | | 0.063 |
| | | (2.48) | | (3.61) | | (1.60) |
| 6. Sulfates | .263 | 0.170 | | 0.097 | | 0.047 |
| | | (2.57) | | (3.96) | | (1.23) |
| | | *Fetal death rate* | | | | |
| 7. Particulates | .434 | 0.274 | 0.004 | 0.171 | | 0.106 |
| | | (2.02) | (2.01) | (5.70) | | (2.11) |
| 8. Sulfates | .434 | 0.171 | 0.004 | 0.181 | | 0.085 |
| | | (1.95) | (1.82) | (5.87) | | (1.71) |

* The *t* statistic: for a one-tailed *t*-test, a value of 1.65 indicates significance at the .05 level. † Total death rate per 10,000: mean, 91.5; S.D., 15.2. Infant death rate (age, <1 year) per 10,000 live births: mean, 255.1; S.D., 36.1. Infant death rate (age, <28 days) per 10,000 live births: mean, 188.0; S.D., 24.4. Fetal death rate per 10,000 live births: mean, 153.9; S.D., 34.4. Suspended particulates ($\mu g/m^3$), minimum reading for a biweekly period: mean, 45.2; S.D., 18.7. Total sulfates ($\mu g/m^3$) (× 10), minimum reading for a biweekly period: mean, 46.9; S.D., 30.6. Persons per square mile: mean, 763.4; S.D., 1387.9. Percentage of nonwhites in population (× 10): mean, 125.2; S.D., 102.8. Percentage of population over 65 (× 10): mean, 84.2; S.D., 21.2. Percentage of families with incomes under $3000 (× 10): mean, 181.6; S.D., 65.7. ‡ The coefficient of determination: a value of .804 indicates a multiple correlation coefficient of .90, and indicates that 80 percent of the variation in the death rate is "explained" by the regression. § Persons per square mile.

particulates would decrease the total death rate by 0.5 percent, the infant death rate by 0.7 percent, the neonatal death rate by 0.6 percent, and the fetal death rate by 0.9 percent. Note that a 10 percent decrease in the percentage of poor families would decrease the total death rate by 0.2 percent and the fetal death rate by 2 percent. A 10 percent decrease in the minimum concentration of sulfates would decrease the total death rate by 0.4 percent, the infant mortality rate by 0.3 percent, the neonatal death rate by 0.4 percent, and the fetal death rate by 0.5 percent.

Each of the relations in Tables 1 and 3 was estimated in alternative ways, including transformation into logarithms, a general quadratic, and a "piecewise" linear form as documented in note 12. The implications about the roles of air pollution and of the socioeconomic variables were unchanged by use of the different functional forms. Another result to be stressed is that, in Table 1, comparable regressions for males and females show almost precisely the same effects for air pollution. This suggests that occupational exposure does not affect these results; the result lends credence to the estimates. A result that we document in note 53 is that it is the minimum level of air pollution that is important, not the occasional peaks. People dealing with this problem should worry about abating air pollution at all times, instead of confining their concern to increased pollution during inversions.

## Some Caveats

In preceding sections we have described a number of studies which quantify the relationship between air pollution and both morbidity and mortality. Is the evidence conclusive? Is it possible for a reasonable man still to object that there is no evidence of a substantial quantitative association? We believe that there is conclusive evidence of such association.[55]

In the studies discussed, a number of countries are considered, and differences in morbidity and mortality rates among different geographical areas, among people within an occupational group, and among children are examined. Various methods are used, ranging from individual medical examinations and interviews to questionnaires and tabulations of existing data. While individual studies may be attacked on the grounds that none manages to provide controls for all causes of ill health, the number of studies and the variety of approaches are persuasive. It is difficult to imagine how factors such as general habits, inher-

55. For a discussion of the limitations of these studies, see B. G. Ferris, Jr. and J. L. Whittenberger (7) and J. R. Goldsmith, *Arch. Environ. Health* 18, 516 (1969).

ited characteristics, and lifetime exercise patterns could be taken into account.

To discredit the results, a critic would have to argue that the relationships found by the investigators are spurious because the level of air pollution is correlated with a third factor, which is the "real" cause of ill health. For example, many studies do not take into account smoking habits, occupational exposure, and the general pace of life. Perhaps city dwellers smoke more, get less exercise, tend to be more overweight, and generally live a more strained, tense life than rural dwellers. If so, morbidity and mortality rates would be higher for city dwellers, yet air pollution would be irrelevant. This explanation cannot account for the relationships found.

Apparently there is little systematic relationship between relevant "third" factors and the level of air pollution. An English study (see note 19) in which smoking habits are examined reveals little evidence of differences by residence. There is evidence in the United States that smoking is more prevalent among lower socioeconomic groups [56] but income or other socioeconomic variables would account for this effect and still leave the pollution coefficient unbiased. More importantly, the correlations between air pollution and mortality are better when one is comparing areas within a city (where more factors are held constant) than when one is comparing rural and urban areas.[57] It is especially hard to believe that the apparent relation between air pollution and ill health is spurious when significant effects are found in studies comparing individuals within strictly defined occupational groups, such as postmen or bus drivers (where incomes and working conditions are comparable and unmeasured habits are likely to be similar).

When there are uncontrolled factors, some studies may show inconclusive or even negative results; only by collecting samples large enough to "average away" spurious effects can dependable results be guaranteed. In the main, each of the studies cited above was based on a substantial sample. It is the body of studies as a whole that we find persuasive.

## *An Examination of Contrary Results*

Uncontrolled factors, together with small samples, are certain to lead to some results contrary to the weight of evidence and to our expectations.

56. See "Smoking and Health Report of the Advisory Committee to the Surgeon General of the Public Health Service," *U.S. Public Health Serv. Publ. No. 1103* (1964), p. 362.

57. This might be explained by noting that farmers tend to be exposed to a high level of pollution in the course of their work (from fertilizers, insecticides, and the exhaust fumes from farm equipment), which causes more deaths from respiratory disease than would be expected from the low level of general air pollution in rural areas.

For example, in some studies [58] no attempt is made to control even for income or social status. From the evidence of studies which did provide such controls, we know that failure to control for income leads to biased results, and so we place little credence in either the positive or negative findings of studies lacking these controls.

Sampling error can be extremely important. For example, Zeidberg *et al.*[59] find mixed results in cross-tabulating respiratory disease mortality with level of air pollution and with income class. In general the relationships are in the expected direction, but they are often insignifiicant. Insignificant results might occur often, if the samples are small, even if air pollution is extremely significant, since sampling errors dominate the explanatory variables.

Another study in which sampling error is important is reported by Ferris and Whittenberger (see note 7). They compared individuals in Berlin, New Hampshire, with residents of Chilliwack (British Columbia), Canada, and—not surprisingly in view of the small samples—failed to find significant differences in the occurrence of respiratory disease. Prindle *et al.*[60] compared two Pennsylvania towns in the same fashion. These two studies are admirable in that individuals were subjected to careful medical examinations. However, only a few hundred individuals were studied, and this means that sampling errors tend to obscure the effects of air pollution. Moreover, there were no controls for other factors, such as smoking. Also, one must be careful to control for a host of other variables if the sample is small. For example, the ethnic origins of the population and their general habits and occupations are known to affect mortality rate. It is exceedingly difficult to control for these factors; use of carefully constructed large samples seems the best answer. Finally, air pollution is measured currently, and it is generally assumed that relative levels have been constant over time and that people have lived at their present addresses for a long period. It is hardly surprising that statistical significance is not always obtained when such assumptions are necessary.

Since investigators are more reluctant to publish negative results than positive ones, and since it is more difficult to get negative results published, it is probable that we are unaware of other studies that fail to find a strong association between air pollution and ill health. We are somewhat reluctant to come to strong conclusions without knowledge of such negative results. However, there seems to be no reasonable alternative to evaluating the evidence at hand and allowing for uncertainty.

58. See, for example, T. Toyama (*16*) and F. L. Petrilli, G. Agnese, S. Kanitz, *Arch. Environ. Health* 12, 733 (1966).

59. L. D. Zeidberg, R. J. M. Horton, E. Landau, *Arch. Environ. Health* 15, 214 (1967).

60. R. A. Prindle, G. W. Wright, R. O. McCaldin, S. C. Marcus, T. C. Lloyd, W. E. Bye, *Amer. J. Public Health* 53, 200 (1963).

Thus, we conclude that an objective observer would have to agree that there is an important association between air pollution and various morbidity and mortality rates.

## The Economic Costs of Disease

Having found a quantitative association between air pollution and both morbidity and mortality, the next question is that of translating the increased sickness and death into dollar units. The relevant question is, How much is society willing to spend to improve health (to lower the incidence of disease)? In other words, how much is it worth to society to relieve painful symptoms, increase the level of comfort of sufferers, prevent disability, and prolong life? It has become common practice to estimate what society is willing to pay by totaling the amount that is spent on medical care and the value of earnings "forgone" as a result of the disability or death (*61*). This cost seems a vast underestimate for the United States in the late 1960's. Society seems willing to spend substantial sums to prolong life or relieve pain. For example, someone with kidney failure can be kept alive by renal dialysis at a cost of $15,000 to $25,000 per year; this sum is substantially in excess of forgone earnings, but today many kidney patients receive this treatment. Another example is leukemia in children; enormous sums are spent to prolong life for a few months, with no economic benefit to society. If ways could be found to keep patients with chronic bronchitis alive and active longer, it seems likely that people would be willing to spend sums substantially greater than the forgone earnings of those helped. So far as preventing disease is concerned, society is willing to spend considerable sums for public health programs such as chest x-rays, inoculation, fluoridation, pure water, and garbage disposal and for private health care programs such as annual physical checkups.

While we believe that the value of earnings forgone as a result of morbidity and mortality provides a gross underestimate of the amount society is willing to pay to lessen pain and premature death caused by disease, we have no other way of deriving numerical estimates of the dollar value of air-pollution abatement. Thus, we proceed with a conventional benefit calculation, using these forgone earnings despite our reservations.

### *Direct and Indirect Costs*

Our figures for the cost of disease are based on *Estimating the Cost of Illness,* by Dorothy P. Rice.[61] Unfortunately, Rice calculated disease costs in quite aggregate terms, and so the category "diseases of the respiratory system" must be broken down. It seems reasonable to assume that both direct and indirect costs would be proportional to the period of hospitalization (total patient-days in hospitals) by disease category.[62]

Rice defines a category of direct disease costs as including expenditures for hospital and nursing home care and for services of physicians, dentists, and members of other health professions. "Other direct costs" (which would add about 50 percent to those just enumerated) consist of a variety of personal and nonpersonal expenditures (such as drugs, eyeglasses, and appliances), school health services, industrial inplant health services, medical activities in federal units other than hospitals, medical research, construction of medical facilities, government public health activities, administrative expenditures of voluntary health agencies, and the net cost of insurance. Since Rice does not allocate "other direct costs" among diseases, we omit it from our cost estimates. However, we conjecture that respiratory diseases represent a substantial portion of this category. Thus, our direct cost estimate is likely to be a substantial underestimate of "true" direct costs (probably more than 50 percent too low).

Estimating indirect cost is an attempt to measure the losses to the nation's economy caused by illness, disability, and premature death. We would argue that such a calculation gives a lower bound for the amount people would be willing to pay to lower the morbidity and mortality rates. These costs are calculated in terms of the earnings forgone by those who are sick, disabled, or prematurely dead.[63]

61. D. P. Rice, "Estimating the Cost of Illness," *Public Health Serv. Publ. No. 947–6* (1966).

62. The category "diseases of the respiratory system" encompasses numbers 470 through 527 of the 1962 International Classification of Diseases, Adapted (ICDA). A report of the Commission on Professional and Hospital Activities, entitled *Length of Stay in Short-Term General Hospitals (1963–1964)* (McGraw-Hill, New York, 1966), gives details on the average lengths of stay and number of patients in 319 U.S. general hospitals for 1963 and 1964 by specific ICDA classifications. From these figures we were able to compute the ratio of total hospitalization by specific disease to total hospitalization for all respiratory diseases. Of the 2,410,900 inpatient days for all respiratory diseases, 232,222 were for acute bronchitis and 177,232 were for "bronchitis, chronic and unspecified." Thus, approximately 17 percent of all inpatient days for respiratory diseases were for some form of bronchitis. On the basis of current hospitalization rates, we find the direct cost of diseases of the respiratory system to be $1581 million annually. An estimated 17 percent of this amount is due to bronchitis; thus, the direct cost of bronchitis is about $268.8 million annually.

63. To calculate the indirect cost of bronchitis, we must do more than take 17 percent of the total indirect cost ($3,305,700) of all diseases of the respiratory system. Almost 50 percent of respiratory disease patients are hospitalized for "hypertrophy of tonsils and adenoids" (ICDA 510). Hospitalization is categorized by age of patient in the Commission on Professional and Hospital Activities report, and we note that 80 percent of these "tonsil and adenoid" patients were under 20

## The Health Cost of Air Pollution

The studies cited earlier in this article show a close association between air pollution and ill health. The evidence is extremely good for some diseases (such as bronchitis and lung cancer) and only suggestive for others (such as cardiovascular disease and nonrespiratory-tract cancers). Not all factors have been taken into account, but we argue that an unbiased observer would have to concede the association. More effort can and should be spent on refining the estimates. However, the point of this exercise is to estimate the health cost of air pollution. We believe that the evidence is sufficiently complete to allow us to infer, roughly, the quantitative associations. We do so with caution, and proceed to translate the effects into dollars. We have attempted to choose our point estimates from the conservative end of the range.

We interpret the studies cited as indicating that mortality from bronchitis would be reduced by about 50 percent if air pollution were lowered to levels currently prevailing in urban areas with relatively clean air. We therefore make the assumption that there would be a 25 to 50 percent reduction in morbidity and mortality due to bronchitis if air pollution in the major urban areas were abated by about 50 percent. Since the cost of bronchitis (in terms of forgone income and current medical expenditures) is $930 million per year, we conclude that from $250 million to $500 million per year would be saved by a 50 percent abatement of air pollution in the major urban areas.

Approximately 25 percent of mortality from lung cancer can be saved by a 50 percent reduction in air pollution, according to the studies cited above. This amounts to an annual cost of about $33 million.

The studies document a strong relationship between all respiratory disease and air pollution. It seems likely that 25 percent of all morbidity and mortality due to respiratory disease could be saved by a 50 percent abatement in air pollution levels. Since the annual cost of respiratory disease is $4887 million, the amount saved by a 50 percent reduction in air pollution in major urban areas would be $1222 million.

There is evidence that over 20 percent of cardiovascular morbidity and about 20 percent of cardiovascular mortality could be saved if air pollution were reduced by 50 percent. We have chosen to put this saving at only 10 percent—that is, $468 million per year.

---

years of age. Thus, it seems clear that the "forgone earnings" of these patients is negligible, and so no indirect costs should be allocated to this group. We therefore excluded the hospitalization of "tonsil and adenoid" patients before computing the percentage of hospitalization due to bronchitis. Thus, we estimated that 20 percent of the indirect cost of respiratory disease can be ascribed to bronchitis.

Finally, there is a good deal of evidence connecting all mortality from cancer with air pollution. It is difficult to arrive at a single figure, but we have estimated that 15 percent of the cost of cancer would be saved by a 50 percent reduction in air pollution—a total of $390 million per year.

Not all of these cost estimates are equally certain. The connection between bronchitis or lung cancer and air pollution is much better documented than the connection between all cancers or all cardiovascular disease and air pollution. The reader may aggregate the costs as he chooses. We estimate the total annual cost that would be saved by a 50 percent reduction in air-pollution levels in major urban areas, in terms of decreased morbidity and mortality, to be $2080 million. A more relevant indication of the cost would be the estimate that 4.5 percent of all economic costs associated with morbidity and mortality would be saved by a 50 percent reduction in air pollution in major urban areas.[64] This percentage estimate is a robust figure; it is not sensitive to the exact figures chosen for calculating the economic cost of ill health.

A final point is that these dollar figures are surely underestimates of the relevant costs. The relevant measure is what people would be willing to pay to reduce morbidity and mortality (for example, to reduce lung cancer by 25 percent). It seems evident that the value used for forgone earnings is a gross underestimate of the actual amount. An additional argument is that many health effects have not been considered in arriving at these costs. For example, relatively low levels of carbon monoxide can affect the central nervous system sufficiently to reduce work efficiency and increase the accident rate.[65] Psychological and esthetic effects are likely to be important, and additional costs associated with the effect of air pollution on vegetation, cleanliness, and the deterioration of materials have not been included in these estimates.[66]

64. There is one bit of evidence that 25 to 50 percent of total morbidity (and therefore mortality) can be associated with air pollution; see L. D. Zeidberg, R. A. Prindle, E. Landau, *Amer. J. Public Health* 54, 85 (1964). If one accepted this evidence as conclusive, it would follow that the annual cost of air pollution, because of health effects, would run between $14 billion and $29 billion.

65. See J. H. Schulte, *Arch. Environ. Health* 7, 524 (1963); A. G. Cooper, "Carbon Monoxide," *U.S. Public Health Serv. Publ. No. 1503* (1966); *Effects of Chronic Exposure to Low Levels of Carbon Monoxide on Human Health, Behavior, and Performance* (National Academy of Sciences and National Academy of Engineering, Washington, D.C., 1969).

66. Another way to estimate the cost of air pollution is to examine the effect of air pollution on property values. See R. J. Anderson, Jr. and T. D. Crocker, "Air pollution and residential property values," paper presented at a meeting of the Econometric Society, New York, December 1969; H. O. Nourse, *Land Econ.* 43, 181 (1967); R. G. Ridker, *Economic Costs of Air Pollution* (Praeger, New York, 1967); R. G. Ridker and J. A. Henning, *Rev. Econ. Statist.* 49, 246 (1967); R. N. S. Harris, G. S. Tolley, C. Harrell, *ibid.* 50, 241 (1968).

25

JACK L. KNETSCH AND
ROBERT K. DAVIS

# *Comparisons of Methods for Recreation Evaluation*

*Jack L. Knetsch is Professor of Economics and Director of the Natural Resources Policy Center at George Washington University, Washington, D.C. Robert K. Davis is a Research Associate at Resources for the Future.*

Evaluation of recreation benefits has made significant headway in the past few years. It appears that concern is increasingly focusing on the hard core of relevant issues concerning the economic benefits of recreation and how we can go about making some useful estimates.

The underlying reasons for this sharpening of focus are largely pragmatic. The rapidly increasing demand for recreation, stemming from the often-cited factors of increasing population, leisure, incomes, mobility, and urbanization, calls for continuing adjustments in resource allocations. This is the case with respect to our land and water resources in general; but more specifically it bears on such matters as the establishment of national recreation areas, setting aside or preserving areas for parks and open spaces in and near expanding urban areas, and clearly on questions of justification, location, and operation of water development projects.

Recreation services have only recently been recognized as products of land and water resource use. As such, they offer problems that do not occur when resolving the conflicting uses of most goods and services—for example, steel and lumber. Conflicting demands for commodities such as these are resolved largely in the market places of the private economy, where users bid against each other for the limited supplies.

Outdoor recreation, however, has developed largely as a non-market commodity. The reasons for this are quite elaborate, but in essence outdoor recreation for the most part is produced and distributed in the absence of a market mechanism, partly because we prefer it that way and have rejected various market outcomes, and partly because many kinds of outdoor recreation experience cannot be packaged and sold by private producers to private consumers. This absence of a market necessitates imputing values to the production of recreation services. Such economic benefits can be taken into account in decisions affecting our use of resources.

## Misunderstandings of Recreation Values

Discussions of values of outdoor recreation have been beset by many misunderstandings. One of these stems from a lack of appreciation that the use of outdoor recreation facilities differs only in kind, but not in principle, from consumption patterns of other goods and services. Another is that the market process takes account of personal and varied consumer satisfactions.

It is, furthermore, the incremental values that are important in making decisions relative to resource allocations. The incremental values of recreation developments of various kinds are a manageable concept which can be used for comparisons, in spite of the very great aggregate value that some may want to attribute to recreation. Nothing is gained—and no doubt a great deal has been lost—by what amounts to ascribing the importance of a total supply of recreation to an added increment, rather than concentrating on the added costs and the added benefits.

A similar difficulty arises with respect to questions of water supply. That man is entirely dependent upon the existence of water is repeatedly emphasized. While true, the point does not matter. Decisions necessarily focus on increments and therefore on the added costs and the added benefits that stem from adding small amounts to the existing total.

Further, no goods or services are priceless in the sense of an infinite price. There is an individual and collective limit to how much we will give up to enjoy the services of any outdoor recreation facility or to preserve any scenic resource. The most relevant economic measure of recreation values, therefore, is willingness on the part of consumers to pay for outdoor recreation services. This set of values is comparable to economic values established for other commodities, for it is the willingness to give up on the part of consumers that establishes values throughout the economy.

Failure to understand these value characteristics results in two types of error. The first is the belief that the only values that are worth considering are those accounted for commercially. A second and related source of error is a belief that outdoor recreation experience is outside the framework of economics, that the relevant values have an aesthetic, deeply personal, and even mystical nature. We believe both of these to be incorrect. In particular, the notion that economic values do not account for aesthetic or personal values is fallacious and misleading. Economically, the use of resources for recreation is fully equivalent to other uses, and the values which are relevant do not necessarily need to be determined in the market place. This last condition does indicate that indirect means of supplying relevant measures of the values produced may be necessary. But this is an empirical problem, albeit one of some considerable dimension, and the primary concern of this paper.

The problem of using imputed values for value determination has been met with a considerable degree of success for some products of water resource development. Procedures have been developed to assess the value of the flood protection, irrigation, and power services produced by the projects, even though in many cases a market does not in fact exist or is inadequate for the actual benefit calculations. Without commenting on the adequacy of these methods, it is generally agreed that such measures are useful in evaluating the output of project services.

## National and Local Benefits

Discussions of these topics have often been further confused by failure to separate two types of economic consequences or benefit. This has led to improper recognition of relevant and legitimate economic interests, and to inferior planning and policy choices.

There are, first, what we may call primary benefits, or national benefits. Second, there are benefits we may refer to as local benefits, or impact benefits. Both sets of values resulting from investment in recreation have economic relevance, but they differ, and they bear differently on decision.

The primary recreation benefits, or values, are in general taken to be expressions of the consumers' willingness to pay for recreation services. These values may or may not register in the commerce of the region or in the commerce of the nation, but this does not make them less real. When appropriately measured, they are useful for guiding social choices at the national level. The other set of accounts is concerned with local expenditure of money for local services associated with recreation. While outdoor recreation is not marketed—in the sense that the services of parks, as such, are not sold to any great extent in any organized market—money does indeed become involved in the form of expenditures for travel, equipment, lodging, and so forth. The amount of money spent in connection with outdoor recreation and tourism is large and growing, making outdoor recreation expenditures of prime concern to localities and regions which may stand to benefit. Our concern is with measuring the more difficult of the two types of benefit just mentioned—national recreation benefits. While these are measured essentially by the consumers' willingness to pay, in some cases the benefits extend to the non-using general public.

## Alternative Measurement Methods

There are obvious advantages to evaluating recreation benefits by market prices in the same manner as their most important resource competitors. However, as we have indicated, past applications have been hampered by disagreement on what are the meaningful values. In spite of growing recognition that recreation has an important economic value, economists and public administrators have been ill-prepared to include it in the social or public calculus in ways that lead to better allocations of resources.

The benefits of recreation from the social or community viewpoint are alleged to be many and varied. Some of the descriptions of public good externalities arising from recreation consumption are gross overstatements of the real values derived from the production of recreation services. But recreation benefits do in fact exist. Where externalities are real—as in cases of recreation in connection with visits to various historic areas or educational facilities, or where preservation of unique ecological units has cultural and scientific values—they should be recognized in assigning values to the development or preservation of the areas. However, it is our view that, by and large, recreation is a consumption good rather than a factor of production, and the benefits to be enjoyed are largely those accruing to the individual consumer participating. This is even more likely to be the case with recreation provided by water projects. The large bulk of primary recreation benefits can be viewed as the value of the output of the project to those who use them. This view stems from the concept that recreation resources produce an economic product. In this sense they are scarce and capable of yielding satisfaction for which people are willing to pay. Finally, some accounting can be made of this economic demand.

As the desirability of establishing values for recreation use of resources has become more apparent over the past few years, a number of methods for measuring or estimating them have been proposed and to some extent used. Some of the measures are clearly incorrect; others attempt to measure appropriate values, but fall short on empirical grounds [ref. 1, 2, 3].*

## *Gross Expenditures Method*

The gross expenditures method attempts to measure the value of recreation to the recreationist in terms of the total amount spent on recreation by the user. These expenditures usually include travel expenses, equipment costs, and expenses incurred while in the recreation area. Estimates of gross recreation expenditures are very popular in some quarters; for one thing, they are likely to produce large figures. It is argued that persons making such expenditures must have received commensurate value or they would not have made them. The usual contention is that the value of a day's recreation is worth at least the amount of money spent by a person for the use of that recreation.

These values have some usefulness in indicating the amount of money that is spent on a particular type of outdoor recreation, but as justification for public expenditure on recreation, or for determining the worth or benefit of the recreation opportunity afforded, they are of little consequence.

The values we seek are those which show not some gross value, but the net increase in value over and above what would occur in the absence of a particular recreation opportunity. Gross expenditures do not indicate the value of the losses sustained if the particular recreation opportunity were to disappear, nor do they show the net gain in value from an increase in a particular recreation opportunity.

## *Market Value of Fish Method*

A proposed method for estimating the recreation benefits afforded by fishing imputes to sport fishing a market value of the fish caught. The main objection to this procedure is the implied definition that the fish alone are the primary objective of the activity.

* References in brackets are identified at the end of the chapter.

### *Cost Method*

The cost method assumes that the value of outdoor recreation resource use is equal to the cost of generating it or, in some extreme applications, that it is a multiple of these costs. This has the effect of justifying any contemplated recreation project. However, the method offers no guide in the case of contemplated loss of recreation opportunities, and allows little or no discrimination between relative values of alternative additions.

### *Market Value Method*

Basic to the market value method measure is a schedule of charges judged to be the market value of the recreation services produced. These charges are multiplied by the actual or expected attendance figures to arrive at a recreation value for the services.

The method is on sound ground in its emphasis on the willingness of users to incur expenses to make choices. However, the market for outdoor recreation is not a commercial one, certainly not for much of the recreation provided publicly and only to a limited extent for private recreation. It is in part because private areas are not fully comparable with public areas that users are willing to pay the fees or charges. It seems, therefore, inappropriate to use charges paid on a private area to estimate the value of recreation on public areas. Also a single value figure or some range of values will be inappropriate for many recreation areas. Physical units of goods and services are not everywhere equally valuable, whether the commodity be sawtimber, grazing, or recreation. Location in the case of recreation affects value greatly. Moreover, differences of quality and attractiveness of recreation areas are not fully comparable or recognized by the unit values.

There are other methods, but few have received much attention. Where does this leave us? The only methods to which we give high marks are based on the concept of willingness to pay for services provided.

## Methods Based on Willingness to Pay

We have alluded to two kinds of problems we face in measuring the benefits of outdoor recreation: the conceptual problems and the measurement problems.

Conceptually, we wish to measure the willingness to pay by consumers of outdoor recreation services as though these consumers were purchasing the services in an open market. The total willingness of consumers to pay for a given amount and quality of outdoor recreation (that is, the area under the demand curve) is the relevant measure we seek. Our conceptual problems are essentially that any measurement of effective demand in the current time period, or even an attempt to project effective demand in future time periods, must necessarily omit from the computation two kinds of demand which may or may not be important. These are option demand and demand generated by the opportunity effect.[1]

Option demand is that demand from individuals who are not now consumers or are not now consuming as much as they anticipate consuming, and who therefore would be willing to pay to perpetuate the availability of the commodities. Such a demand is not likely to be measured by observance or simulation of market phenomena. The opportunity effect derives from those unanticipated increases in demand caused by improving the opportunities to engage in a recreational activity and thereby acquainting consumers with new and different sets of opportunities to which they adapt through learning processes. To our knowledge no methods have been proposed which might be used to measure those two kinds of demand for a good.

Notwithstanding the undoubted reality of these kinds of demand, our presumption is that effective demand is likely to be the predominant component of the aggregate demand for outdoor recreation of the abundant and reproducible sorts we have in mind. We further presume that this quantity can be estimated in a useful way, although by fairly indirect means, for we have no market guide of the usual sort. Two methods —a direct interview, and an imputation of a demand curve from travel cost data—currently appear to offer reasonable means of obtaining meaningful estimates.

## *Interview Methods*

The essence of the interview method of measuring recreation benefits is that through a properly constructed interview approach one can elicit from recreationists information concerning the maximum price they would pay in order to avoid being deprived of the use of a particular area for whatever use they may make of it. The argument for the existence of something to be measured rests on the conception that the recreationist is engaged in the utility maximizing process and has made a rational series of allocations of time and money in order to participate in the recreation being evaluated. Since the opportunity itself is avail-

1. These concepts are developed by Davidson, Adams, and Seneca in "The Social Value of Water Recreational Facilities Resulting from an Improvement in Water Quality: The Delaware Estuary," Allen V. Kneese and Stephen C. Smith, eds., *Water Research*, Johns Hopkins Press, Baltimore, 1966.

able at zero or nominal price, the interview provides the means for discovering the price the person would pay if this opportunity were marketed, other things being equal.

The chief problem to be reckoned with in evaluating interview responses is the degree of reliability that can be attached to the information the respondent provides the interviewer. Particularly on questions dealing with matters of opinion, the responses are subject to many kinds of bias.

One such bias of particular interest to economists stems from the gaming strategy that a consumer of a public good may pursue on the theory that, if he understates his preference for the good, he will escape being charged as much as he is willing to pay without being deprived of the amount of the good he now desires. This may be a false issue, particularly when it comes to pursuing recreation on private lands or waters, because the consumer may be well aware that the owner could, through the exercise of his private property rights, exclude the user from the areas now occupied. An equally good case can be made that, on state and national park lands to which there is limited access, particularly when at the access points the authority of the state is represented by uniformed park patrolmen, recreationists would have no trouble visualizing the existence of the power to exclude them. This being the case, it is not unreasonable to expect the recreationist to be aware of some willingness to pay on his part in order to avoid being excluded from the area he now uses.

Counterbalancing the possibility that the recreationist may purposely understate his willingness to pay in order to escape charges is the possibility that he may wish to bid up his apparent benefits in order to make a case for preserving the area in its current use, a case equally appropriate on private or public lands and waters.

The problem, to continue the argument, is narrowed to one of phrasing the question in such a way that the recreationist is not asked to give his opinion on the propriety of charging for the use of recreation areas.

It has become something of a principle in survey methodology that the less hypothetical the question, the more stable and reliable the response. By this principle, the respondent ought to be a consumer of the product rather than a potential consumer, thus distinguishing the data collected as pertaining to effective demand rather than to option or potential demand. It may also be preferable to impose the conditions on the interview that it occur at a time when the respondent is engaged in the activity. This may contribute to the accuracy of the responses by reducing the requirement that he project from one situation to another. (Admittedly, it is desirable to experiment with the methodology on this question, as well as others, in order to determine its sensitivity to such variations.)

In sum then, we expect to discover the consumer's willingness to pay through a properly constructed interview, and further, we expect that this measure will be the same quantity as would be registered in an organized market for the commodity consumed by the respondent. In other words, we hold a deterministic view that something exists to be measured, and is a sufficiently real and stable phenomenon that the measurement is useful.

*The Interview Procedure* The willingness to pay of a sample of users of a forest recreation area in northern Maine was determined in interviews on the site [ref. 4]. The interviews included a bidding game in which respondents could react to increased costs of visiting the area. Bids were systematically raised or lowered until the user switched his reaction from inclusion to exclusion or vice versa. At the beginning of the interview rapport was established with the respondents largely through objective questions inquiring into their recreation activities on the area, on other areas, and the details of their trips. The bidding questions were interspersed with a series of propositions for which the respondent was to indicate his opinion in the form of a positive, negative, or neutral reaction. His reactions to increased expenses connected with the visit constituted the essence of the bidding game. Personal questions regarding income, education, and the like were confined to the end of the interview.

The sampling procedure amounted to cluster sampling, since the procedure followed was to locate areas of use such as campgrounds and to systematically sample from the available clusters of users. The interviews were conducted from June through November by visiting areas in the privately owned forests of northern Maine and in Baxter State Park.

The data from the interviews is pooled to include hunters, fishermen, and summer campers. This pooling is defended largely on the grounds that no structural differences between identifiable strata were detected in a multiple regression analysis of the responses.

The procedure imputes a discontinuous demand curve to the individual household which may be realistic under the time constraints faced particularly by vacation visitors and other non-repeating visitors. This rectangular demand curve (Figure 1) reflects a disposition either to come at the current level of use or not to come at all if costs rise above a limiting value. Its realism is supported by a number of respondents whose reaction to the excluding bid was precisely that they would not come at all. It seems reasonable to view the use of remote areas such as northern Maine as lumpy commodities which must be consumed in five- or six-day lumps or not all. Deriving an aggregate demand function from the individual responses so characterized is simply a matter of taking the distribution function of willingness to pay cumulated on a less-than basis. This results in a continuous demand schedule which can be

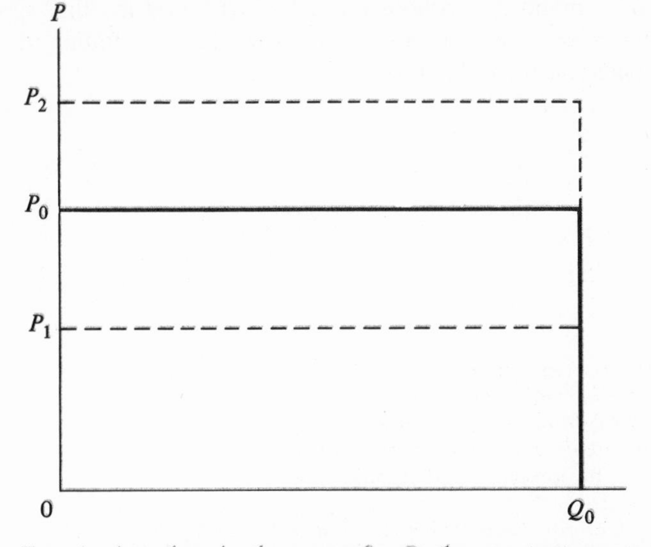


FIG. 1. At prices in the range $0-P_0$ the constant amount $Q_0$ will be demanded. Above $P_0$ demand will fall to zero. The individual may be in one of three states depending on the reigning price. Consider three individual cases with market price at $P_0$: The user paying $P_1$ is excluded; $P_0$ is associated with the marginal user; and $P_2$ is the willingness to pay of the third user who is included at the reigning price, $P_0$.

interpreted for the aggregate user population as a conventional demand schedule.

For the sample of 185 interviews, willingness-to-pay-per-household-day ranges from zero to $16.66. Zero willingness to pay was encountered in only three interviews. At the other extreme, one or two respondents were unable to place an upper limit on their willingness to pay. The distribution of willingness to pay shows a marked skewness toward the high values. The modal willingness to pay occurs between $1.00 and $2.00 per day per household.

Sixty per cent of the variance of willingness to pay among the interviews is explained in a multiple regression equation with willingness-to-pay-per-household-visit a function of income of the household, years of experience by the household in visiting the area, and the length of the stay in the area. (See Equation 1.) While the large negative intercept of this equation necessitated by its linear form causes some difficulties of interpretation, the exhibited relation between willingness to pay, and income, experience, and length of stay appears reasonable. The household income not only reflects an ability to pay, but a positive income elasticity of demand for outdoor recreation as found in other studies. It is also significant that an internal consistency was found in the responses to income-related questions.

| | $R^2$ | |
|---|---|---|
| $W = -48.57 + 2.85\,Y + 2.88E + 4.76L$ <br> $\quad\quad\quad\quad (1.52) \quad (0.58) \quad (1.03)$ | .5925 | (1) |
| $W = .74L^{.76}\,E^{.20}\,Y^{.60}$ <br> $\quad\quad (.13)\;(.07)\;(.17)$ | .3591* | (2) |

Standard errors of regression equations: (1) 39.7957; (2) 2.2007.
Standard errors of coefficients are shown in parentheses.
$W$ = household willingness to pay for a visit.
$E$ = years of acquaintance with the area visited.
$Y$ = income of the household in thousands of dollars.
$L$ = length of visit in days.
$F$ = ratios of both equations are highly significant.
* Obtained from arithmetic values of residual and total variances. ($R^2$ of the logarithmic transformation is .4309.)

The significance of years of experience in returning to the area may be interpreted as the effect of an accumulated consumer capital consisting of knowledge of the area, acquisition of skills which enhance the enjoyment of the area, and in some cases use of permanent or mobile housing on the area.

The significance of length of stay in the regression equations is that it both measures the quantity of goods consumed and also reflects a quality dimension suggesting that longer stays probably reflect a greater degree of preference for the area.

Colinearity among explanatory variables was very low. The general economic consistency and rationality of the responses appear to be high. Respondents' comments indicated they were turning over in their minds the alternatives available in much the same way that a rational shopper considers the price and desirability of different cuts or kinds of meat. Both the success in finding acceptable and significant explanatory variables and a certain amount of internal consistency in the responses suggest that considerable weight can be attached to the interview method.

*The Simulated Demand Schedule* While providing an adequate equation for predicting the willingness to pay of any user, the results of the interviews do not serve as direct estimates of willingness to pay of the user population, because the income, length of stay, and years' experience of the interviewed sample do not accurately represent the characteristics of the population of users. Fortunately, it was possible to obtain a reliable sample of the users by administering a questionnaire to systematically selected samples of users stopped at the traffic checking stations on the private forest lands. A logarithmic estimating equation, although not as well fitting, but free of a negative range, was used to compute the willingness to pay for each household in the sample. (See Equation 2.) The observations were then expanded by the sampling fraction to account for the total number of users during the recreation season.

The next step in the analysis consists of arraying the user population by willingness to pay, and building a cumulative distribution downward from the upper limit of the distribution. Table 1 shows the resulting demand and benefit schedule. The schedule accounts for the total of about 10,300 user households estimated to be the user population in a 450,000-acre area of the Maine woods near Moosehead Lake, known as the Pittston area.

TABLE 1. Demand and Benefit Schedules for Pittston Area Based on Alternative Estimates of Willingness to Pay

| | Interview results | | Willingness to drive (interview method) | | Willingness to drive (travel cost method) | |
|---|---|---|---|---|---|---|
| Price | Household visits | Benefits [1] | Household visits | Benefits [1] | Household visits | Benefits [1] |
| $70.00 | 0 | 0 | | | | |
| 60.00 | 11.36 | $ 747.77 | | | | |
| 50.00 | 15.35 | 983.56 | | | | |
| 40.00 | 44.31 | 2,281.46 | | | | |
| 30.00 | 150.22 | 6,003.19 | 11.36 | $ 384.79 | 165 | $ 3,800 |
| 26.00 | 215.80 | 7,829.71 | | | | |
| 22.00 | 391.07 | 12,027.89 | | | | |
| 20.00 | 536.51 | 15,099.31 | 76.96 | 1,890.12 | 422 | 12,134 |
| 18.00 | 757.86 | 19,275.95 | | | | |
| 16.00 | 1,069.01 | 24,607.81 | | | | |
| 14.00 | 1,497.75 | 31,027.17 | 392.29 | 7,287.06 | | |
| 12.00 | 1,866.41 | 35,802.70 | | | | |
| 10.00 | 2,459.70 | 42,289.68 | 2,157.91 | 28,921.93 | 1,328 | 26,202 |
| 8.00 | 3,100.99 | 48,135.01 | | | | |
| 6.00 | 4,171.89 | 55,794.64 | | | | |
| 4.00 | 5,926.94 | 64,436.36 | 5,721.06 | 53,531.68 | 3,459 | 44,760 |
| 2.00 | 7,866.02 | 70,222.66 | | | | |
| 0.00 | 10,333.22 | 71,460.94 | 10,339.45 | 63,689.99 | 10,333 | 69,450 |

1. Benefits are computed as the integral of the demand schedule from price maximum to price indicated. Willingness to drive computations are based on an assumed charge of 5¢ per mile for the one-way mileage.

The demand schedule is noticeably elastic from the upper limit of $60.00 to about $6.00, at which point total revenues are maximized. The interval from $60.00 to $6.00 accounts for the estimated willingness to pay of nearly half of the using households. Total benefits at $6.00 are $56,000. The price range below $6.00 accounts for the other half of the using households, but only for $15,000 in additional benefits. Benefits are estimated as the cumulative willingness to pay or the revenues available to a discriminating monopolist.

*Willingness to Drive vs. Willingness to Pay* An alternative expression of the willingness of recreationists to incur additional costs in order to continue using an area may be found in their willingness to drive additional distances. This measure was first proposed by Ullman and Volk [ref. 5] although in a different version than is used here. (See also [ref. 6].)

Willingness to drive additional distances was elicited from respondents by the same technique used to elicit willingness to pay. If there are biases involving strategies to avoid paying for these recreation areas, then certainly willingness to drive is to be preferred over willingness to pay as an expression of value. Analysis of the willingness to drive responses shows that a partly different set of variables must be used to explain the responses. Equation 3 shows willingness to drive extra miles to be a function of length of stay and miles driven to reach the area.

$$Wm = 41.85 + \underset{(3.03)}{20.56L} + \underset{(.04)}{.15M} \qquad (R^2 = .3928) \tag{3}$$

$Wm$ = willingness to drive additional miles.
$L$ = length of visit in days.
$M$ = miles traveled to area.

The respondents thus expressed a willingness to exert an additional driving effort, just as they expressed a willingness to make an additional money outlay if this became a requisite to using the area. Moreover, there is a significant correspondence between willingness to pay and willingness to drive. The simple correlation coefficient between these two variables is .5. Because of the correlation with length of stay, the reduction in unexplained variance produced by adding either variable to the equation in which the other variable is the dependent one is not very high. However, willingness to pay was found to increase about 5¢ per mile as a function of willingness to drive additional miles. This result gives us a basis for transforming willingness to drive into willingness to pay.

We may now construct a demand schedule for the Pittston area on the basis of willingness to drive, and compute a willingness to pay at 5¢

per mile. The resulting demand and benefit schedules appear in Table 1. The estimated $64,000 of total benefits is very close to that developed from the willingness to pay interview. While one may quibble about the evaluation of a mile of extra driving and about the treatment of one way versus round-trip distance, the first approximation using the obvious value of 5¢ and one-way mileage as reported by the respondents produces a result so close to the first result that we need look no further for marginal adjustments. The initial result strongly suggests that mileage measures and expenditure measures have equal validity as a measure of benefits in this particular case at least.

There are some differences between the respective demand schedules worth noting. The much lower price intercept on the willingness to drive schedule reflects the effect of the time constraint in traveling as well as our possibly erroneous constant transformation of miles to dollars when an increasing cost per mile would be more reasonable. The travel schedule is also elastic over more of its range than the dollar schedule—also perhaps a result of the constant transformation employed.

This initial success with alternative derivations of the benefits schedule now leads us to examine an alternative method for estimating the willingness to drive schedule.

## *Travel-Cost Method of Estimating User-Demand Curve*

The direct interview approach to the estimate of a true price-quantity relationship, or demand curve, for the recreation experience is one approach to the benefit calculations based on willingness to pay. An alternative approach has received some recognition and has been applied in a number of limited instances with at least a fair degree of success. This uses travel-cost data as a proxy for price in imputing a demand curve for recreation facilities. [Ref. 7, 8, 9, 10.] As with the direct interview approach, we believe that estimates derived from this approach are relevant and useful for measuring user benefits of outdoor recreation.

The travel-cost method imputes the price-quantity reactions of consumers by examining their actual current spending behavior with respect to travel cost. The method can be shown by using a simple, hypothetical example. Assume a free recreation or park area at varying distances from three centers of population given in Table 2.

TABLE 2. Visits to a Hypothetical Recreation Area

| City | Population | Cost of visit | Visits made | Visits/1,000 pop. |
|---|---|---|---|---|
| A | 1,000 | $1.00 | 400 | 400 |
| B | 2,000 | 3.00 | 400 | 200 |
| C | 4,000 | 4.00 | 400 | 100 |

The cost of visiting the area is of major concern and would include such items as transportation, lodging, and food cost above those incurred if the trip were not made. Each cost would vary with the distance from the park to the city involved. Consequently, the number of visits, or rather the rate of visits per unit total population of each city, would also vary.

The visits per unit of population, in this case per thousand population, may then be plotted against the cost per visit. A line drawn through the three points of such a plot would have the relationship given by the equation of $C=5-V$, or perhaps more conveniently $V=5-C$, where $C$ is cost of a visit and $V$ is the rate of visits in hundreds per thousand population. This information is taken directly from the tabulation of consumer behavior. The linear relationship assumed here is for convenience. Actual data may very well show, for example, that $1.00 change in cost might have only a slight effect on visit rate where the visit is already high in cost, and a large effect on low-cost visits.

The construction of a demand curve to the recreation area, relating number of visits to varying cost, involves a second step. Essentially, it derives the demand curve from the equation relating visit rates to cost, by relating visit rates of each zone to simulated increases in cost and multiplying by the relative populations in each zone. Thus we might first assume a price of $1.00, which is an added cost of $1.00 for visits to the area from each of the three different centers used in our hypothetical example. This would have the expected result of reducing the number of visitors coming from each of the centers. The expected reduction is estimated from the visit-cost relationship. The total visits suggested by these calculations for different prices or differing added cost are given as:

| Price (added cost) | Quantity (total visits) |
|---|---|
| $0.00 | 1,200 |
| 1.00 | 500 |
| 2.00 | 200 |
| 3.00 | 100 |
| 4.00 | 0 |

These results may then be taken as the demand curve relating price to visits to the recreation area. While this analysis takes visits as a simple function of cost, in principle there is no difficulty in extending the analysis to other factors important in recreation demand, such as alternative sites available, the inherent attractiveness of the area in question or at least its characteristics in this regard, and possibly even some measure of congestion.

A difficulty with this method of benefit approximation is a consistent bias in the imputed demand curve resulting from the basic assumption that the disutility of overcoming distance is a function only of money cost. Clearly this is not so. The disutility is most likely to be the sum of at least three factors: money cost, time cost, and the utility (plus or minus) of driving, or traveling. The total of these three factors is demonstrably negative. but we do not know enough about the significance of the last two components. In all likelihood their sum—that is, of the utility or disutility of driving and the time cost—imposes costs in addition to money. To the extent that this is true the benefit estimate will be conservatively biased, for, as has been indicated, it is assumed that the only thing causing differences in attendance rates for cities located at different distances to a recreation area will be the differences in money cost. The method then postulates that if money cost changes are affected, the changes in rates will be proportional. What this bias amounts to is, essentially, a failure to establish a complete transformation function relating the three components of overcoming distance to the total effect on visitation rates. The resulting conservative bias must be regarded as an understatement of the recreation benefits which the approach is designed to measure.

*Application to Pittston Area* The travel-cost method was applied to the same area as that used to illustrate the interview method of recreation benefit estimation. The same data were utilized to allow at least a crude comparison of the methods. In all, 6,678 respondents who said the Pittston area was the main destination of their trip were used in the analysis.

Visit rates of visitors from groups of counties near the area and from some states at greater distances were plotted against distance. The results were fairly consistent considering the rough nature of the approximations used in estimating distance. A curve was drawn through the points, giving a relationship between visit rates and distance. The demand curve was then calculated, giving a price-quantity relationship based on added distance (or added toll cost) and total visits. It was assumed initially that travel cost would be 5¢ per mile, using one-way distance to conform with our earlier analysis of travel cost by the interview method.

The results at this point were not comparable to the interview method because of a difference in the number of users accounted for. It will be recalled that in the analysis we are now describing only those respondents were used who had specifically stated that the visit to the Pittston area was the main destination of the trip. In order to make this number comparable to the total number of users accounted for in the interview estimate, we counted at half weight the 1,327 respondents who said that Pittston was *not* the primary destination of the trip, and also included in this group the non-response questionnaires and others with incomplete information. In this way we accounted for the same number of users as in the interview estimate. This very crude approximation points out the problems of the multiple-destination visit, but perhaps adequately serves the present purpose.

On the basis of these approximations, the benefit estimates on an annual basis were $70,000, assuming 5¢ per mile one-way distance. While the assumptions made throughout this analysis are subject to refinement, the exercise does seem to illustrate that the procedure is feasible from a practical standpoint and does produce results that are economically meaningful.

## Comparison between Travel-Cost and Interview Methods

Having demonstrated that fairly close results are obtained from both the interview and imputation methods of estimating recreation benefits on the basis of reactions to travel costs, and further that the interview method of directly estimating willingness to pay agrees closely with both estimates based on travel costs, we can now begin to assess the meaning of these results. In some ways the task would be easier if the results had not agreed so closely, for the three methodologies may imply different things about the users' reactions to increased costs. At least, it is not obvious without further probing as to why the agreement is so close.

The interview and imputation methods of estimating benefits on the basis of willingness to incur additional travel costs do not, for example neatly imply the same relationship between distance traveled and willingness to incur additional travel costs. The estimating equation derived from the interviews (Equation 3) suggests that the farther one has traveled, the greater additional distance he will travel. Yet the imputation procedure implies that the willingness to drive by populations in the respective zones does not vary consistently with distance. Furthermore, according to the interviews, responses to the monetary measure of willingness to pay do not attribute any variance in willingness to pay to the distance factor, nor is an indirect relationship obvious. It seems relevant to inquire into the implied effects of these factors to discover why the alternative procedures appear to imply substantially different determinants of willingness to pay.

The superficial agreement in results may be upheld by this kind of further probing, but there are also some methodological issues which should not be overlooked. The travel-cost methods are obviously sensitive to such matters as the weighting given to multiple-destination visits and to the transformation used to derive costs from mileage values. Both methods are sensitive to the usual problems of choosing an appropriately fitting equation for the derivation of the demand schedule. The interview method has a poorly understood sensitivity to the various methodologies that might be employed in its use. Moreover, even the minimal use of interviews in studies of recreation benefits makes the method far more costly than the imputation method based on travel costs.

There are, however, complementarities in the two basic methods which may prove highly useful. In the first place, the two methods may serve as checks on each other in applied situations. One is certainly in a better position from having two methods produce nearly identical answers than if he has to depend on only one. There are also interesting possibilities that interviews may be the best way of resolving the ambiguities in the travel-cost method concerning the treatment of multiple-destination cases and for finding the appropriate valuation for converting distance into dollars. Much can be said for letting the recreationist tell us how to handle these problems.

In sum, we have examined three methods of measuring recreation benefits. All three measure recreationists' willingness to pay. This, we argue, is the appropriate measure of primary, or national, benefits. Furthermore, the measures are in rough agreement as to the benefits ascribable to an area of the Maine woods. This may be taken as evidence that we are on the right track. There are, however, some rough spots to be ironed out of each of the methods—an endeavor we believe to be worthy of major research effort if benefit-cost analysis is to contribute its full potential in planning decisions affecting recreation investments in land and water resources.

## References

[1] Lerner, Lionel. "Quantitative Indices of Recreational Values," in *Water Resources and Economic Development of the West*. Report No. 11. Proceedings, Conference of Committee on the Economics of Water Resources Development of Western Agricultural Economics Research Council with Western Farm Economics Association. Reno: University of Nevada, 1962.

[2] Merewitz, Leonard. "Recreational Benefits of Water-Resource Development." Unpublished paper of Harvard Water Program, 1965.

[3] Crutchfield, James. "Valuation of Fishery Resources," *Land Economics,* Vol. 38, No. 2 (1962).

[4] Davis, Robert K. "The Value of Outdoor Recreation: An Economic Study of the Maine Woods." Ph.D. thesis, Harvard University, 1963.

[5] Ullman, Edward, and Volk, Donald. "An Operational Model for Pre-

dicting Reservoir Attendance and Benefits: Implications of a Location Approach to Water Recreation," *Proceedings Michigan Academy of Sciences,* 1961.

[6] Meramec Basin Research Project. "Recreation," Chap. 5 in *The Meramec Basin,* Vol. 3, St. Louis: Washington University, December 1961.

[7] Clawson, Marion. *Methods of Measuring the Demand for and Value of Outdoor Recreation.* RFF Reprint No. 10. Washington: Resources for the Future, Inc., 1959.

[8] Knetsch, Jack L. "Outdoor Recreation Demands and Benefits," *Land Economics,* Vol. 39, No. 4 (1963).

[9] ——. "Economics of Including Recreation as a Purpose of Water Resources Projects," *Journal of Farm Economics,* December 1964. Also RFF Reprint No. 50. Washington: Resources for the Future, Inc.

[10] Brown, William G., Singh, Ajner, and Castle, Emery N. *An Economic Evaluation of the Oregon Salmon and Steelhead Sport Fishery.* Technical Bulletin 78. Corvallis: Oregon Experiment Station, 1964.

# 26

WASSILY LEONTIEF

# *Environmental Repercussions and the Economic Structure: An Input-Output Approach*

*Wassily Leontief is Henry Lee Professor of Economics at Harvard University.*

## I

Pollution is a by-product of regular economic activities. In each of its many forms it is related in a measurable way to some particular consumption or production process: The quantity of carbon monoxide released in the air bears, for example, a definite relationship to the amount of fuel burned by various types of automotive engines; the discharge of polluted water into our streams and lakes is linked directly to the level of output of the steel, the paper, the textile and all the other water-using industries and its amount depends, in each instance, on the technological characteristics of the particular industry.

Input-output analysis describes and explains the level of output of each sector of a given national economy in terms of its relationships to the corresponding levels of activities in all the other sectors. In its more complicated multi-regional and dynamic versions the input-output approach permits us to explain the spatial distribution of output and consumption of various goods and services and of their growth or decline —as the case may be—over time.

Frequently unnoticed and too often disregarded, undesirable by-products (as well as certain valuable, but unpaid-for natural inputs) are linked directly to the network of physical relationships that govern the day-to-day operations of our economic system. The technical interdependence between the levels of desirable and undesirable outputs can be described in terms of structural coefficients similar to those used to trace the structural interdependence between all the regular branches of production and consumption. As a matter of fact, it can be described and analyzed as an integral part of that network.

It is the purpose of this report first to explain how such "externalities" can be incorporated into the conventional input-output picture of a national economy and, second, to demonstrate that—once this has been done—conventional input-output computations can yield concrete replies to some of the fundamental factual questions that should be asked and answered before a practical solution can be found to problems raised by the undesirable environmental effects of modern technology and uncontrolled economic growth.

## II

Proceeding on the assumption that the basic conceptual framework of a static input-output analysis is familiar to the reader, I will link up the following exposition to the numerical examples and elementary equations presented in chapter 7 of my book entitled *"Input Output Economics"* (Oxford University Press, N.Y. 1966).

Consider a simple economy consisting of two producing sectors, say, Agriculture and Manufacture, and Households. Each one of the two industries absorbs some of its annual output itself, supplies some to the other industry and delivers the rest to final consumers—in this case represented by the Households. These inter-sectoral flows can be conveniently entered in an input-output table.

The magnitude of the total outputs of the two industries and of the two different kinds of inputs absorbed in each of them depends on (1) the amounts of agricultural and manufactured goods that had to be delivered to the final consumers, i.e., the Households and (2) the input requirements of the two industries determined by their specific technological structures. In this particular instance Agriculture is assumed to

TABLE 1. Input-Output Table of a National Economy (in physical units)

| From | Into: Sector 1 Agriculture | Sector 2 Manufacture | Final demand Households | Total output |
|---|---|---|---|---|
| Sector 1 Agriculture | 25 | 20 | 55 | 100 bushels of wheat |
| Sector 2 Manufacture | 14 | 6 | 30 | 50 yards of cloth |

require 0.25 (=25 / 100) units of agricultural and 0.14 (=14 / 100) units of manufactured inputs to produce a bushel of wheat, while the manufacturing sector needs 0.40 (=20 / 50) units of agricultural and 0.12 (=6 / 50) units of manufactured product to make a yard of cloth.

The "cooking recipes" of the two producing sectors can also be presented in a compact tabular form:

TABLE 2. Input Requirements per Unit of Output

| From | Into: Sector 1 Agriculture | Sector 2 Manufacture |
|---|---|---|
| Sector 1 Agriculture | 0.25 | 0.40 |
| Sector 2 Manufacture | 0.14 | 0.12 |

This is the "structural matrix" of the economy. The numbers entered in the first column are the technical input coefficients of the Agriculture sector and those shown in the second are the input coefficients of the Manufacture sector.

## III

The technical coefficients determine how large the total annual outputs of agricultural and of manufactured goods must be if they are to satisfy not only the given direct demand (for each of the two kinds of goods) by the final users, i.e., the Households, but also the intermediate demand depending in its turn on the total level of output in each of the two productive sectors.

These somewhat circular relationships are described concisely by the following two equations:

$$X_1 - 0.25X_1 - 0.40X_2 = Y_1$$
$$X_2 - 0.12X_2 - 0.14X_1 = Y_2$$

or in a rearranged form,

$$\begin{aligned} 0.75X_1 - 0.40X_2 &= Y_1 \\ -0.14X_1 + 0.88X_2 &= Y_2 \end{aligned} \tag{1}$$

$X_1$ and $X_2$ represent the unknown total outputs of agricultural and manufactured commodities respectively; $Y_1$ and $Y_2$ the given amounts of agricultural and manufactured products to be delivered to the final consumers.

These two linear equations with two unknowns can obviously be solved, for $X_1$ and $X_2$ in terms of any given $Y_1$ and $Y_2$.

Their "general" solution can be written in form of the following two equations:

$$\begin{aligned} X_1 &= 1.457Y_1 + 0.662Y_2 \\ X_2 &= 0.232Y_1 + 1.242Y_2. \end{aligned} \tag{2}$$

By inserting on the right-hand side the given magnitudes of $Y_1$ and $Y_2$ we can compute the magnitudes of $X_1$ and $X_2$. In the particular case described in table 1, $Y_1 = 55$ and $Y_2 = 30$. Performing the necessary multiplications and additions one finds the corresponding magnitudes of $X_1$ and $X_2$ to be, indeed, equal to the total outputs of agricultural (100 bushels) and manufactured (50 yards) goods, as shown in table 1.

The matrix, i.e., the square set table of numbers appearing on the right-hand side of (2),

$$\begin{bmatrix} 1.457 & 0.662 \\ 0.232 & 1.242 \end{bmatrix} \tag{3}$$

is called the "inverse" of matrix,

$$\begin{bmatrix} 0.75 & -0.40 \\ -0.14 & 0.88 \end{bmatrix} \tag{4}$$

describing the set constants appearing on the left-hand side of the original equations in (1).

Any change in the technology of either Manufacture or Agriculture, i.e., in any one of the four input coefficients entered in table 2, would entail a corresponding change in the structural matrix (4) and, consequently, of its inverse (3). Even if the final demand for agricultural ($Y_1$) and manufactured ($Y_2$) goods remained the same, their total outputs, $X_1$

and $X_2$, would have to change, if the balance between the total outputs and inputs of both kinds of goods were to be maintained. On the other hand, if the level of the final demands $Y_1$ and $Y_2$ had changed, but the technology remained the same, the corresponding changes in the total outputs $X_1$ and $X_2$ could be determined from the same general solution (2).

In dealing with real economic problems one takes, of course, into account simultaneously the effect of technological changes and of anticipated shifts in the levels of final deliveries. The structural matrices used in such computations contain not two but several hundred sectors, but the analytical approach remains the same. In order to keep the following verbal argument and the numerical examples illustrating it quite simple, pollution produced directly by Households and other final users is not considered in it. A concise description of the way in which pollution generated by the final demand sectors can be introduced—along with pollution originating in the producing sectors—into the quantitative description and numerical solution of the input-output system is relegated to the Mathematical Appendix.

## IV

As has been said before, pollution and other undesirable—or desirable—external effects of productive or consumptive activities should for all practical purposes be considered part of the economic system.

The quantitative dependence of each kind of external output (or input) on the level of one or more conventional economic activities to which it is known to be related must be described by an appropriate technical coefficient and all these coefficients have to be incorporated in the structural matrix of the economy in question.

Let it be assumed, for example, that the technology employed by the Manufacture sector leads to a release into the air of 0.20 grams of a solid pollutant per yard of cloth produced by it, while agricultural technology adds 0.50 grams per unit (i.e., each bushel of wheat) of its total output.

Using $\overline{X}_3$ to represent the yet unknown total quantity of this external output, we can add to the two original equations of output system (1) a third,

$$\begin{aligned} 0.75X_1 - 0.40X_2 \qquad &= Y_1 \\ -0.14X_1 + 0.88X_2 \qquad &= Y_2 \\ 0.50X_1 + 0.20X_2 - \overline{X}_3 &= 0 \end{aligned} \tag{5}$$

In the last equation the first term describes the amount of pollution produced by Agriculture as depending on that sector's total output, $X_1$, while the second represents, in the same way, the pollution originating in Manufacture as a function of $X_2$; the equation as a whole simply states that $\overline{X}_3$, i.e., the total amount of that particular type pollution generated by the economic system as a whole, equals the sum total of the amounts produced by all its separate sectors.

Given the final demands $Y_1$ and $Y_2$ for agricultural and manufactured products, this set of three equations can be solved not only for their total outputs $X_1$ and $X_2$ but also for the unknown total output $\overline{X}_3$ of the undesirable pollutant.

The coefficients of the left-hand side of augmented input-output system (5) form the matrix,

$$\left\{\begin{matrix} 0.75 & -0.40 & 0 \\ -0.14 & 0.88 & 0 \\ 0.50 & 0.20 & -1 \end{matrix}\right\} \qquad (5a)$$

A "general solution" of system (5) would in its form be similar to the general solution (2) of system (1); only it would consist of three rather than two equations and the "inverse" of the structural matrix (4) appearing on the right-hand side would have three rows and columns.

Instead of inverting the enlarged structural matrix one can obtain the same result in two steps. First, use the inverse (4) of the original smaller matrix to derive, from the two-equation system (2), the outputs of agricultural ($X_1$) and manufactured ($X_2$) goods required to satisfy any given combination of final demands $Y_1$ and $Y_2$. Second, determine the corresponding "output" of pollutants, i.e., $\overline{X}_3$, by entering the values of $X_1$ and $X_2$ thus obtained in the last equation of set (5).

Let $Y_1 = 55$ and $Y_2 = 30$; these are the levels of the final demand for agricultural and manufactured products as shown on the input-output table 1. Inserting these numbers on the right-hand side (5), we find—using the general solution (2) of the first two equations—that $X_1 = 100$ and $X_2 = 50$. As should have been expected they are identical with the corresponding total output figures in table 1. Using the third equation in (5) we find, $X_3 = 60$. This is the total amount of the pollutant generated by both industries.

By performing a similar computation for $Y_1 = 55$ and $Y_2 = 0$ and then for $Y_1 = 0$ and $Y_2 = 30$, we could find out that 42.62 of these 60 grams of pollution are associated with agricultural and manufactured activities contributing directly and indirectly to the delivery to Households of 55 bushels of wheat, while the remaining 17.38 grams can be imputed to productive activities contributing directly and indirectly to final delivery of the 30 yards of cloth.

Had the final demand for cloth fallen from 30 yards to 15, the amount of pollution traceable in it would be reduced from 17.38 to 8.69 grams.

## V

Before proceeding with further analytical exploration, it seems to be appropriate to introduce the pollution-flows explicitly in the original table 1:

TABLE 3. Input-Output Table of the National Economy with Pollutants Included (in physical units)

| | Into | | | |
|---|---|---|---|---|
| From | Sector 1 Agriculture | Sector 2 Manufacture | Households | Total output |
| Sector 1 Agriculture | 25 | 20 | 55 | 100 bushels of wheat |
| Sector 2 Manufacture | 14 | 6 | 30 | 50 yards of cloth |
| Sector 3 Air pollution | 50 | 10 | | 60 grams of pollutant |

The entry at the bottom of the first column in table 3 indicates that Agriculture produced 50 grams of pollutant or 0.50 grams per bushel of wheat. Multiplying the pollutant-output-coefficient of the manufacturing sector with its total output we find that it has contributed 10 to the grand total of 60 grams of pollution.

Conventional economic statistics concern themselves with production and consumption of goods and services that are supposed to have in our competitive private enterprise economy some positive market value. This explains why the production and consumption of DDT is, for example, entered in conventional input-output tables while the production and the consumption of carbon-monoxide generated by internal combustion engines is not. Since private and public bookkeeping, that constitutes the ultimate source of the most conventional economic statistics, does not concern itself with such "non-market" transactions, their magnitude has to be estimated indirectly through detailed analysis of the underlying technical relationships.

Problems of costing and of pricing are bound, however, to arise as soon as we go beyond explaining and measuring pollution toward doing something about it.

## VI

A conventional national or regional input-output table contains a "value-added" row. It shows, in dollar figures, the wages, depreciation charges, profits, taxes and other costs incurred by each producing sector in addition to payments for inputs purchased from other producing sectors. Most of that "value-added" represents the cost of labor, capital, and other so-called primary factors of production, and depends on the physical amounts of such inputs and their prices. The wage bill of an industry equals, for example, the total number of man-years times the wage rate per man-year.

In table 4 the original national input-output table is extended to include a labor input or total employment row.

TABLE 4. Input-Output Table with Labor Inputs Included (in physical and in money units)

| | Into | | | |
|---|---|---|---|---|
| From | Sector 1 Agriculture | Sector 2 Manufacture | Households | Total output |
| Sector 1 Agriculture | 25 | 20 | 55 | 100 bushels of wheat |
| Sector 2 Manufacture | 14 | 6 | 30 | 50 yards of cloth |
| Labor inputs (value-added) | 80 ($80) | 180 ($180) | | 260 man-years ($260) |

The "cooking recipes" as shown in table 2 can be accordingly extended to include the labor input coefficients of both industries expressed in man-hours as well as in money units.

TABLE 5. Input Requirements per Unit of Output (including labor or value-added)

| | Into | |
|---|---|---|
| From | Sector 1 Agriculture | Sector 2 Manufacture |
| Sector 1 Agriculture | 0.25 | 0.40 |
| Sector 2 Manufacture | 0.14 | 0.12 |
| Primary input-labor in man-hours (at $1 per hour) | 0.80 ($0.80) | 3.60 ($3.60) |

In section III it was shown how the general solution of the original input-output system (2) can be used to determine the total outputs of agricultural and manufactured products ($X_1$ and $X_2$) required to satisfy any given combination of deliveries of these goods ($Y_1$ and $Y_2$) to final Households. The corresponding total labor inputs can be derived by multiplying the appropriate labor coefficients ($k_1$ and $k_2$) with each sector's total output. The sum of both products yields the labor input $L$ of the economy as a whole.

$$L = k_1 X_1 + k_2 X_2. \tag{6}$$

Assuming a wage rate of \$1 per hour we find (see table 5) the payment for primary inputs per unit of the total output to be \$0.80 in Agriculture and \$3.60 in Manufacture. That implies that the prices of one bushel of wheat ($p_1$) and of a yard of cloth ($p_2$) must be just high enough to permit Agriculture to yield a "value-added" of $v_1$ ($=0.80$) and Manufacture $v_2$ ($=3.60$) per unit of their respective outputs after having paid for all the other inputs specified by their respective "cooking recipes."

$$\begin{aligned} p_1 - 0.25p_1 - 0.14p_2 &= v_1 \\ p_2 - 0.12p_2 - 0.40p_1 &= v_2 \end{aligned}$$

or in a rearranged form,

$$\begin{aligned} 0.75p_1 - 0.14p_2 &= v_1 \\ -0.40p_1 + 0.88p_2 &= v_2 \end{aligned} \tag{7}$$

The "general solution" of these two equations permitting us to compute $p_1$ and $p_2$ from any given combination of values-added, $v_1$ and $v_2$, is

$$\begin{aligned} p_1 &= 1.457v_1 + 0.232v_2 \\ p_2 &= 0.662v_1 + 1.242v_2. \end{aligned} \tag{8}$$

With $v_1 = \$0.80$ and $v_2 = \$3.60$ we have, $p_1 = \$2.00$ and $p_2 = \$5.00$. Multiplying the physical quantities of wheat and cloth entered in the first and second rows of table 4 with appropriate prices, we can transform it into a familiar input-output table in which all transactions are shown in dollars.

## VII

Within the framework of the open input-output system described above any reduction or increase in the output level of pollutants can be traced either to changes in the final demand for specific goods and services, changes in the technical structure of one or more sectors of the economy, or to some combination of the two.

The economist cannot devise new technology, but, as has been demonstrated above, he can explain or even anticipate the effect of any given technological change on the output of pollutants (as well as of all the other goods and services). He can determine the effects of such a change on sectoral, and, consequently, also on total demand for the "primary factor of production." With given "values-added" coefficients he can, moreover, estimate the effect of such a change on prices of various goods and services.

After the explanations given above, a single example should suffice to show how any of these questions can be formulated and answered in input-output terms.

Consider the simple two-sector economy whose original state and structure are described in tables 3, 4, 5 and 6. Assume that a process has been introduced permitting elimination (or prevention) of pollution and that the input requirements of that process amount to two man-years of labor (or $2.00 of value-added) and 0.20 yards of cloth per gram of pollutant prevented from being discharged—either by Agriculture or Manufacture—into the air.

TABLE 6. Structural Matrix of a National Economy with Pollution Output and Anti-Pollution Input Coefficients Included

| | Output Sectors | | |
|---|---|---|---|
| Inputs and Pollutant Output | Sector 1 Agriculture | Sector 2 Manufacture | Elimination of pollutant |
| Sector 1 Agriculture | 0.25 | 0.40 | 0 |
| Sector 2 Manufacture | 0.14 | 0.12 | 0.20 |
| Pollutant (output) | 0.50 | 0.20 | |
| Labor (value-added) | 0.80 ($0.80) | 3.60 ($3.60) | 2.00 ($2.00) |

Combined with the previously introduced sets of technical coefficients this additional information yields the following complex structural matrix of the national economy.

The input-output balance of the entire economy can be described by the following set of four equations:

$$\begin{aligned} 0.75X_1 - 0.40X_2 &= Y_1 && \text{(wheat)} \\ -0.14X_1 + 0.88X_2 - 0.20X_3 &= Y_2 && \text{(cotton cloth)} \\ 0.50X_1 + 0.20X_2 - X_3 &= Y_3 && \text{(pollutant)} \\ -0.80X_1 - 3.60X_2 - 2.00X_3 + L &= Y_4 && \text{(labor)} \end{aligned} \tag{9}$$

Variables:

$X_1$: total output of agricultural products
$X_2$: total output of manufactured products
$X_3$: total amount of eliminated pollutant
$L$ : employment
$Y_1$: final demand for agricultural products
$Y_2$: final demand for manufactured products
$Y_3$: total uneliminated amount of pollutant
$Y_4$: total amount of labor employed by Household and other "final demand" sectors.[1]

Instead of describing complete elimination of all pollution, the third equation contains on its right-hand side $Y_3$, the amount of uneliminated pollutant. Unlike all other elements of the given vector of final deliveries it is not "demanded" but, rather, tolerated.[2]

The general solution of that system, for the unknown $X$'s in terms of any given set of $Y$'s is written out in full below

$$\begin{aligned}
X_1 &= 1.573 Y_1 + 0.749 Y_2 - 0.149 Y_3 + 0.000 Y_4 && \text{Agriculture} \\
X_2 &= 0.449 Y_1 + 1.404 Y_2 - 0.280 Y_3 + 0.000 Y_4 && \text{Manufacture} \\
X_3 &= 0.876 Y_1 + 0.655 Y_2 - 1.131 Y_3 + 1.000 Y_4 && \text{Pollutant} \\
L &= 4.628 Y_1 + 6.965 Y_2 - 3.393 Y_3 + 0.000 Y_4 && \text{Labor}
\end{aligned} \tag{10}$$

The square set of coefficients (each multiplied with the appropriate $Y$) on the right-hand side of (10) is the inverse of the matrix of constants appearing on the left-hand side of (9). The inversion was, of course, performed on a computer.

The first equation shows that each additional bushel of agricultural product delivered to final consumers (i.e., Households) would require (directly and indirectly) an increase of the total output of agricultural sector ($X_1$) by 1.573 bushels, while the final delivery of an additional yard of cloth would imply a rise of total agricultural outputs by 0.749 bushels.

The next term in the same equation measures the (direct and indirect) relationship between the total output of agricultural products ($X_1$) and the "delivery" to final users of $Y_3$ grams of uneliminated pollutants.

The constant $-0.149$ associated with it in this final equation indicates that a reduction in the total amount of pollutant delivered to final

1. In all numerical examples presented in this paper $Y_4$ is assumed to equal zero.
2. In (5), which describes a system that generates pollution, but does not contain any activity combating it, the variable $\overline{X}_3$ stands for the total amount of uneliminated pollution that is represented by $Y_3$ in system (9).

consumers by one gram would require an increase of agricultural output by 0.149 bushels.

Tracing down the column of coefficients associated with $Y_3$ in the second, third and fourth equations we can see what effect a reduction in the amount of pollutant delivered to the final users would have on the total output levels of all other industries. Manufacture would have to produce 0.280 additional yards of cloth. Sector 3, the anti-pollution industry itself, would be required to eliminate 1.131 grams of pollutant to make possible the reduction of its final delivery by 1 gram, the reason for this being that economic activities required (directly and indirectly) for elimination of pollution do, in fact, generate some of it themselves.

The coefficients of the first two terms on the right-hand side of the third equation show how the level of operation of the anti-pollution industry ($X_3$) would have to vary with changes in the amounts of agricultural and manufactured goods purchased by final consumers, if the amount of uneliminated pollutant ($Y_3$) were kept constant. The last equation shows that the total, i.e., direct and indirect, labor input required to reduce $Y_3$ by 1 gram amounts to 3.393 man-years. This can be compared with 4.628 man-years required for delivery to the final users of an additional bushel of wheat and 6.965 man-years needed to let them have one more yard of cloth.

Starting with the assumption that Households, i.e., the final users, consume 55 bushels of wheat and 30 yards of cloth and also are ready to tolerate 30 grams of uneliminated pollution, the general solution (10) was used to determine the physical magnitudes of the intersectoral input-output flows shown in table 7.

The entries in the third row show that the agricultural and manufactured sectors generate 63.93 (=52.25 + 11.68) grams of pollution of which 33.93 are eliminated by the anti-pollution industry and the remaining 30 are delivered to Households.

## VIII

The dollar figures entered in parentheses are based on prices the derivation of which is explained below.

The original equation system (7), describing the price-cost relationships within the agricultural and manufacturing sectors has now to be expanded through inclusion of a third equation stating that the price of "eliminating one gram of pollution" (i.e., $p_3$) should be just high enough to cover—after payment for inputs purchased from other industries has been met—the value-added, $v_3$, i.e., the payments to labor and other primary factors employed directly by the anti-pollution industry.

$$\begin{aligned} p_1 - 0.25p_1 - 0.14p_2 &= v_1 \\ p_2 - 0.12p_2 - 0.40p_1 &= v_2 \\ p_3 \qquad\qquad - 0.20p_2 &= v_3 \end{aligned}$$

or in rearranged form,

$$\begin{aligned} 0.75p_1 - 0.14p_2 \qquad &= v_1 \\ -0.40p_1 + 0.88p_2 \qquad &= v_2 \\ -0.20p_2 + p_3 &= v_3 \end{aligned} \tag{11}$$

The general solution of these equations—analogous to (8) is

$$\begin{aligned} p_1 &= 1.457v_1 + 0.232v_2 \\ p_2 &= 0.662v_1 + 1.242v_2 \\ p_3 &= 0.132v_1 + 0.248v_2 + v_3 \end{aligned} \tag{12}$$

Assuming as before, $v_1 = 0.80$, $v_2 = 3.60$ and $v_3 = 2.00$, we find,

$$\begin{aligned} p_1 &= \$2.00 \\ p_2 &= \$5.00 \\ p_3 &= \$3.00 \end{aligned}$$

The price (=cost per unit) of eliminating pollution turns out to be \$3.00 per gram. The prices of agricultural and manufactured products remain the same as they were before.

Putting corresponding dollar values on all the physical transactions shown on the input-output table 7 we find that the labor employed by the three sectors adds up to \$361.80. The wheat and cloth delivered to final consumers cost \$260.00. The remaining \$101.80 of the value-added earned by the Households will just suffice to pay the price, i.e., to defray the costs of eliminating 33.93 of the total of 63.93 grams of pollution generated by the system. These payments could be made directly or they might be collected in form of taxes imposed on the Households and used by the Government to cover the costs of the privately or publicly operated anti-pollution industry.

The price system would be different, if through voluntary action or to obey a special law, each industry undertook to eliminate, at its own expense, all or at least some specified fraction of the pollution generated by it. The added costs would, of course, be included in the price of its marketable product.

Let, for example, the agricultural and manufacturing sectors bear the costs of eliminating, say, 50 percent of the pollution that, under prevailing technical conditions, would be generated by each one of them. They

TABLE 7. Input-Output Table of the National Economy (Surplus pollution is eliminated by the Anti-Pollution Industry)

| Inputs and Pollutants Output | Output Sectors | | | Final Deliveries to Households | Totals |
|---|---|---|---|---|---|
| | Sector 1 Agriculture | Sector 2 Manufacture | Anti-Pollution | | |
| Sector 1 Agriculture (bushels) | 26.12 ($52.24) | 23.37 ($46.74) | 0 | 55 ($110.00) | 104.50 ($203.99) |
| Sector 2 Manufacture (yards) | 14.63 ($73.15) | 7.01 ($35.05) | 6.79 ($33.94) | 30 ($150.00) | 58.43 ($292.13) |
| Pollutant (grams) | 52.25 | 11.68 | −33.93 | 30 ($101.80 paid for elimination of 33.93 grams of pollutant) | |
| Labor (man-years) | 83.60 ($83.60) | 210.34 ($210.34) | 67.86 ($67.86) | 0 | 361.80 ($361.80) |
| Column Totals | $208.99 | $292.13 | $101.80 | $361.80 | |

$p_1 = \$2.00$, $p_2 = \$5.00$, $p_3 = \$3.00$, $p_k = \$1.00$ (wage rate).

may either engage in anti-pollution operations on their own account or pay an appropriately prorated tax.

In either case the first two equations in (11) have to be modified by inclusion of additional terms: the outlay for eliminating 0.25 grams and 0.10 grams of pollutant per unit of agricultural and industrial output respectively.

$$\begin{aligned} 0.75p_1 - 0.14p_2 - 0.25p_3 &= v_3 \\ -0.40p_1 + 0.88p_2 - 0.10p_3 &= v_2 \\ -0.20p_2 + \quad p_3 &= v_3 \end{aligned} \qquad (13)$$

The "inversion" of the modified matrix of structural coefficients appearing on the left-hand side yields the following general solution of the price system:

$$\begin{aligned} p_1 &= 1.511v_1 + 0.334v_2 + 0.411v_3 \\ p_2 &= 0.703v_1 + 1.318v_2 + 0.308v_3 \\ p_3 &= 0.141v_1 + 0.264v_2 + 1.062v_3 \end{aligned} \qquad (14)$$

With "values-added" in all the three sectors remaining the same as they were before (i.e., $v_1 = \$.80$, $v_2 = \$3.60$, $v_3 = \$2.60$) these new sets of prices are as follows:

$$p_1 = \$3.234$$
$$p_2 = \$5.923$$
$$p_3 = \$3.185$$

While purchasing a bushel of wheat or a yard of cloth the purchaser now pays for elimination of some of the pollution generated in production of that good. The prices are now higher than they were before. From the point of view of Households, i.e., of the final consumers, the relationship between real costs and real benefits remains, nevertheless, the same; having paid for some anti-pollution activities indirectly they will have to spend less on them directly.

## IX

The final table 8 shows the flows of goods and services between all the sectors of the national economy analyzed above. The structural characteristics of the system—presented in the form of a complete set of technical input-output coefficients—were assumed to be given; so was the vector of final demand, i.e., quantities of products of each industry delivered to Households (and other final users) as well as the uneliminated amount of pollutant that, for one reason or another, they are prepared to "tolerate." Each industry is assumed to be responsible for elimination of 50 percent of pollution that would have been generated in the absence of such counter measures. The Households defray—directly or through tax contributions—the cost of reducing the net output of pollution still further to the amount that they do, in fact, accept.

On the basis of this structural information we can compute the outputs and the inputs of all sectors of the economy, including the anti-pollution industries, corresponding to any given "bill of final demand." With information on "value-added," i.e., the income paid out by each sector per unit of its total output, we can, furthermore, determine the prices of all outputs, the total income received by the final consumers and the breakdown of their total expenditures by types of goods consumed.

The 30 grams of pollutant entered in the "bill of final demand" are delivered free of charge. The $6.26 entered in the same box represents the cost of that part of anti-pollution activities that was covered by Households directly, rather than through payment of higher prices for agricultural and manufactured goods.

TABLE 8. Input-Output Table of a National Economy with Pollution-Related Activities Presented Separately

| | Agriculture | | | Manufacture | | | | | |
|---|---|---|---|---|---|---|---|---|---|
| | Wheat | Anti-pollution | Total | Cloth | Anti-pollution | Total | Anti-pollution | Final deliveries to households | National totals |
| Agriculture | 26.12 ($84.47) | 0 | 26.12 ($84.47) | 23.37 ($75.58) | 0 | 23.37 ($75.58) | 0 | 55 ($177.87) | 105.50 ($337.96) |
| Manufacture | 14.63 - ($86.65) | 5.23 ($30.98) | 19.86 ($117.63) | 7.01 ($41.52) | 1.17 ($6.93) | 8.18 ($48.45) | .39 ($2.33) | 30 ($177.69) | 58.43 ($346.07) |
| Pollutant | 52.25 | −26.13 | 26.12 | 11.69 | −5.85 | 5.84 | −1.97 | 30 ($6.26 paid for elimination of 1.97 grams of pollutant | |
| Labor (value-added) | 83.60 ($83.60) | 52.26 ($52.26) | 135.86 ($135.86) | 210.34 ($210.34) | ($11.70) ($11.70) | ($222.04) ($222.04) | ($3.93) ($3.93) | | 361.8 ($361.80) |
| Total Costs | ($254.72) | ($83.24) | ($337.96) | ($327.44) | ($18.63) | ($346.07) | ($6.26) | ($361.80) | |

$p_1 = \$3.23$, $p_2 = \$5.92$, $p_3 = \$3.19$.
$v_1 = \$0.80$, $v_2 = \$3.60$, $v_3 = \$2.00$.

The input requirements of anti-pollution activities paid for by the agricultural and manufacturing sectors and all the other input requirements are shown separately and then combined in the total input columns. The figures entered in the pollution row show accordingly the amount of pollution that would be generated by the principal production process, the amount eliminated (entered with a minus sign), and finally the amount actually released by the industry in question. The amount (1.97) eliminated by anti-pollution activities not controlled by other sectors is entered in a separate column that shows also the corresponding inputs.

From a purely formal point of view the only difference between table 8 and table 7 is that in the latter all input requirements of Agriculture and Manufacture and the amount of pollutant released by each of them are shown in a single column, while in the former the productive and anti-pollution activities are described also separately. If such subdivision proves to be impossible and if, furthermore, no separate anti-pollution industry can be identified, we have to rely on the still simpler analytical approach that led up to the construction of table 3.

## X

Once appropriate sets of technical input and output coefficients have been compiled, generation and elimination of all the various kinds of pollutants can be analyzed as what they actually are—integral parts of the economic process.

Studies of regional and multi-regional systems, multi-sectoral projections of economic growth and, in particular, the effects of anticipated technological changes, as well as all other special types of input-output analysis can, thus, be extended so as to cover the production and elimination of pollution as well.

The compilation and organization of additional quantitative information required for such extension could be accelerated by systematic utilization of practical experience gained by public and private research organizations already actively engaged in compilation of various types of input-output tables.

# Mathematical Appendix

## *Static-Open Input-Output System with Pollution-Related Activities Built In*

### Notation

*Commodities and Services*

$1, 2, 3, \ldots i \ldots j \ldots m, m+1, m+2, \ldots g \ldots k \ldots n$

useful goods — pollutants

*Technical Coefficients*

$a_{ij}$—input of good $i$ per unit of output of good $j$ (produced by sector $j$)

$a_{ig}$—input of good $i$ per unit of eliminated pollutant $g$ (eliminated by sector $g$)

$a_{gi}$—output of pollutant $g$ per unit of output of good $i$ (produced by sector $i$)

$a_{gk}$—output of pollutant $g$ per unit of eliminated pollutant $k$ (eliminated by sector $k$)

$r_{gi}$, $r_{gk}$—proportion of pollutant $g$ generated by industry $i$ or $k$ eliminated at the expense of that industry.

*Variables*

$x_i$—total output of good $i$

$x_g$—total amount of pollutant $g$ eliminated

$y_i$—final delivery of good $i$ (to Households)

$y_g$—final delivery of pollutant $g$ (to Households)

$p_i$—price of good $i$

$p_g$—the "price" of eliminating one unit of pollutant $g$

$v_i$—"value-added" in industry $i$ per unit of good $i$ produced by it

$v_g$—"value-added" in anti-pollution sector $g$ per unit of pollutant $g$ eliminated by it.

*Vectors and Matrices*

$A_{11} = [a_{ij}]$ $\quad i, j = 1, 2, 3, \ldots, m$

$A_{21} = [a_{gi}]$ $\quad i = 1, 2, 3, \ldots m$

$A_{12} = [a_{ig}]$ $\quad g = m+1, m+2, m+3, \ldots, n$

$A_{22} = [a_{gk}]$ $\quad g, k = m+1, m+2, m+3, \ldots, n$

$Q_{21} = [q_{gi}]$ $\quad i = 1, 2, \ldots m$

$\quad g = m+1, m+2, \ldots n$

$Q_{22} = [q_{gk}]$ $\quad g, k = m+1, m+2, \ldots, n$

where $q_{gi} = r_{gi} a_{gi}$

$q_{gk} = r_{gk} a_{gk}$

$$X_1 = \begin{Bmatrix} x_1 \\ x_2 \\ \cdot \\ \cdot \\ \cdot \\ x_m \end{Bmatrix} \qquad Y_1 = \begin{Bmatrix} y_1 \\ y_2 \\ \cdot \\ \cdot \\ \cdot \\ y_m \end{Bmatrix} \qquad V_1 = \begin{Bmatrix} v_1 \\ v_2 \\ \cdot \\ \cdot \\ \cdot \\ v_m \end{Bmatrix}$$

$$X_2 = \begin{Bmatrix} x_{m+1} \\ x_{m+2} \\ \cdot \\ \cdot \\ \cdot \\ x_n \end{Bmatrix} \qquad Y_2 = \begin{Bmatrix} y_{m+1} \\ y_{m+2} \\ \cdot \\ \cdot \\ \cdot \\ y_n \end{Bmatrix} \qquad V_2 = \begin{Bmatrix} v_{m+1} \\ v_{m+2} \\ \cdot \\ \cdot \\ \cdot \\ v_n \end{Bmatrix}$$

*Physical Input-Output Balance*

$$\begin{bmatrix} I - A_{11} & -A_{12} \\ A_{21} & -I + A_{22} \end{bmatrix} \begin{bmatrix} X_1 \\ X_2 \end{bmatrix} = \begin{bmatrix} Y_1 \\ Y_2 \end{bmatrix} \tag{15}$$

$$\begin{bmatrix} X_1 \\ X_2 \end{bmatrix} = \begin{bmatrix} I - A_{11} & -A_{12} \\ A_{21} & -I + A_{22} \end{bmatrix}^{-1} \begin{bmatrix} Y_1 \\ Y_2 \end{bmatrix} \tag{16}$$

*Input-Output Balance between Prices and Values-added*

$$\begin{bmatrix} I - A'_{11} & -Q'_{21} \\ -A'_{12} & I - Q'_{22} \end{bmatrix} \begin{bmatrix} P_1 \\ P_2 \end{bmatrix} = \begin{bmatrix} V_1 \\ V_2 \end{bmatrix} \tag{17}$$

$$\begin{bmatrix} P_1 \\ P_2 \end{bmatrix} = \begin{bmatrix} I - A'_{11} & -Q'_{21} \\ -A'_{12} & I - Q'_{22} \end{bmatrix}^{-1} \begin{bmatrix} V_1 \\ V_2 \end{bmatrix} \tag{18}$$

Supplementary Notation and Equations
Accounting for Pollution Generated
Directly by Final Consumption

Notation

*Technical Coefficients*

$a_{gy,(i)}$—output of pollutant generated by consumption of one unit of commodity $i$ delivered to final demand.

*Variables*

$y_g$*—sum total of pollutant $g$ "delivered" from all industries to and generated within the final demand factor,

$x_g^*$—total gross output of pollutant $g$ generated by all industries and in the final demand sector.

$$A_y = \begin{Bmatrix} a_{m+1,y(1)} & a_{m+1,y(1)} & \cdots & a_{m+1,y(m)} \\ a_{m+2,y(2)} & a_{m+2,y(2)} & \cdots & a_{m+2,y(m)} \\ \cdot & \cdot & & \\ \cdot & \cdot & & \\ \cdot & \cdot & & \\ a_n\, y_1 & a_n\, y_2 & \cdots & a_n\, y_m \end{Bmatrix}$$

$$Y_2^* = \begin{Bmatrix} y^*_{m+1} \\ y^*_{m+2} \\ \cdot \\ \cdot \\ \cdot \\ y_n^* \end{Bmatrix} \qquad x_g^* = \begin{Bmatrix} x^*_{m+1} \\ x^*_{m+2} \\ \cdot \\ \cdot \\ \cdot \\ x_n^* \end{Bmatrix}$$

In case some pollution is generated within the final demand sector itself, the vector $Y_2$ appearing on the right-hand side of (15) and (16) has to be replaced by vector $Y_2 - Y_2^*$,
where

$$Y_2^* = A_y Y_1. \tag{19}$$

The price-values added equations (17), (18) do not have to be modified.

Total gross output of pollutants generated by all industries and the final demand sector does not enter explicitly in any of the equations presented above; it can, however, be computed on the basis of the following equation,

$$\mathrm{X}^* = [\mathrm{A}_{21} \,:\, \mathrm{A}_{22}] \begin{bmatrix} X_1 \\ \cdots \\ X_2 \end{bmatrix} + Y_2^*. \tag{20}$$

# FURTHER READINGS ON ECONOMICS OF THE ENVIRONMENT

## Section I *An Overview*


Black, R., A. Muhieh, et al. *The National Solid Waste Survey*. Washington, D.C.: U.S. Department of Health, Education and Welfare, October 24, 1968.

Council on Environmental Quality. Second Annual Report, *Environmental Quality*. Washington, D.C., U.S. Government Printing Office, August 1971.

Dales, J. H. *Pollution, Property and Prices, An Essay in Policymaking and Economics*. Toronto: University of Toronto Press, 1968.

D'Arge, Ralph C., "Essay on Economic Growth and Environmental Quality," *Swedish Journal of Economics, 73* (March 1971), 25–41.

Dolan, Edwin G. *TANSTAAFL, The Economic Strategy for Environmental Crisis*. New York: Holt, Reinhart and Winston, Inc., 1971.

Herfindahl, Orris C., and Allen V. Kneese. *Quality of the Environment: An Economic Approach to Some Problems in Using Land, Water and Air*. Baltimore: Johns Hopkins Press, 1965.

Krutilla, John. "Conservation Reconsidered," *The American Economic Review, 57* (September 1967), 777–786.

Krutilla, John. "Some Environmental Effects of Economic Development," *Daedalus* (Fall 1970), 1058–1070.

Landsberg, Hans H. "The U.S. Resource Outlook, Quantity and Quality," *Daedalus* (Fall 1967), 1034–1057.

Mishan, E. J. *Technology and Growth, the Price We Pay*. New York: Frederick A. Praeger, 1970.

National Academy of Sciences, The Committee on Pollution. Report to the Federal Council for Science and Technology, *Waste Management and Control*. Washington, D.C.: National Research Council, 1966.

President's Science Advisory Committee, Environmental Panel, J. W. Tukey, Chairman. *Restoring the Quality of the Environment*. Washington, D.C.: White House, 1965.

The Study of Critical Environmental Problems Sponsored by the Massachusetts Institute of Technology. Report, *Man's Impact on the Global Environment*. Cambridge, Mass.: MIT Press, 1970.

Tsuru, Shigeto, ed. *Environmental Disruption: Proceedings of the International Symposium, Tokyo, March, 1970,* Tokyo: Asahi Evening News, 1970. Symposium held under the sponsorship of the International Social Science Council and the Japan Science Council.

*The "materials balance" approach, discussed in Kneese's paper, is explored more extensively in the three references below:*

Ayres, Robert U., and Allen V. Kneese. "Pollution and Environmental Quality," in Harvey S. Perloff, ed., *The Quality of the Urban Environment.* Baltimore: Johns Hopkins Press, 1968, 35–71.

Ayres, Robert U., and Allen V. Kneese. "Production, Consumption and Externalities," *The American Economic Review, 59* (June 1969), 282–297.

Kneese, Allen V., Robert U. Ayres, and Ralph C. D'Arge. *Economics and the Environment: A Materials Balance Approach.* Baltimore: Johns Hopkins Press, 1970.

## Section II *Formal Analysis*

*In most of the references listed below the issues examined in Section II are pursued at a more advanced level of theoretical analysis.*

Bator, Francis M. "The Anatomy of Market Failure," *The Quarterly Journal of Economics, 72* (August 1958), 351–379.

Buchanan, James M. and William C. Stubblebine. "Externality," *Economica, 29* (November 1962), 371–384.

Buchanan, James M. "Politics, Policy and the Pigovian Margins," *Economica,* N.S. *29* (February 1962), 17–28.

Davis, O. A., and A. Whinston. "On Externalities, Information and the Government-Assisted Invisible Hand," *Economica,* N.S. *33* (August 1966), 303–318.

Dolbear, F. T. "On the Theory of Optimal Externality," *The American Economic Review, 57* (March 1967), 90–103.

Kneese, Allen V., and Blair T. Bower. *Managing Water Quality: Economics, Technology, Institutions.* Baltimore: Johns Hopkins Press, 1968.

Jarrett, Henry, ed. *Environmental Quality.* Baltimore: Johns Hopkins Press, 1966. See in particular Ralph Turvey, "Side Effects of Resource Use," pp. 47–60, and a comment by Roland McKean, "Some Problems of Criteria and Adequacy of Information," pp. 61–65.

Mishan, E. J. "Reflections on Recent Developments in the Concept of External Effects," *Canadian Journal of Economics, 31* (February 1965), 3–34.

Mishan, E. J. "Welfare Criteria for External Effects," *American Economic Review, 51* (September 1961), 594–613.

Nutter, G. W. "The Coase Theorem on Social Cost," *Journal of Law and Economics, 11* (October 1968), 503–507.

Pigou, A. C. *The Economics of Welfare.* 4th ed. Section II. London: Macmillan and Co., Ltd., 1932.

Rothenberg, Jerome. "The Economics of Congestion and Pollution," *The American Economic Review, 60* (May 1970), 114–121.

Samuelson, Paul A. "Diagrammatic Exposition of a Theory of Public Expenditures," *Review of Economics and Statistics, 37* (November 1955), 350–356.

Scitovsky, Tibor. "External Diseconomies in the Modern Economy," *The Western Economic Journal, 4* (Summer 1966), 197–202.

Tulloch, Gordon. "Problems in the Theory of Public Choice, Social Cost and Government Action," *The American Economic Review, 59* (May 1969), 189–197.

Wellisz, Stanislaw. "On External Diseconomies and the Government-Assisted Invisible Hand," *Economica,* N.S. *31* (November 1964), 345–362.

## Section III *Policies for Environmental Protection*

Baumol, William J., and Wallace E. Oates. "The Use of Standards and Prices to Protect the Environment," *The Swedish Journal of Economics, 73* (March 1971), 42–54.

Burrows, Paul. "On External Costs and the Visible Arm of the Law," *Oxford Economic Papers, 22* (March 1970), 39–56.

Crowe, Beryl L. "The Tragedy of the Common Revisited," *Science, 166* (November 28, 1969), 1103–1107.

Demsetz, H. "The Exchange and Enforcement of Property Rights," *Journal of Law and Economics, 7* (October 1964), 11–26.

Hardin, Garrett. "The Tragedy of the Common," *Science, 162* (December 13, 1968), 1243–1248.

Haveman, Robert H., and Julius Margolis, eds. *Public Expenditures and Policy Analysis*. Chicago: Markham Publishing Co., 1970. Contains many of the papers, some in revised form, which originally appeared in the Joint Economic Committee's three-volume compendium, *The Analysis and Evaluation of Public Expenditures: The PPB System.*

Kneese, Allen V. "Environmental Economics and Policy," *The American Economic Review, 61* (May 1971), 153–166.

Mishan, E. J. "Pangloss on Pollution," *The Swedish Journal of Economics, 73* (March 1971), 111–120.

Mishan, E. J. "Pareto Optimality and the Law," *Oxford Economic Papers, 19* (November 1967), 255–287.

Roberts, Marc J. "Organizing Water Pollution Control: The Scope and Structure of River Basin Authorities," *Public Policy, 19* (Winter 1971), 79–141.

Reich, Charles A. "The New Property," *Yale Law Journal, 73* (April 1964), 733–787.

Teller, Azriel. "Air Pollution Abatement: Economic Rationality and Reality," *Daedalus* (Fall 1967), 1082–1098.

## Section IV *The Roots of Environmental Degradation*

Commoner, Barry. *The Closing Circle, Nature, Man and Technology*. New York: Alfred A. Knopf, 1971.

Dahmen, Erik. "Environmental Control and Economic Systems," *The Swedish Journal of Economics, 75* (March 1971), 67–75.

Schelling, Thomas C. "On the Ecology of Micromotives," *The Public Interest,* No. 25 (Fall 1971), 59–98.

White, Lynn, Jr. "The Historical Roots of Our Ecological Crisis," *Science, 155* (March 10, 1967), 1203–1207.

## Section V *Measuring Costs and Benefits*

Davidson, Paul, F. Gerard Adams, and Joseph Seneca. "The Social Value of Water Recreational Facilities Resulting from an Improvement in Water Quality: The Delaware Estuary," in Kneese, Allen V. and Stephen C. Smith, *Water Research* (Baltimore: Johns Hopkins Press, 1966), 175–211.

Eckstein, Otto. "A Survey of the Theory of Public Expenditures Criteria," in National Bureau of Economic Research, *Public Finances: Needs, Sources and Utilization*. Princeton, N.J.: Princeton University Press, 1961, pp. 439–504.

Fabricant, Neil, and Robert M. Hallman. *Toward a Rational Power Policy, Energy, Politics and Pollution, a Report of the Environmental Protection Agency of the City of New York*. New York: George Braziller, 1971.

Lave, Lester B., and Eugene P. Seskin. "Health and Air Pollution," *The Swedish Journal of Economics, 73* (March 1971), 76–95.

Maas, Arthur. "Benefit Cost Analysis—Its Relevance for Public Investment Decisions," *The Quarterly Journal of Economics, 80* (May 1966), 208–226. See also "Comment" by Robert Haveman in same journal, *81* (November 1967), 695–699.

McKean, Roland. "The Unseen Hand in Government," *The American Economic Review, 55* (June 1965), 496–507.

Prest, A. R., and Ralph Turvey. "Cost-Benefit Analysis: A Survey," *The Economic Journal, 74* (December 1965), 683–735.

Ridker, Ronald G. *Economic Costs of Air Pollution*. New York: Praeger, 1967.